MW01049926

Total Man

Total Man

Stan Gooch

Holt, Rinehart and Winston
NEW YORK CHICAGO SAN FRANCISCO

Library of Congress Catalog Card Number: 73-182750
ISBN: 0-03-091383-7
First published in the United States in 1973

Printed in the United States of America

Contents

Part Five: The Rise to Tyranny of Western Consciousness

Part Six: Evolutionary Prospect: The Three-Dimensional Personality

As to the indistinctness of dreams, that is a characteristic like any other. We cannot dictate to things their characteristics.

SIGMUND FREUD

Acknowledgements

My appreciation is due firstly to a number of individuals who, without in every case being even aware that they were so doing, helped me towards the initial formulation of the theory of personality presented here – not least by their demonstration that the word 'academic' need not be synonymous with 'narrow' or 'partisan': Professor Ronald Fletcher, Professor Brian Foss, Dr Ilse Grahame, Mrs F. M. Hodgson, Professor R. S. Peters and Professor Edna Purdie.

Special thanks are due to Sir Cyril Burt and Professor Dan Miller for their critical reading of the book in manuscript and for their many constructive suggestions. Here I must also record my gratitude to Adolf Wood for his careful editing of the final text and his wholehearted concern for its welfare.

Finally I would like to thank Mr Edward Luttwak for, among other items, the title of the book.

My thanks are due to the publishers of the following works for permission to quote:

Dylan Thomas – *Collected Poems*, published by J. M. Dent & Sons Ltd and the Trustees for the Copyrights of the late Dylan Thomas
(published in USA by New Directions Inc.)

R. D. Laing – *The Divided Self*, published by Tavistock Publications Ltd
(published in USA by Pantheon Books Inc.)

Translated by Richard Wilhelm/Cary F. Baynes – *I Ching or The Book of Changes*, published by Routledge & Kegan Paul Ltd
(published in USA and Canada by Bollingen Foundation at Princeton University Press)

C. G. Jung – *Synchronicity: An A-Causal Connecting Principle*, from *Collected Works of C. G. Jung*, published by Routledge & Kegan Paul Ltd
(published in USA and Canada by Bollingen Foundation at Princeton University Press)

Sigmund Freud – *Introductory Lectures on Psycho-Analysis*, published by George Allen & Unwin Ltd
(published in USA by Liveright Publishing Co.)

C. T. Morgan/E. Stellar – *Physiological Psychology*, published by McGraw-Hill Book Co.

Bruno Bettelheim – *Joey: A 'Mechanical Boy'*, published by Scientific American Inc.

R. S. Woodworth/H. Schlosberg – *Experimental Psychology*, copyright held by Laurence Pollinger Ltd

Preface

Contact with colleagues and editors rapidly, and rightly, shatters every author's belief that his book speaks so clearly for itself that no introduction is necessary. He eventually comes to accept that a few words of preface are not merely advisable, but essential.

The book should be thought of in first place more as the scaffold which makes the building possible than the building itself – the outline or projection of a theory more than a finished theory. The genesis of this theory dates from my attempts as a psychology undergraduate twelve years ago to reconcile the approach of learning and conditioning theorists with, for instance, that of the psychoanalysts. In the years which followed my conclusions expanded to take in such fields of human activity as politics, art, religion, mythology, anthropology and still others.

Were the book then anything other than an outline, the fifteen chapters would necessarily have become at least fifteen volumes. I am myself only too aware of how much has been left unsaid and unconsidered. I have necessarily omitted important writers – some on the one hand who would be inimical to my position, but equally others who would be supportive. Even where a writer is in fact considered, I can only be said to have sampled, not exhausted, his views. My deliberate policy throughout has been to stop short in the pursuance of any theme or argument whenever I felt a first case had been established – a sufficient brief to justify bringing the matter before the court, if not to make the result of the trial a foregone conclusion.

The subject of the book is personality. This term is used in its broadest possible sense to refer to the totality of man's psychological, emotional and intellectual life. As the word 'body' includes a man's total physical characteristics, so the term 'personality' should be understood here to embrace his entire psychological being.[1]

The coin of the book is generalization. In stating this I hope I have disarmed those who would wish to use that fact as a criticism. One should hardly complain if the villages of England do not appear on a map of the world.

Essentially, the view I have taken of other 'theories of personality', such as those of Marx, Christianity, Freud, Pavlov, Nietzsche and many others, is not that all, or all but one, are wrong, but on the contrary that all are correct. I was, and am still, unable to escape the implications of the fact that each of the theorists in question was concerned with one and the same human being. Instead of rejecting all such views, I found myself essentially rejecting none. This shift of emphasis is, I believe, crucial. The task is therewith changed from one of selection, to one of assembly.

The arguments of the book are certainly speculative. No data has been considered too broad or too extreme to warrant exclusion on those grounds alone. Additionally, since all barriers between areas of study are of man's, and only of man's, making, I have dismantled these whenever it seemed useful to do so. Established or generally accepted fact of course also plays a major part in the book: and I have throughout been at pains to distinguish between such fact and my own speculations. Specific reference to particular authorities is made where I felt doubt might arise as to the authenticity of the evidence cited, or where my own interpretation of an author's views alone might seem unduly biased. Occasionally it has been necessary to draw on

1. A theory of personality, in psychology, is one which attempts to accommodate all the differing moods and functions – feeling, thinking, learning, dreaming, continuity of identity and so on, the changes and permanences of human psychological life – in terms of some overall, unifying structure.

some learned source not readily available to the general reader, but wherever possible, I have deliberately given preference to popular accounts, literary works, magazine articles, encyclopedias and so forth – not simply on the grounds of their availability, but because I believe that these more reflect the general mood in society than do narrowly professional or academic journals. Why, one might ask, am I concerned with the more general, rather than the academic, mood?

The social scientist, erroneously, as I believe, has adopted many of the practices of the physical scientist on the implicit, often explicit, assumption that psychology and sociology *are* sciences. I myself on the other hand, together with some other psychologists, consider the wholesale application of the methods of the physical sciences to the study of human behaviour to be among the major disasters of our time. This does not mean, however, that I believe those methods have no place at all in behavioural studies – though I have not space here to outline my precise position. The point I do wish to make very briefly – a slightly different one – is this. Because of the fact that we ourselves are the objects of the psychologist's and sociologist's study, we cannot grant the psychologist the same automatic authority that we grant the professional physicist or chemist. Rather, the position resembles that which pertains in democracies in respect of Parliament. The people elect representatives to govern them – individuals whom they consider particularly suited to do so – granting them by such election a mandate to run the affairs of the country as they think best. It is when a point is reached that government behaves in ways deeply unacceptable to the people that that mandate is withdrawn. So it must be, I think, with the social sciences. We in a sense grant, or have granted, a mandate. But we do not thereby lose our inalienable right as human beings – the objects of the psychologist's study – to reject not merely his findings but, if necessary, even his methods.

My personal feeling, then, is that the focus of the psychological instrument has currently been adjusted to a level *below* that necessary for an adequate view of man to emerge. For example, I find present accounts of the human psyche and human society by social scientists, with rare exceptions, to be in almost every way inferior to those of the novelist, the poet and the playwright. This is of course my personal view. But if matters *are* so, how might the position have arisen?

The situation is best approached from the standpoint that psychology and sociology, whatever else they may be (and along with all the physical sciences), are themselves *human behaviours*. Objectivity itself is likewise not some property of the universe, but also only a human behaviour. We must not hesitate to study it and them as just that. Indeed, to take the position to its extreme, I suggest that the universe may profitably be regarded as a galactic Rohrschach test, on to which we project and in which we perceive our own, and only our own, inner processes. Here one includes, as emphasized, the so-called objective processes. What we, then (with an odd pride at having apparently excluded ourselves from the equation), call the laws of the universe, are at most the laws of our interaction with it. So, for example, while the universe manifests something termed wavelengths of light, it is itself no colour at all, not even black.

One or two final comments can be made. If reactions obtained to the book while in manuscript are typical, it seems that it and the central arguments would have enjoyed a far less turbulent passage, had I chosen to confine them to the area of general psychology. Approval of my main conclusions at this level has been voiced by several authorities, among them Sir Cyril Burt. (This statement, of course, in no sense necessarily implies the agreement of those authorities with any specific conclusion.) I am extremely grateful to the people concerned for their generosity and encouragement, particularly in view of the fact that the book can hardly be said to flatter the image of the professional social scientist. However, my conclusions do not stop short at psychology. On the contrary they have led and indeed forced me into far-reaching physiological speculations. It is here that criticism has been marked, and it must certainly go on record that the views I express are wholly novel – that is to say, are not currently envisaged by any professional physiologist. That, however, is not quite the same as saying that no hints towards my position exist in the professional literature – and I have quoted extensively from the writings of physiologists in support of my speculations. It remains to be seen how these will fare.

With this I have probably sufficiently prepared the ground for the book which follows. Its development will be found to be idiosyncratic. That is, it is selective towards its own ends. It asks of the reader, though not unreasonably, a willingness to follow a line of argument where the argument happens to lead.

Part One:

The Restless Dead: The Personality in Literature, Legend, Magic and Religion

Duality in Literature and Legend 1

Zwei Seelen wohnen, ach! in meiner Brust,
Die eine will sich von der andern trennen

Two spirits dwell within my breast
Each seeks therein its separate existence

J.W.V.GOETHE, *Faust*, *Part One*, 1808

In its barest outline the Faust legend is that of a learned scholar who in return for worldly pleasures and certain magical powers sells his soul to the Devil.

There actually existed a relatively small-time, though not unintelligent, trickster named either Georgius or Johannes Faust. It is possible that two, or more, men of this name were involved. Numerous references to the surname are made by officials and clerics up and down sixteenth-century Germany. Johannes Faust, at any rate, received the equivalent of a B.A. degree at Heidelberg University in 1509.

In 1587, following the death of this individual or individuals, an anonymous author published a *Faust* biography in Frankfurt-am-Main. This was a not specially well-written, and indeed speculative, venture, which far from giving a true picture of any real Faust attributed to him all manner of far-fetched and scandalous activities from folk-anecdotes then in general circulation.

Such stories, already centuries old, were regularly reattributed to the magician, wizard or trickster of the moment. On this occasion, however, the labels stuck, many of these elements thenceforth becoming the property of Faust and no other. They were dramatized by Marlowe in England – later by Goethe – and the permanent form of the legend was now established.

This first known written version, the *Volksbuch vom Doktor Faustus*, travelled rapidly through Europe. It went to eighteen editions in ten years, appearing almost immediately in translation in England, France and Holland. It was one of these translations that Marlowe came upon. Versions of the Faust drama which Marlowe wrote in 1604 were then performed regularly throughout Europe for the next two hundred years. Later the play continued its career in the puppet-theatre. It was in this form that Goethe made its acquaintance at the end of the eighteenth century.

It is already of interest that such a man or men whose abilities in 'magic' were well below those of many of their contemporaries, and whose scholarship was certainly inferior (one of whom, moreover, indulged in 'the most dastardly form of lewdness' with the boys in his charge), should have been chosen or at least have emerged as a composite, representative figure of the age. Further, that a badly written book about this figure should have seized the public imagination. Third, that out of these two circumstances should have come a universal theme which from then on, as we shall see, occupies the attention of the major artistic talents of each succeeding generation.

In more detail, the plot of Goethe's own version of the Faust story is as follows. An elderly scholar, realizing that life has passed him by and that his single-minded devotion to study has brought him no nearer any *real* knowledge, attempts to conjure up a devil. He meets with indifferent success, but is later visited by Mephistopheles, who in return for Faust's soul at death promises him youth, boundless pleasures and time to find his ultimate answers. The devil is true to his word – except that Faust for his part discovers only fresh disillusion. In particular he seduces a young girl, Gretchen, who, having borne his child out of wedlock, murders it and is executed for infanticide. The tragic Faust is carried away by Mephistopheles.

Some only of those who both earlier and subsequently concerned themselves with the Faust story include: Marlowe (*The Tragicall History of Dr Faustus*, 1604), Lessing (*Faustfragmente*, 1784), Goethe (*Faust*, *Parts One and Two*, 1808 and 1832), Berlioz (*Damnation de Faust*, 1846), Heine (*Doktor Faust*, *Ein Tanzpoem*, 1851), Gounod (*Faust*, 1859), Valéry (*Mon Faust*,

1946) and Thomas Mann (*Doktor Faustus*, 1947). One final recent example is the Richard Burton film of *Doctor Faustus* (1968).[1]

English readers – less well acquainted than they might be with Goethe's work – may not appreciate that his talents were not confined to literature. He was poet, writer, painter, impresario, educationalist, scientist and statesman – the universal man. Some of his scientific contributions stand to the present day. He had completed a version of *Faust, Part One* in 1775 when he was twenty-six. *Faust, Part Two* was completed fifty-seven years later in 1832, the year of his death. Work on the second part had occupied him throughout his life.

If we are looking for some gauge of the significance of the myriad events which make up our lives and the weave of history, may we not do far worse than consider the themes which have continually engaged the attention of great artists of each generation? Such, indeed, is the starting point of this book.

With this we turn to a story which will be more familiar to English readers, that of *The Strange Case of Dr Jekyll and Mr Hyde*, published in 1886. If the change of level seems disconcerting, it may become less so. R. L. Stevenson's story concerns an elderly brilliant physician, Dr Jekyll, a man continuously preoccupied with the question of good and evil, who undertakes a study of the dark side of the mind. He eventually distils a liquor which when taken releases a man's hidden personality – his dark self. Jekyll is himself the subject of his experiments. His *alter ego*, Mr Hyde, is even a physically different personality. He is shorter, younger, coarser, with matted hair on the backs of his hands, a totally unbridled creature of uncontrollable anger and depraved appetites. Gradually Hyde comes to assume control of Jekyll even without the potion, and finally takes over almost completely. Among other incidents he has perpetrated an atrocious murder, and on the point of being arrested, with the last

1. Of Berlioz the *Encyclopædia Britannica* comments, 'Among his dramatic works, two became internationally known, the *Damnation de Faust* and *L'Enfance du Christ.*' Gounod the same work describes as 'a French composer whose fame rests chiefly on his opera *Faust*'.

vestiges of conscious control, Jekyll commits suicide. In the somewhat better-known screen versions of this story Jekyll is usually a young man, often with a fiancée, a pure girl whom he as Hyde forsakes for the company of prostitutes.

The plot is melodramatic, and Stevenson was a prolific writer. Yet it was *Jekyll and Hyde* which established his reputation with the ordinary reader. In some ways the story may seem a far remove from Goethe's *Faust*. But are not, for instance, the lines 'two spirits dwell within my breast, each seeks therein its separate existence' still more applicable to Jekyll than to Faust?

Whether he put any actual credence in the matter or not, Goethe felt at least artistically obliged to externalize the devil as a separate character. Stevenson's devil is wholly internalized – a difference we may choose to consider significant.

Jekyll and Hyde has been called one of the screen's most popular themes; at least seventeen films have been based on the story. The first film version was made in 1908, the most recent, apparently, in 1963. The theme has inspired some great screen performances by such actors as John Barrymore, Conrad Veidt, Spencer Tracy, Boris Karloff, Paul Massie and Frederic March – who received an Academy Award for the role.

In the events surrounding *Jekyll and Hyde* do we not see, in miniature, and over a shorter time-span, something of the circumstances of the Faust story? The parallels go further. Just as elements in Marlowe's *Faustus* were quickly expanded to produce popular stage versions at the level of knockabout farce, so we shortly find knockabout versions of Stevenson's theme involving Abbot and Costello, Jerry Lewis, Bernard Bresslaw and others. There is even what might be called a puppet version, played by Sylvester the cartoon cat.[2] The evolution of these themes from tragic to comic is not so much one of debasement, but rather one of acceptance. (This is a feature of importance that will be considered again.)

From the dual vantage-point of Faust and Dr Jekyll we are in fact in touch with an extensive continuum of legend and folk-

2. *Dr Jekyll's Hide*, 1958.

lore in existence for hundreds, in fact thousands, of years. The form changes from time to time and the similarities are not always obvious at first sight. As a next step some of the further forms of this major theme are now reviewed, and the interconnections between versions indicated.

Central to the theme is a state of possession, which may be *internal* or *external*, though the distinction is in any case not absolute. While Jekyll was internally possessed, Faust is seen more as controlled by an external agency. In all cases, however, the person is somehow taken over by some other intelligence or force. This take-over may be with or without his permission, and may have the apparently paradoxical effect of either increasing or diminishing his powers. What always suffers in one way or other is freedom of action. Internal possession occurs in a variety of forms – the frenzy of witch-doctors or medicine men; the tongues of fire which filled Christ's disciples; the trance of the spiritualist medium; in sleep-walking, epilepsy and actual madness. At the other end of this scale comes the zombie of Haiti. Here the individual has abdicated his own volition, his own body, to the point of being dead – although he is in fact only sometimes described as dead. At other times it is said that the 'conjure man' has 'stolen the living man's soul'. In whichever event, the practitioner here has fully taken charge of the other person's body and uses it entirely for his own ends. The hypnotist, too, exercises a similar, though perhaps more limited power.

From the zombie it is only a short, if perhaps unexpected, jump to the puppet. The puppet, and the zombie also, represent the most extreme forms of external possession. The puppet literally has no life of its own except that imparted by its master – or so it would seem.

In innumerable forms the puppet has figured centrally, and in many areas still figures, in the religious rituals and entertainment of all races. The statement holds as true for classical Greece as for India, China, Mexico and ancient Egypt. Puppet theatres flourished not merely in rural areas, but in urban areas where there was no shortage of human entertainers and in countries where it was no disgrace to be an actor. The art, as we know, was practised in Goethe's day. As for antiquity, complete puppet theatres have been found in Egyptian tombs, and single puppets are regularly found with the dead throughout the ancient world. There are, perhaps, grounds for suggesting that man's interest in puppets is greater than can solely be accounted for in rational terms.

Consider a moment the ventriloquist and the ventriloquist's dummy. Is our sole interest in the fact that the ventriloquist speaks without moving his lips? Or is our interest rather with this strange creature which is alive and not alive, which in some ways is more limited, yet in other ways less limited, than a human being: which in a way is under our control and does what we tell it, but which in another sense does things against our will and has an almost macabre life of its own?

Probably the best known of the possessors of recent times is Count Dracula, from the novel by Bram Stoker. Of it a commentator writes:

> This strange tale of vampires and werewolves has worn surprisingly well. It has been presented on the stage and its principal character, Dracula, has become a well-known figure of literary reference.[3]

Surprisingly, perhaps, the tale was written as recently as 1897. Fourteen film versions have been made, showing that again the situation of the *Faustbuch* and *Dr Jekyll* has recurred – a rather far-fetched, melodramatic idea, incorporated in an apparently one-shot story does not merely survive, but continues to extend its influence.

The basis of the Dracula story is that certain individuals when dead are not truly dead: by night they come to life and search out living victims whose blood they drink. Those so bitten eventually die and become vampires themselves. By day the vampire sleeps in a coffin or box, which must contain soil from his native land. One of the few ways of destroying a vampire is to drive a stake through his heart during this period. Of interest for our purposes is the increase in the vampire's body-hair, and the fact not

3. *Masterpieces of World Literature*, Ed. F. N. Magill.

so much that he comes to life by night as that he sleeps all day.

Vampirism is a local European variant of a world-wide phenomenon, lycanthropy. This term (from Greek *lykos*, a wolf) is today applied to a psychiatric disorder in which the patient imagines himself to be an animal. The belief that all men could and that certain men did become animals is universal and dates from prehistoric times.

The trail from vampires and werewolves leads naturally to the hordes of demons, wraiths and spectres which jostle for elbow-room in the world's folk-lore. These were waiting in all kinds of places to spring upon the unwary human being, given half a chance. Apocryphal writings, in particular the *Book of Enoch*, state that demons are the offspring of unions between fallen angels and human mothers. The fallen angels themselves are devils and include the Devil himself. One recalls that the crimes of the fallen angels were self-assertion and the wrongful revelation of divine secrets: in this connection one notes, too, that sorcerers like Faust conjured up devils to gain knowledge and revelation, rather than to pave the way to a life of depravity.

Angels as such are significant for several reasons. First, though often mentioned in the Old Testament, they come into their own principally in the New Testament. Second, they have human form and are either masculine or sexless – female angels making their appearance only with the Renaissance. Third, angels come from above, not from below. Fourth, Greek *aggelos* means a messenger. At a later stage, from one or two remarks by Christ, the idea is evolved that each human being has his own guardian angel. This notion becomes confused with the belief that human beings have a detachable soul, which is an exact counterpart of themselves. This spiritual self could separate from the material self and wander abroad – the so-called *Doppelgänger*. It was death to meet your own doppelganger. Norwegian folklore has a similar figure in the Follower. This Scandinavian manifestation often took the form of an animal having the characteristics of the person concerned. Thus, this myth in turn links back to were-wolves. Further echoes are found in the belief of the ancient

Egyptians, for example, that every living thing (men, animals, plants) had two forms – the visible individual, and an invisible double or second self.

In connection with the general theme of possession and the concept of the doppelganger one cannot escape the link also with the legend of Amphitryon. In this story Zeus, chief of the gods, is consumed with a passion for the mortal Alcmene, wife of Amphitryon. While Amphitryon is away at war, Zeus changes himself into a double of the husband, the only way he can seduce the virtuous wife. With this story we come to still another of the perennial legends. The earliest extant literary treatment of the theme is the comedy by Plautus (died 185 B.C.). This, however, appears to be based on still older plays. The basic plot has come to be one of the most frequently treated in world comedy. Like *Faust*[4] it has caught the attention of outstanding talents. Specifically, one has Molière's *Amphitryon* (1668), H. v.Kleist's *Amphitryon* (1806), Jean Giradoux's *Amphitryon 38* (1926) – the number thirty-eight being a joking reference to the many earlier versions of the play; and, more obliquely, Shakespeare's *Comedy of Errors.* The sub-plot of, for instance, Kleist's drama is equally significant in the present context. Here Mercury, the servant of Zeus, impersonates Amphitryon's servant Sosia, and appears to the servant as himself.

In progressing from the puppet to Dracula and beyond we branched away from another road, which leads to the mechanical doll. There is, first, the isolated but significant tale of *Pinocchio*, the puppet who by an act of moral courage earns the right to become a real boy. The main route takes us on to the wind-up or clockwork doll. Such figures are the subjects, for instance, of Delibes's *Coppélia* (1870), Stravinsky's *Petrouchka* (1911) and Rossini's *La Boutique Fantasque* (1919). In *Coppélia* a real girl substitutes herself for the life-size mechanical doll which the toymaker has made, and creates havoc in the shop. In Rossini's

4. *Faust*, *Jekyll*, *Dracula* and *Amphitryon*, despite the many differences of detail, are considered to be essentially treating the same subject – in what sense the later argument will show.

similar ballet a large number of mechanical figures come to life of their own volition, and run wild when the shop is shut for the night. The theme of toys which come to life at night is very frequently found in children's tales.

Allied here also is the story (written, astonishingly, over 150 years ago) by Mary Shelley, of the man-made monster *Frankenstein*. The toys which come to life of their own accord are daunting enough. Worse still is a creation of our mind which becomes entirely independent of us, and is moreover almost totally destructive. With *Frankenstein* the macabre as opposed to the amusing elements are explicit. In the sense of an escape from control, however, all these stories are linked. Essentially this is the process, too, of *Jekyll and Hyde*.

Apart from their fascination for the artistic imagination – a consideration we shall ultimately return to – the main element which unites all the diverse material so far reviewed is 'two-ness'. There is one thing, and something else – a possessor and a possessed. We must temporarily delay facing the question of precisely what or who is possessing whom or what. So far there are three possibilities. One, a human being is *externally* possessed; two, a human being is *internally* possessed; three, a human being is *himself* the external possessor – as with the puppet-master, the toy-maker and the monster-maker. In this third case, however, the situation is inherently unstable – very soon the tables are wholly or partly turned and the puppet-master finds himself dancing to his own creation. This fact highlights the absence in this literature of the fourth possibility, namely the *human being* as *internal* possessor. This is a feasible but for some reason neglected alternative. May one ask whether what is really being dealt with, what is really feared, is *being possessed*?

Even this question stops short of the truth. Is not the real fear the possibility that the possessor is always within us, and somehow *is not us*?

The almost ludicrous question which is posed is – are we each two creatures? Are we each ourselves *and* a doppelganger? And if this is true, is the other then a friend or an enemy? And worse still, which of the two is the *senior* – are *we* the real householder in this our house, or are *we* the upstart, the lodger, the unwelcome guest, who may at any moment be turned out of his estates?

To lower the temperature at this point – though still vigorous as we have seen, the themes and situations we have examined all somehow undeniably bear the stamp of the past. Whatever may or may not be the real issues hidden in the material cited (and, certainly, its wide if former provenance appears to demand some kind of explanation) are we nevertheless perhaps looking only at a historical phenomenon? Or, may we perhaps find these issues as vigorous as ever elsewhere in a slightly different guise? Were that to be so, would there not then be a situation *very much* requiring an explanation – the more so perhaps because so patently neglected in any current writings on mankind?

In search of some present position, we turn now to a consideration of a twentieth-century phenomenon, science fiction.

Attention need hardly be drawn to the rapid growth of this literary genre in the years since the war – at which point it went public following a period of underground growth. One saw then the appearance of the SF novel in the best-seller lists, for example John Wyndham's *The Day of the Triffids* (1951).[5] The effective capture by SF not merely of the 'way-out' but of the average adult mind is perhaps best seen today in the emergence of the big-budget SF film for adults – Kubrick's *2001*, Vadim's *Barbarella*, and so on.

Let us consider a few themes from current SF writing.

(1) A party of scientists arrives from outer space; they examine Earth's ecology, operate on the brains of some anthropoid apes and depart, leaving the changed apes to evolve into *homo sapiens*.

5. Science Fiction, without that label, was actually already no stranger to the reading public – so, Thomas More's *Utopia* (1516), Swift's *Gulliver's Travels* (1726), Verne's *From the Earth to the Moon* (1865), or Wells's *Time Machine* (1895).

(2) Another version, with a grimmer twist. A party of outer-space scientists arrives on Earth, ostensibly to observe anthropoid apes. One of the group, a sociologist, is vaguely puzzled by his inclusion in the party. During the night he is drugged; he awakes to find himself in the body of an ape, the victim of a brain transplant. The space-ship takes off as the tragic ape-man turns to his stinking companions and begins teaching them to build huts.

(3) A family of short, incredibly ugly, immensely strong, smelly, magical-mystical, creative and imaginative hill-billy junk merchants prove to be the last survivors and direct descendants of Neanderthal man, who preceded and were destroyed by *homo sapiens.*

(4) A party of space explorers crashes, and they are obliged to make their lives on Earth. Fairly rapid degeneration sets in over a few generations, with the resultant loss of skills and knowledge. In the course of many generations the upward climb to civilization begins again, leading to present-day man with his dim memories of wheels in the sky, chariots of fire, forgotten 'magical' powers, and a yearning to return to space.

(5) The first space astronaut finds on leaving Earth's gravity that his mind has been released from a prison. All the mental powers mankind has dreamed of and tried to grasp – telepathy, telekinesis, true understanding, creativity – are suddenly his. An Earth-bound, non-physical parasite which preys on every human being from birth has been forced to abandon its parasitic grip at the entry to free space.

(6) Parents find themselves unable to have any more children, and the present generation of children ceases physical growth. Man proves to be the primitive, larval stage of a higher life-form, which now abandons Earth.

Reading these outlines, some may have experienced for the first time the 'charge' or sense of exhilaration which many SF readers obtain from the genre. This is perhaps the same sensation described by Goethe in his autobiography on first seeing the

Faust puppet-play: 'The significant puppet-fable re-echoed and hummed in many tones in my head.' It is possible also that the original listeners to Greek legends experienced this same excitement – a sense of 'this is where it's at'.

While some of the further details of the content of these outlines must be left for a later stage, in terms of the similarities between the old and the new material one notes again the clear and persisting emphasis on 'two-ness'. Significantly, we have now for the first time the human being as internal possessor. This we see in the first outline, but more clearly in the second and fourth. A human being, instead of being overwhelmed by some alien force or agency, finds himself *rooted in* some undesirable, basically captive position (where, nonetheless, he is in a limited sense master). A similar theme is seen in the third outline, where man is shown as a displacer, a pusher-aside of what has gone before. In (5) and (6) he is an escaper from the undesirable or lowly-captive situation. The last two outlines are, perhaps, less real or poignant and the 'escapes' contain an element of wish-fulfilment.

Of the *dis*similarities between the old and new material, one, for the moment, is emphasized. Whereas the older material looks mainly backward and inward, the recent material looks forward and outward. In a sense also these two divisions look respectively downward (into graves, dungeons, or whatever) and upward (into the sky and outer space). In those terms angels would not be the last of the vampires but the first of the spacemen.

These SF themes, and many others beside them, represent the new, vigorous growth of the twentieth century. It is of significance that this growth takes place not in some overlooked corner, as a kind of lunatic fringe, but nearer than ever to the nerve centre of the scientific establishment.[6]

6. Older themes, incidentally, are meantime not neglected. Space permits the mention of only a very few examples. In respect of possession in the old sense one would cite Nigel Kneale's outstanding *Quatermass* serials, the already mentioned *Day of the Triffids*, John Christopher's *The Possessors*, the

At this point we can perhaps look again, rather more seriously, at the earlier forms of these absurdly persistent ideas.

We turn, then, to a more detailed examination of the older material, with a view to isolating some of the less immediately obvious aspects of the major theme – which are really sub-themes in their own right. Among these are loss of freedom – clearly related to the theme of possession – loss of innocence, and the destructive nature of knowledge. These last two may not seem to relate directly either to possession, or to each other, in any obvious sense. However, one notes in passing that, for instance, to possess is one way of describing a man having sexual intercourse with a girl; that the original meaning of innocent is 'not harmful'; and that in the Bible to know a woman is, again, to have intercourse with her.

We examine first the loss of innocence. Of the many protagonists considered, who loses his or her innocence? Faust is one. At the opening of the drama he is discontented, but still virtuous – still a virgin, in fact – a respected, prayerful, middle-class teacher. His search for knowledge brings him Mephistopheles. Under this being's tutelage Faust then becomes a drunkard, a thief and a libertine. Much the same holds for Jekyll. (Although the latter, apparently, had one or two wild times as a youngster, he has resolutely put that behind him, and since lived a blameless life dedicated to healing.) He, like Faust, is a bachelor. Preoccupation with the problem of good and evil – the search for knowledge – leads him to Mr Hyde and the road taken by Faust.

The crime of the fallen angels, as noted, is given variously as self-assertion and the unlawful revelation of divine secrets. This

---

B.B.C.'s *Dr Who* (now in its seventh year, an endless belt of variations on the possession theme, itself producing the Dalek fever which gripped the nation's children), Richard Mathieson's disconcerting reversal of the vampire theme *I Am Legend*, and Walter Miller's revival of the good-evil, spiritual-temporal conflict in *A Canticle for Leibowitz*.

much, then, is similar but there seems here to be no sex element. Yet in the apocryphal writings we are told that demons are the offspring of fallen angels and human mothers. Thus sex *is* somehow involved.

Can we include Dracula? The answer is yes, but one must point out that one is here dealing with a very ancient level of awareness, where meanings are much less straightforward. First, consider the vampire's original crime by which he became what he is. This fits readily enough. Either he is a suicide (a form of overstepping prescribed boundaries) or he has practised the forbidden arts in his lifetime, that is, sought the forbidden knowledge. But what of the *innocent* victims of the vampire who at death also assume the vampire form? The solution to this apparent difficulty is that on the first occasion the vampire may only enter *by the invitation or connivance* of the victim. By this action the victim takes on the guilt of the vampire. Second, biting the victim symbolizes the sexual act. If this claim seems a trifle rash, let us remember that mild bites are a feature of normal love-play, and more importantly that biting figures prominently in pornographic writing. It is also standard knowledge that the female domestic cat (and the female rat, guinea-pig, rabbit, etc.) will only permit the male to have intercourse with her after he bites, or while he is biting, the back of her neck. In other animal species the female cooperates only after the male has lacerated and actually drawn blood from her stomach.[7]

Consideration of loss of innocence and the acquisition of forbidden knowledge brings us inevitably to the best-known legend of the Western world, the story of Adam and Eve. This, as we shall see, is a collector's item, an immensely compacted legend containing all the elements so far discussed. Before turning to it, however, a point has been reached where a little more must be said of the general characteristics and nature of legends and allied material.

First, the nature of legends is symbolic. Precisely what a

7. These matters are discussed further in connection with inherited behaviour-patterns in Chapter 11.

symbol is, and what a symbol does, is not easy to describe, since the subject cannot be approached intellectually or logically, in the narrow sense of those terms, with any real profit. The low estate into which symbols have fallen in the West, not only in terms of value but of understanding, is, I believe, due to wholly mistaken attempts at intellectualization. It will be possible to say more about symbolism elsewhere – but for the moment let us consider an actual instance.

The soil of his native land – the earth – to which the vampire must return each night symbolizes or represents the 'female principle' – in this instance specifically his mother. What grounds do we have for this claim? One might perhaps consider first the many obvious parallels in language between a woman producing children and the earth producing crops. The words fertile, barren and seed, for example, apply both to cropping and child-bearing. This kind of 'evidence' however, may not particularly impress, especially at this stage of the argument. But let us turn aside for the moment to consider the story of the struggle between Hercules and Antaeus, the giant son of the Earth Goddess. In this conflict Hercules found that every time he seized Antaeus and dashed him to the ground, his opponent sprang up with renewed vigour. In time Hercules realized that each contact with his mother, the Earth, gave Antaeus fresh strength. With this realization Hercules raised his adversary clear of the ground and was able to crush the rapidly weakening giant to death.

In the late 1950s H. F. Harlow at the Primate Laboratory of the University of Wisconsin initiated a programme of research into the relationship of the infant rhesus monkey with its early environment. As part of this programme baby monkeys were reared with a variety of mother substitutes – a wire model or a cloth-covered wire model – some of which fed the baby from a nipple and some which did not. The 'models', incidentally, were merely cylinders of wire, lacking arms, with stylized, non-moving heads. It is not possible to summarize here all the findings of this research programme. What emerged clearly, however, is that baby monkeys *in simple physical contact* with a wholly inert, cloth-covered wire 'mother' developed normally in many respects. The feeding aspect was irrelevant. Offered the choice between a plain wire, feeding mother and a non-feeding cloth mother, the infant preferred the latter.[8]

Still more meaningful, perhaps, is the fact that those monkey infants which were prevented from touching the substitute mother, or who were reared in the complete absence of such a mother, grew into adults that were entirely psychologically, socially and sexually inadequate.

One may not, of course, argue without further consideration straight from monkeys to human beings. The observation of human children, however, has in general borne out the relevance of these findings.[9] The cuddling of teddy-bears or a favourite blanket seems in particular to have the same comforting properties for the young human child. Mothers confirm also how often a young child returns to them momentarily in the course of play or exploration.

In the light of Harlow's findings the insight of the Antaeus and Dracula legends, particularly the former, is staggering. They do not merely highlight the importance of touching, they show it to be paramount over all else. If a recent scientific insight was apparently already understood in Greece several thousand years ago, does this not lead one to speculate what other facts about human existence may be contained in legend that we have as yet not rediscovered? Might not therefore a non-dismissive examination of legendary material spare us a good

8. A brief excerpt from Harlow's actual account:

'We exposed our monkey infants to the stress of fear with strange objects, for example a mechanical bear which moved forward beating a drum. The infant would cling to its cloth mother, rubbing its body against hers. Then with its fear assuaged through intimate contact with the mother, it would turn to look at the previously terrifying bear without the slightest sign of alarm. Indeed, the infant would sometimes even leave the protection of the mother and approach the object that a few moments before had reduced it to abject terror.' (H. F. Harlow, *Scientific American*, June, 1959).

9. See for example John Bowlby's *Child Care and the Growth of Love*, Penguin, 1953.

deal of effort in other areas of study, notably in psychology and psychiatry?

Legends in the telling and re-telling may be thought of as stones which have been in the sea for many years and by its action become rounded and smoothed; or as liquids repeatedly distilled so that only essence remains. Much of what Freud has said of dreams also applies to legends – depending partly, however, on the age of the legend – particularly his insights into the mechanisms of condensation and displacement. These produce respectively (a) a high degree of concentration, so that even the smallest detail is important and may summarize a complex point, and (b) a shifting of emphasis – so that what is played down or apparently merely included by chance may sometimes carry the real weight and meaning of the story. One particular item may do duty for a dozen others, including its own opposite. One must in general not be surprised to discover the front at the back, the back at the front, the parts scattered or sometimes wholly absent, as perhaps in the manner of cubist painting. The links tend to be associative rather than logical – that is, a tree may stand for a man because both are tall, or a whirlpool indicate the seductive and allegedly treacherous qualities of woman, since both attract and pull you down. One will here look in vain for ordered, scientific reasoning or for any justification of assumptions made. Unlike logical and scientific thought, which attempts to persuade and demonstrate, with a dream you must take it or leave it – and the dream does not care which you do.[10]

With this brief introduction, we can again turn to the consideration of Adam and Eve. Eve is in effect two people – Eve the virgin (Eve I) and Eve the sexually aware woman (Eve II). Indeed, all the figures in these various narratives may be thought of as two people, Before and After. They begin by being 'innocent' (unharmful?). They end by acquiring knowledge and guilt. This holds certainly for Faust and Jekyll. Similarly at the outset of the biblical story Adam is innocent. Like Faust and Jekyll

10. Compare in this respect the cryptic, once-for-all utterances of the Delphic Oracle.

he is a (*de facto*) bachelor, and in regular employment. The appearance of the serpent represents the intrusion of knowledge, but symbolizes also the 'male principle' – in fact, the penis. The serpent in this sense may be thought of as Eve's aroused interest in sexual matters with dawning womanhood.[11] She fancies Adam's penis. Moved by such feelings she offers Adam the apple – representing again knowledge, but symbolizing more particularly the roundness and ripeness of the adult female form. Of great interest here is that both the male and female symbols in the story represent knowledge.

Adam, together with Eve, was expelled from the Garden of Eden because he knew too much. The phrase reminds us that in gangster stories former associates are killed because they possess knowledge which could lead to the arrest and execution of their one-time bosses. Did Adam have knowledge which could lead to the arrest and execution of God?

In essence the stories of Adam and Eve, Faust and Jekyll are the same. Their story-lines can be summarized in the last analysis as (1) gaining knowledge (2) having sex.[12] Certainly there are differences of detail, but the broad outlines coincide as follows:[13]

---

11. In one way indeed the story of Adam and Eve can be regarded as little more than an allegory of a boy and a girl growing from childhood into puberty and adolescence. Eve, correctly, reaches puberty first.

12. Here one notes a sharp difference from the later SF material. Implicitly, though to an extent also explicitly, these later stories reject sex. This is seen, for example, in the disgust with which the 'transplanted' space man views his animal body, and in the displacement and destruction by modern man of the sexual and earthy (an interesting word, in the light of our recent discussion) Neanderthal. In general, however, SF is not so much anti-sexual as a-sexual. While extraordinarily gripping in terms of its ideas, anyone in search of sexual titillation or stimulation is in general wasting his time with SF. This is not an accident. Also, the human beings portrayed in SF stories are notoriously wooden and two-dimensional. In the country of science fiction, the Idea and the Robot rule.

13. The reader will, of course, decide for himself, both here and elsewhere, whether or how far these outlines constitute oversimplifications, or distortions, of the originals.

(1) Adam, a simple gardener, lives a blameless and celibate life in the company of Eve I. Eve I is tempted by, and listens to, the devil. The now Eve II tempts Adam with the forbidden act and the forbidden knowledge, which includes knowledge of good and evil. With the loss of innocence the state of contentment and one-ness with God is lost. Both Adam II and Eve II are delivered into the hands of death.

(2) The traditional Faust is a respected teacher, living a blameless, celibate, but now not altogether contented life. He has observed Gretchen, a young innocent girl of the neighbourhood. Seeking to extend his knowledge beyond what is permitted, he contacts the devil. A pact is made and Faust meets the wanton Helen of Troy. With the devil's help he is subsequently able to seduce Gretchen. Her former contentment is turned to misery and death, and Faust himself passes into the hands of Mephistopheles to await destruction.

(3) Jekyll is a respected doctor, living a blameless, celibate, though now not altogether contented life. He is preoccupied with the question of good and evil. In pursuing it he investigates his dark side, and in finding the hidden knowledge becomes its prey. As Hyde he is the companion of prostitutes, destroying both the happiness of the girl who loves him and finally himself.

Among the many important points here, one that is often played down is that the woman is as surely destroyed by contact with the male as the male with her. Adam *and* Eve are cast out of paradise. Gretchen dies forsaken, a murderess. Nevertheless, what is usually emphasized in these stories is the fall of man by contact with woman (perhaps because men have always dominated the means of literary production and distribution).

Many other legends attest to this fall of man. In *Tannhäuser*, for instance, the knight of the same name in the course of his wanderings enters the cave of the Lady Venus. Here he abandons himself to a life of sensual pleasure, forgetful of knightly duties. Eventually he is overcome with remorse and invokes the aid of the Virgin Mary to return to the outside world. Later he journeys to the Pope to entreat forgiveness of his sins.

Some of the many well-known tales having an essentially similar form and the same implications as *Tannhäuser* are those of Samson and Delilah, of Circe, of Scylla and Charybdis, and of the Lorelei. All of these emphasize above all the dangers of approaching the situation in question. The situation, in most cases explicitly, in others implicitly, is sexual in nature. The reader is entitled to disagree here, but to the present writer the sexual implications of Charybdis, the female whirlpool, are only too clear.

Tannhäuser is interesting partly by reason of its directness. As we move from the distant past nearer to the present day, stories and legends tend to become more explicit and less subtly symbolic. Here a much older legend has been overlaid by a later Christian ethic. We see at its clearest so far the identification of evil with Venus, who has known men carnally, and good with the Virgin, who has not. *Tannhäuser* raises also in particular the question of the final outcome, when this is not to be automatic and irrevocable destruction. In connection with outcomes we look again at *Faust* and the Bible.

In a sense the first and second parts of Goethe's *Faust* represent the Old and New Testaments respectively. Those familiar with these works might like to reflect for themselves in what sense this might be true. Those who are not may find the following acceptable. In what follows, the Old and New Testaments are taken to be the story of man, and man to be the single, central character.

At the beginning of the Old Testament man loses his early innocence and his special relationship with God through over-eagerness for knowledge, becoming in the process carnally involved with woman. He is then cast out and wanders the earth seeking answers, but finds only false gods and non-solutions, alternately wallowing in the joys, and suffering the agonies, of the flesh. In the New Testament God himself at last intervenes – through his son – and a reprieve is granted. With this man has regained the possibility of one-ness with God and of re-entering paradise. This is a hard bargain, however, involving sexual abstinence and bodily death.

In Goethe's *Faust*, *Part One* the central character, seeking too much knowledge, finds instead only the lusts of the flesh and no real answers. The gratification of lust brings no happiness or contentment, but only pain. In *Faust*, *Part Two* we find the hero working for the good of others (financing and directing irrigation projects, etc.). On his death he is snatched away from the devil up into heaven, despite the terms of the pact, partly as a result of the intercession of Gretchen in the form of an angel. The solution involves the plea that sin is an inevitable consequence of striving, and since striving is good, the sin is ultimately forgiveable.

Up to the time of his suicide Jekyll, too, follows the general pattern of the Old Testament and *Faust*, *Part One*. In his case, however, there is nothing corresponding to the New Testament and *Faust*, *Part Two*. A hint of a possible solution is nevertheless given, as we shall see.

Goethe wrote the first part of his *Faust*, the statement of the problem, while still a young man. The second part, the solution to the problem, occupied the rest of his long life. This second part has never achieved the popularity of the first. Perhaps that fact alone can be accepted as the world's judgement of the validity of the solution. Thus the man of genius, devoting his whole life to the problem, was nonetheless unable to offer us an acceptable answer. Many, indeed, have found Faust's escape from the final consequences of his bargain totally unacceptable. They find the two key statements '*es irrt der Mensch*, *solang er strebt*' (erring is an inevitable consequence of striving) and '*das Ewig-Weibliche zieht uns hinan*' (the eternal-feminine always leads us on to greater achievement) an inadequate justification for the forgiveness.

The New Testament solution was on the face of it a better one, seeming to many for centuries to be the final answer. Unhappily, it appears not to be. In fairness to Goethe also, one has to point out that his, despite the celestial props, was an attempt at a rational solution, while that of the New Testament remains in essence magical. Making Christ a supernatural figure, while at the same time presenting him as one of us, was an excellent each-way bet, and paid dividends. This is perhaps an unfair comment, in the sense that it implies deliberate engineering on the part of the early founders of the Church. What is true, however, is that these men were content to pass off *symbolically* true events as *literally* true events on a largely defenceless proletarian population. Had there not been some truth at one or other level Christianity, of course, could and would not have survived.

The message of the New Testament is essentially that man cannot go it alone. By his own efforts he is not capable of finding a solution. If he is to be saved it is only through divine intervention.

Christ was able to place the Kingdom of Heaven within man, though not apparently the Kingdom of Hell, while Goethe to some extent, and Stevenson wholly, could locate also the devil and his temptations within. Jekyll therefore writes in his confession: 'I saw that of the two natures that contended in the field of my consciousness, even if I could be rightly said to be either, it was only because I was radically both.' Both figures, certainly, are within. But like Goethe, Stevenson can only think of a *solution* in terms of separation. Jekyll goes on:

> I had learned to dwell with pleasure, as on a beloved day-dream, on the thought of the separation of these two elements. If each, I told myself, could be housed in separate identities, life would be relieved of all that was unbearable.

Despite his intellectual and literary gifts, then, Stevenson could not even attempt a practical solution to his hero's dilemma. This fact, together with the great failed attempts of Christ and Goethe, should convince us of the magnitude of the task – whatever it may be. That there *is* some kind of 'task' confronting mankind would seem to be clear. Perhaps one should say, more conservatively, that at least a first case has been made out for its existence. One of our present concerns would then be to translate or describe that 'task' in more straightforward psychological language.

With this the review of certain aspects of literary and legendary narrative is concluded.[14] It leaves us with many loose ends and a number of recurrent themes. The most important of these is the pervading and persistent preoccupation with 'two-ness'. The instances given have far from exhausted the potential supply. One could have considered whether our interest in, for example, Siamese twins, ordinary twins, forgeries, robots, wax-works, photographs of oneself, the understudy, fancy-dress parties, mistaken identity, the phenomena of polarity, the public and private face, and many, many related subjects is wholly to be explained in terms of the obvious or conscious.

At both ends so to speak of the 'two-ness' theme there is some suggestion of an original one-ness (i.e. happiness, unity with God, Garden of Eden, Atlantis, or whatever); and there are hints of a need or desire to re-achieve that one-ness, at least in the Christian tradition, through reconciliation with God. Although that aspect was not highlighted, one could include here the stories of Shangri-la, Utopia, Nirvana, El Dorado and the Promised Land; perhaps too such items as the look of peace which comes over the face of the staked vampire before he crumbles to dust.[15]

Other themes noted were that man does well enough, until he becomes involved with woman – and vice-versa. Similarly, that

14. Among other important legends that of Dionysus could have been profitably included. Dionysus (or Bacchus) was worshipped in frenzied, orgiastic ceremonies in which raw flesh, and sometimes human flesh, was consumed. The god himself often took bestial form. He was further noted for being *twice* born, and for dying and rising again. Principally since Nietzsche, Dionysus – the ecstatic, emotional, dark – is contrasted with Apollo – the orderly, rational, shining. So many elements of this legendary material are germane not only here but throughout the book that I have not earmarked it for any one chapter. Instead reference is made at particular scattered points.

15. Christ said that unless a man died and was born again, he should not see the Kingdom of Heaven.

man was once innocent but has since become corrupt. This last, incidentally, is by no means only a religious theme. The idea is, for example, a cornerstone of the philosophy of Jean-Jacques Rousseau and the progressive education movement. It recurs again in the Communist view that social and psychological ills are not endemic or natural to man, but are produced by the Capitalist system, which corrupts him.

A further related theme is that knowledge, especially carnal knowledge, is dangerous and destructive. One might in that connection have touched on the story of Prometheus, who stole fire and the secret of civilization from the gods. For this Prometheus was punished by having his liver pecked out for ever by an eagle – while mankind in general was punished by evil, toil and disease. The Book of Genesis pronounces the same punishment on man for the crime of eating the apple of understanding.

Above all, there was the dominant theme of possession and of being possessed. In this connection two overall, opposing tendencies were noted – that of looking inward and back, and that of looking forward and out.

Inevitably, these opening chapters must be expositional rather than explanatory. The next chapter will serve rather to complicate than to clarify the situation, in that it takes us still further into the area of 'expendable' material, the beliefs and practices of medieval and earlier man – usually considered to have a certain interest for the historian and the archaeologist, but hardly for the psychologist.

Dualism in Religion and Magic 2

Magical thinking is not random, it has its own laws and its own logic, but it is poetic rather than rational. It leaps to conclusions which are usually scientifically unwarranted, but which often seem poetically right. It is a type of thinking which has been prevalent all through the history of Europe, which lies behind huge areas of our religion, philosophy and literature, and which is a major guide-post to the regions of the spiritual and the supernatural, the regions of which science has nothing to say. There is no necessity to accept it, but it rings many a far-away, summoning bell in the depths of the mind.
RICHARD CAVENDISH, *The Black Arts*.

The material of legends, fairy-stories and religious myths is far from being a kind of psychological junk-heap, a midden as it were of former mental contents, now happily discarded. On the contrary I suggest it offers the only hope of making sense of important features of our psychological make-up and the nature of our present society. The detailed defence of this view is one of the tasks which face this book. For the moment the following may suffice.

It can be seen first that a child in the course of growing up recapitulates in at least some respects the earlier physical, behavioural and psychological history of our species. The first of these three claims is beyond dispute. For example, the human embryo (and the embryo of all mammals) shows during the first weeks of its development rudimentary fish-gills, so briefly recapitulating a period when our very distant ancestors lived under water. Around the end of the first month the embryo also possesses a well-marked tail. Very occasionally a full-term human baby is born with such a tail, or a complete coat of fur, and so on. There is an impressive list of such features.[1] An

1. It should, however, be borne in mind that while such observations provide convincing evidence of our non-human ancestry, these traces themselves have been subject to evolutionary change – so that what we see in the present-day embryo is almost certainly not identical in form to the earlier actuality. This is an important point, which may apply equally at the behavioural-psychological level.

instance of *behavioural* recapitulation is the inward or grasping flexion of the foot when the sole of the very young baby's foot is tickled. The growing youngster, too, later passes through a stage of ambidexterity before settling for the dominance of one or other hand. Since other primates (apes, monkeys, etc.) do not develop this hand-dominance, it can be assumed that the human child is here briefly repeating an earlier phase of primate evolution. On the *psychological* level, children pass through a stage of firm belief in magic and the supernatural – a belief once common both among adult societies of former times,[2] as it is among adult members of primitive tribes to this day. Further, children and the primitive tribesman of today (and our own early ancestors) preserve links with the natural environment (witness, perhaps, the close attachment of children to their pets and indeed to all animals) which we as modern adults in general lose. Other instances of apparent psychological recapitulation – a notion, I must stress, to which by no means all psychologists subscribe – will be indicated at more appropriate stages in the course of this book.

Why are we interested in making such comparisons at this point?

Hospitalized neurotics and psychotics as a group demonstrate behaviours which on closer examination are once again extremely reminiscent of some of the behaviours of children and primitive peoples – for example, omnipotent-magical beliefs. Much of the disturbance and madness of the mental patient, so apparently deviant from the norms of our present society, can also be shown to be much closer to the norms of the behaviour of children and primitive man. It is not for nothing that the clinical psychologist (wittingly or otherwise) speaks of 'regression' (literally, going back), 'primitivation', 'retardation' (being held back) or 'arrest'; of immaturity, childishness and so on. In the extreme forms of mental illness do we not find, too, something suspiciously like the monsters and terrors of myth-

2. Anthropology and archaeology provide us with some information about these – though inevitably not as much as we should like.

ology re-activated and abroad once more in the mind of the patient – roused perhaps like vampires from their un-dead graves?

One has provisionally four possibly 'related' groups, namely: children; modern primitive man; early man; and mental patients.

In prior defence of our forthcoming excursion into magical material there remains finally the suggestion made in the last chapter that the legends and the allied subject-matter we are now considering may contain, in disguised or unperceived form, facts about human existence which we have not yet re-discovered along our present, twentieth-century route. One could be more specific, and still more extravagant, by saying instead 'facts about the structure and functioning of the human nervous system and the physiological basis of personality, as yet unknown to modern physiology'.[3]

The remainder of the present chapter is expositional, and extremely speculative in character – to what extent justifiably the eventual outcome will show. The reader need not be too much at pains at this stage to retain all details either of the exposition or the tentative attempts at interpretation, since all the matters reviewed are taken up again on a number of subsequent occasions in later chapters.

If we were able to experience the world as children experience it but at the same time to think as mature adults, there are certain questions we might ask ourselves. Such as why are gnomes, dwarfs, trolls, brownies, goblins and leprechauns always short, masculine, ugly, strong and irritable, and usually malevolent? Why, too, is Punch short, ugly, strong and bad-tempered? And is it for the same reason that giants, in addition to being always *tall*, are similarly masculine, ugly, strong and bad-tempered? Why are fairies, though small, principally *feminine*, graceful and

3. The term personality, as used in this book, is defined in the Preface.

often, though not invariably, kind? Why are there male fairies but no female goblins or leprechauns? Why have dwarfs no one to make love to? Male fairies do not, of course, literally 'make love' to female fairies (at least not in the fairy-stories of our Western childhood), although they court them and reign with them as king. On examination, the male fairy indeed does not seem to be a particularly masculine figure.[4] He has, for instance, no beard, unlike the bearded dwarf or the shaggy giant. It would seem also true to claim that the male fairy is subordinate in many ways to the female fairy.

An interesting point is that despite their clear feminity, female fairies are not noted for the size of their breasts. Yet they are not above sucking other people's, and also those of cows. In all parts of Europe the traditional dish to set out for the wee folk is a saucer of milk.

If the males and females of the fairy world are not particularly sexual, the same is not true of those other figures of that world, the witch and the wizard. Not only are these of normal size, but they are sometimes extremely, and overtly, sexual. This is actually much less true of the wizard, though true of the wizard-as-devil. (The transformation into 'devil' can take place fairly readily, as we see in a sense with Drs Faust and Jekyll.)

Unlike her counterpart the fairy, the witch is commonly old, ugly or even hideous, and possibly deformed. She may have hairs growing from her chin. She is almost invariably cruel. In these characteristics she resembles the goblin or dwarf more than she resembles the fairy.

The wizard is different again. He too will be old, but though sometimes stooped a little with age he will not be deformed. Quite on the contrary, he is often commanding and upright, tall, and even noble. He is given to sternness and unapproachability rather than cruelty. Indeed, he may be essentially kind, like Merlin or, much later, Tolkien's Gandalf the Grey in *The Lord of the Rings*. The wizard has *authority*, particularly

4. It is probably no coincidence that male homosexuals sometimes refer to each other as fairies.

over the witch. In this respect there is a parallel with the female fairy's authority over the male fairy.

While the picture is by no means complete, what seems to emerge from the foregoing is this. The fairy is young, attractive, and the dominant figure *vis-à-vis* the male fairy, who is in any case rather un-masculine: the (female) fairy also rules effectively over dwarfs, gnomes, goblins, etc. *These* are hairy and ugly and usually have temperaments to match. Moreover, dwarfs tend to live and toil in darkness, below ground, while fairies revel and dance at the surface in moonlit glades.

The opposite positions hold for our older group. Here it is the *female* who is misshapen and ugly, rather masculine, and clearly subordinate to the male. He is now tall, commanding and by no means unattractive. Moreover, he lives in a castle on a mountain, or in some spacious cave. It is now the witch who inhabits a rude hovel, in the depths of the dark forest.

Some reversal of roles appears to have occurred here. Is it perhaps of significance that the (young) female has the best role in the diminutive world, and the (old) male the same good part in the life-size – though still of course magical – world? These considerations lead shortly to the question of who is writing these scripts and for whom.

Perhaps enough has already been said to suggest that fairies are a 'childish', pre-pubertal, pre-adolescent or – to use the appropriate Freudian term – a *pre-genital* phenomenon. One points to the diminutive size of the protagonists, to the non-sexual nature of the relationships and to the association, for instance, with milk – the food of infants. Conversely, there is some evidence to suggest that the world of witches and wizards is *post*-genital. The normal-size, and overt sexuality, of the protagonists are pointers to this.

If one is prepared to look non-dismissively at the magic world, one is very much struck by its completeness – that is by its self-contained nature. It does not appear to feel the need of any of the highly-prized features or values of everyday Western society. That statement requires some amplification.

Certainly there are *equivalents* of, for example, money (gold and jewels), factories and mines (dwarfs working and toiling underground) and transport (fairy chariots, magical swans and such creatures). Do we not see, however, that these are symbolic forms – and that such normally allied matters as bank rates, safety regulations and driving licences are wholly irrelevant? The world of magic operates on different premises. As in dreams, what is there is there, and the way things are is the way things are. It is a wholly self-justifying world. The question which may quite legitimately be asked, however, especially since this world does not 'really' exist, is who sees it (or who first saw it), who describes it, and how are its characteristics agreed upon?[5]

One cannot at this stage of the argument provide a full answer to these questions. Instead the matter is approached somewhat obliquely. First we call to the witness-box an old nursery rhyme.

> What are little girls made of?
> Sugar and spice and all things nice . . .
> What are little boys made of?
> Frogs and snails and puppy-dogs' tails . . .

Of this we ask, which of the persons described most resembles a fairy and which a dwarf? The answer of course is that little girls (as described) are more like fairies, while the nastiness of the boys recalls the nastiness of the goblin brood.

There is a further similarity between fairies and little girls, and that is in their fondness for babies. Fairies are said to steal them, and there are few things little girls like better than dolls. Recalling what was said earlier, and with this additional piece of information in mind, one asks the question, if one *had* to choose one of the following as the author or authoress of the fairy world,

5. 'Agreed' is not to deny, of course, that there are differences between, say, the magical world of the ancient Egyptians and the magical world of medieval Europe. It is simply that these differences are wholly outweighed by the consistency of the principles involved.

which of these would one choose: a little boy; a little girl; a woman; or a man? The answer, on the basis of best fit, would be a little girl.

What is the relation of witches to babies? They also steal them. But they steal them in order to kill them or eat them. Witches are firmly anti-baby and anti-marriage. Who, in real life, tends to be more against marriage and babies, men or women?

Recalling also what was said earlier, and with this further snippet in mind, let us pose the question, if one *had* to choose the author or authoress of the witch-wizard world from the following, who would one choose: a little boy; a little girl; a woman; or a man? The answer, *faute de mieux*, has to be a man.[6]

Two points now. First, what does all this mean in more rational terms? To deal with this point one must raise another – are there actually two 'viewers' involved, or are these merely different aspects of one viewer?

Clearly, one is not really saying that, for instance, fairy-stories were actually invented by little girls. What one is saying is that the attributes of little girls – i.e. of pre-genital females – appear in fairy-stories to represent the most prized and 'effective' qualities. In the post-genital world of witches and wizards the most desirable and effective attributes appear to be those of the adult male.

6. If the phenomena of the two sections of the magic world are respectively the views of a little girl and those of a man, what becomes of the views of the little boy and the woman? The short answer is that in general the little boy goes along with the little girl's view; and the woman goes along with that of the man. Nevertheless, there *is* also a separate boy's view and a woman's view, though these are perhaps in something of a minor key.

In less patently anthropomorphic terms we have three interacting variables involved here. There is an age-size factor (young-small versus old-big), a gender factor (male versus female) and a sexuality factor (for versus against sex).

Table I at the end of Section 2 of this chapter considers all possible permutations of perceiver and perceived and presents some of the figures which result at each point. These figures are then considered further in a later chapter.

The answer to the second question – are we speaking of two personalities or of two aspects of one personality – is both.

Taking the single-personality question first, it would appear that the child is governed by female aspects of the personality – with the male aspects in abeyance and not prized. The hairiness and ugliness of the dwarf and the fact that he is kept out of sight shows the (pre-genital) female's distaste for the (potentially sexual) male. The figure of the giant perhaps expresses the young female's *fear* of this potential.

The adult is apparently more ruled by masculine aspects of the personality. *Now* it is the female who is held in distaste, who is ugly and deformed and banished to the nether regions. The male part of the personality has realized its sexual potential (and perhaps in some ways in the wizard even gone beyond that). The female is also now sexual. But whereas before sexual maturity the embodied female figure was free and in command, after sexual maturity, it seems, she becomes man's vassal. Is it then only sex which causes the female's downfall and through sex that she loses her independence?

The view that the single, full-term individual personality is constructed of two major parts or periods is, firstly, not merely in line with the speculations and findings of our first chapter, but that these parts are here named the 'female' and the 'male' is very much in line also with Jung's psychoanalytic theory. Jung's view is that every male has a female shadow side termed the *anima*, and that every female has a male shadow side termed the *animus*. The male, according to Jung, is consciously man but unconsciously woman, while the female is consciously woman and unconsciously man. Although the present book will not in fact be able to meet him wholly on this interpretation, Jung's view that the opposite facet of personality is unconscious accords very well with our observation that among the diminutive-magical and 'feminine' people, the *dwarfs* work underground, while among the full-size-magical and 'masculine' people the *witch* lives in the darkness of the forest.

If the young of our species – that is, if we as children – are governed by certain 'feminine' considerations and attitudes, can it be tentatively proposed that both early man and primitive man (leaving aside for the moment the mentally ill) are also in some sense governed or directed by these? And if the adult of the species is, as alleged, governed by a certain 'maleness', has this anything to tell us, for instance, of the nature of Western society? These speculations we shall leave for the moment in abeyance.

What of the second possibility, that in magic what we see are two quite different personalities? If this is the case, then it would begin to look as if the personalities of the male and female are radically different in some way or ways. For the female begins up but ends down: while the male begins down but finishes up. This neat schema will unfortunately prove a little too simple.

A possible derivation of the terms 'witch' and 'wizard' is from the Old English *witan*, meaning to know. In that case we would seem to have here a link with one of the themes of Chapter 1. Does the fairy who 'knows' become a witch, and does the dwarf who 'knows' become a wizard? If we take 'knowledge' to mean carnal knowledge this would take care of the witch rather neatly. For the male, however, one term will not do. It appears to be necessary to think of carnal knowledge on the one hand and academic knowledge on the other. The dwarf with carnal knowledge is perhaps the devil: the dwarf with academic knowledge is possibly the wizard. This division of knowledge into two parts (a step whose rightness will be confirmed from other directions) would enable us to understand both the sexuality of the Devil, and the relative a-sexuality of the wizard. It appears a possibility that woman is not able to achieve the divorce of sexual and a-sexual knowledge.

Magical and legendary material is certainly akin to that of dreams and in at least one major respect – in its almost inexhaustible compactness and richness. As we raise it to the light and inspect it, it reflects like a diamond, continually new insights and unsuspected depths. To repeat what we said earlier of dreams, no detail of a fairy-tale or legend is insignificant,

Moreover, no item would disgrace itself by having less than half a dozen meanings. This is, of course, in direct conflict with logical or scientific exposition, where the aim and wish is to have one precise meaning for each respective item. The multiplicity and deviousness of the dream, legend and fairy-tale is, however, in no sense a free-for-all, much less a meaningless free-for-all. As the internal structure and external cut of a diamond together determine the quality and kind of light emitted, so the (fixed) inner logic and (chosen) outer form of the legend combine to produce its discernible meaning. While the attempt to transpose symbols into a more rational coin is perhaps not itself entirely in error, one must never lose sight of the fact that the essential meaning *is* symbolic, and must in the final analysis be evaluated in these, and not in more narrowly logical terms.

With this we turn to one final aspect of the behaviour of magical people, that of flying. Our examination is for the moment solely expositional and explanation is deferred. However, Freud's view that flying in dreams always indicated sexual activity may usefully be recalled.

In the tradition we were discussing fairies fly and dwarfs, goblins and so on do not: of the two, fairies are the more 'magical'. The fairy carries a wand, while the dwarf (and incidentally the male fairy) does not. Loss of her wand by a fairy usually means loss of her ability to fly. The wand we may regard at its most basic level as the male organ – but in the present context it may be thought of, too, as a symbol of general sexuality. The wand, in the fairy's hand, is clearly under her control.

The witch has a broomstick which she places between her legs, and which enables her to fly vigorously. *This* symbolism will probably require no explanation! In one sense the broomstick is the witch's servant – but in another, perhaps, it is her master.[7] Unlike the fairy wand, the broomstick may be per-

7. One could enlarge on the witch's habitual, bad-tempered nastiness, and suggest that this is perhaps the result of sexual frustration – compare here the behaviour of the 'office spinster'. The witch, sexually rampant, is frustrated possibly in that she cannot get *enough* of the male principle.

sonified to the extent that it can speak and sometimes move of its own volition: personification, where such exists, is masculine.

The wizard has a wand or staff and he too can fly. This, however, is not the first behaviour one thinks of in connection with wizards. He seldom uses a broomstick but prefers, say, a steed that gallops like the wind, or simple magical appearing and disappearing. The best fliers of all (from another context admittedly) are angels – those a-sexual though basically male figures, who carry no wand. Not only are angels of course first-class fliers, but they actually *live* in the sky.

What can we make of these scattered observations? The dwarf does not fly at all; the wizard flies, or can fly, reasonably well; the angel flies very well. And as far as dwelling-places go, there is a similar progression – under the ground, on the mountain, in the sky. The fairy, with her wand, flies quite well. (Actually fairies flit as much as fly and tend perhaps not to venture too far from their grottoes or dwellings.) The witch, with her larger broomstick, flies very well – indeed, broomstick flying is perhaps her most typical characteristic. She will undertake long journeys – e.g., into the far Harz mountains to celebrate *Walpurgisnacht*. It seems, then, that the sexual witch flies better than the a- or anti-sexual fairy. While the wizard (with the sex symbol in his hand) flies less well and is also in other senses more earthbound than the completely a-sexual angel. Again we seem to have some kind of reversal. Sexuality enables the female to fly better (a kind of freedom in captivity) but the same sexuality prevents the male flying as well as he might.

With these rather puzzling observations, this brief review of the fairy world is completed. Those familiar with this field will consider the review inadequate if not actually wholly unfair. One has examined the specifically English tradition of folk-magic and in particular only the recent phase of that. At best it might be said that the tradition of Europe had been considered. And what one sees in Europe in recent times is in any case itself only the last degeneration of a once powerful, serious and sinister tradition. On the other hand, the decay of a once great or tragic theme to the level of farce is a process we have already observed elsewhere.

A further, more serious criticism is that of special or contrived pleading – not always easy to counter in the type of areas under discussion. The critic might point out, for example, that there are many similarities between fairies and angels and that no doubt some free-for-all rigmarole could be fairly readily devised to link them in the 'explanation' of some non-existent 'event'. Let us consider this submission.

It is, firstly, quite true that there are obvious similarities between these figures. It has already been pointed out that both are sexless or a-sexual and both fly. The *dis*similarities are nonetheless clear, persistent and significant. Firstly, the fairy flits rather than flies. She flies about the surface of the earth, at most into the hills. She is rarely seen in the sky. Her revelries take the form of dancing on the ground, usually in a clearing in the forest. One of her main tasks in the English tradition is to make plants and flowers grow. And while she is in a sense a-sexual, she is nevertheless essentially *female*.

The angel flies, soars or rises in the air. The sky is its natural habitat whence it comes and to which it returns. Angels are frequently identified with light and may be seen by day. The fairy, however, is a creature of the night and her luminosity is much rather that of the marsh-light, the glow-worm or the moon. The revelries of angels take the form of singing in the heavens. When *it* transgresses, it is said to *fall*. To where does it fall – of course to the Earth or through to the nether-regions. (Earth, as discussed in the previous chapter, represents and symbolizes the female element.) And finally, though in a sense a-sexual, the angel is nevertheless essentially *masculine*.

A question the reader might ponder again is how and where all these legends and stories begin, and more importantly, why they survive. To demonstrate here the extraordinary vitality and power of the material (and at the same time the hopeless inadequacy of the rationalistic explanations usually given) the following extract from the *Encyclopaedia Britannica* is offered, not in respect of a fairy-story, but in connection with the legend of Scylla and Charybdis.

Scylla, daughter of Phorcys and Cratais, is a supernatural creature with twelve feet and six heads on long, snaky necks, each head having a triple row of sharklike teeth, while her loins are girt with the heads of baying dogs. From her lair in a cave she devours whatever ventures within reach, and takes toll of six of Odysseus' companions. Charybdis, lurking under the fig-tree a bow-shot away on the opposite shore, drinks down and belches forth the waters thrice a day and is fatal to shipping. The ship-wrecked Odysseus barely escapes her clutches by clinging to a tree until the improvised raft which she has swallowed floats to the surface again after many hours. Both Scylla and Charybdis give poetic expression to the dangers confronting Greek mariners when they first ventured into the perilous and uncharted waters of the Western Mediterranean.

Can we really so lightly accept that a mariner, even a superstitious one, would perceive a sea-scape, though unknown, essentially no different from those of his home waters, as a six-headed woman with the heads of baying dogs about her loins? Does this horrendous vision really fall within the normal bounds of poetic muse? Or do we not, reading the above, feel the vigorous life of the legend and the triviality of the accompanying academic explanation?

The material of the section above, together with that of the first chapter, enables certain tentative conclusions to be drawn. These are as follows. All mythical creatures of the earth are female, or they are captive or fallen males. All mythical creatures of the air are male, or they are liberated-captive females. A further duo of conclusions may seem somewhat less justified at this stage. All creatures of the *night* are female and/or they are sexual.[8] All creatures of the *day* are masculine, or they are

8. Dwarfs, giants, brownies and so on are seen as at least *potentially* sexual, by reason for instance of their body-hair.

a-sexual. In support of these latest conclusions one might perhaps consider the domestic cat. Cats are night creatures, and as a species we regard them as female. The male cat as such is noted for his randiness.

The principal subject of this present section, the Devil, is a creature of considerable antiquity, especially if one considers also his forebears. Christianity, and related early religions and Christian heresies (Gnosticism, Manichaeism, etc.), arose within a context of pagan religions, in most if not all of which elements of nature-worship were to be found. Whenever a new religion ousted an older one, it was fairly standard for the 'God' of the old religion to become the 'devil' of the new one, for obvious reasons. One of the gods so ousted by Christianity in Europe was Pan.

In Pan one sees clearly the animal ancestry of our own later devil. This forerunner is truly half animal, being a goat from the waist down. Above the waist he is human, but sports two horns on his head, has also a goat-like beard and the pointed ears of that or some similar animal. Pan, as then conceived, was not so much evil as excessively randy, devious, mischievous and irresponsible.

Medieval pictures of our own devil often show him with cloven feet, a long, barbed tail, a beard and horns. These (together with the trident or spear which he sometimes carries) may be taken both literally as their animal parts and symbolically as the male appendage. He is also occasionally portrayed as other animals.

The two horns of the goat-Pan-Devil have particular significance. Today, in the form of two raised fingers, they represent still in England one of the strongest personal insults – while in Shakespeare's time two horns on the head indicated that a man was being deceived by his wife. More importantly, however, these horns probably represent the two-ness brought about by evil, as contrasted with the (original) one-ness of God. This idea has, of course, clear links with the subject-matter of our first chapter. Christianity, while admitting the existence of evil, saw it as an offshoot of God, *permitted* by him to exist, for his own purposes. The Gnostic heretics, on the other hand, as part of mainstream dualism, saw evil as *irreducibly separate* from good. This is principally why Gnosticism is considered to be a heresy, and the two-horn symbol a blasphemy.

Digressing briefly to the 'popular culture' of the twentieth century, as far as he can be said to exist there at all the Devil is reduced to a kind of unpleasant petty criminal, an outsider. Presented on stage or film he is urbane, perhaps in evening dress, but without of course any real breeding. The horns are vestigial or non-existent, but he retains a small beard or a pointed moustache. It is of interest that he is dark-haired. A blond devil is almost unthinkable. At this point he has merged with, or perhaps is, the *Villain*.[9] The Villain is seen at his most exaggerated in melodrama, although he pervades very much larger areas of our literature. This generalized character is the swine, the bastard, the playboy and so on.

It is very pertinent to ask at this point why, until very recently, the devil and the villain were almost invariably dark while the hero, conversely, usually fair. One *might* suggest that this is because Christianity and modern civilization are primarily a European, even a north European phenomenon (though this is not really true). It would then be in a sense natural to find undesirable traits assigned to 'inferior' types of peoples – namely the dark-skinned races of the South, of Africa and Latin America.[10]

9. See the extended versions of Table 1 (p. 61) on pp. 435 and 436.

10. One of the several weak points in this rational explanation is that for a number of negro and dark-skinned races the colour of religion is white. Such beliefs pre-date the actual contact with white peoples. One of the reasons why the white conquerer had sometimes so little difficulty invading newly-discovered countries was because the inhabitants thought the conquerors to be their gods come to earth. In Polynesia and other lands brides are sometimes kept for months in darkened rooms to make their skins lighter. Certainly, then, there is more to the association of whiteness or lightness with good, and black with the Devil, than meets the logical eye.

Whiteness appears also to be akin to some extent with smoothness. The

The Christian Devil, to revert to our general argument, has as is well known other, more noble, antecedents than those of Pan and the goat. In Christian mythology the Devil, as Lucifer the archangel, is at the outset virtual co-ruler of the universe together with God himself. Before the Fall, Lucifer was a being of great splendour and power. At first he was great even in defeat. (By the Middle Ages in Europe, however, he is a far meaner and more lowly figure. Possibly the last time he is perceived as having any real stature is in Milton's *Paradise Lost.* A mere hundred years later Goethe could only with difficulty maintain him as a supernatural figure at all.)

Linked surprisingly with this question of his mixed antecedents is that other curious feature of the Devil's make-up, his close association with knowledge, with wisdom and with intellect. This we consider below.

It may as well be suggested at this particular juncture, however, that 'knowledge' is really a blanket term having more than one component. The principal knowledge which the Devil purveys and is linked with, is not really academic-scientific knowledge. For, as we know, as long as both Faust and Jekyll concerned themselves with that they came to no harm, and met with no temptations. Nor, however, is the knowledge we are speaking of merely carnal or sexual knowledge – though that is a part of it. It is knowledge of the self – self-knowledge – which is really at issue here. This, certainly, is externalized or projected to the extent that it is spoken of as knowledge of mysteries, or in Christian jargon, of the Mystery. And undeniably, too, it tends to lead, one way or another, to sexuality, as we shall see. (Objective, scientific knowledge, however, does not.) Perhaps we should best think of a continuum with Illumination (self-knowledge) at one end and Sex (lewdness and depravity) at the

---

priests and initiates of religious orders from England to China shave their heads, as do nuns and the wives of orthodox Jews. Solomon thought the Queen of Sheba was a witch because she had hairy legs. Again, the devil has a beard.

other. Lucifer and the fallen angels had no difficulty in moving all the way from one extreme to the other (just as light can become darkness with no break of continuity). It is this second form of knowledge, which in later chapters will be referred to as Knowledge II, that is principally under discussion in the remainder of this section, and much less academic-scientific knowledge (or Knowledge I). The matter is actually still more complicated, but this will serve.

The devil's connection with forms of knowledge (Knowledge II, that is) reveals itself at every turn. We have mentioned already the probable derivation of the words 'witch' and 'wizard'. Magic and magician in turn are from *magus*, a wise man or priest. It will be recalled that three Magi followed the star to Christ's birthplace. During the Middle Ages one of the main reasons for raising a devil was to gain knowledge and information. In the Book of Genesis the serpent is described as 'more subtle [meaning wise] than any other creature', and it is with the promise of knowledge (specifically forbidden by God) that he tempts Eve.

Manichaeism and Gnosticism (from the Greek word *gnosis*, knowledge) were religious movements of late antiquity that preceded and for a time ran parallel with Christianity – to some extent fertilizing it. As its name implies, Gnosticism was very much concerned with knowledge and knowing. The aspect particularly relevant here is its fundamentally dualist position – the division of the universe into the two absolute, irreducible principles of good and evil. These two last terms are synonymous with light and darkness, spiritual and material. The world of matter (of darkness) belongs to the Devil and he alone is its ruler. Gnosticism teaches that this was his from the outset, while the similar dualism of the Christian mainstream states that the Devil *became* its ruler, under licence. This difference of opinion on one side, the terms Lord, King and Prince of Darkness are common to both religions, and the *de facto* ownership of the material world by the Devil is common to both.

In Manichaeism, as in Gnosticism (both important religious

movements), we have the following account. The Devil came by chance upon the boundary between the kingdoms of darkness and light – saw, desired and invaded the realm of light. At this point light was swallowed by matter, and light lost awareness of its own nature. Matter for its part now grew dependent on the swallowed light and sought to retain it.

What does this story recall – the theme of possession perhaps? Suppose one wrote instead: 'Jekyll at this point is swallowed by Hyde, who engulfs him, and Jekyll loses awareness of his own nature.' The style is not especially good, but the meaning holds up. Or suppose one wrote: 'The witch, for her part, now grows dependent on the swallowed sexual (or male) element and seeks to retain the broomstick.' The syntax has deteriorated further, but does not the sense remain possible? Finally, for good measure, how about: 'The brain of the travelling space scientist is swallowed at this point by the body of the terrestrial ape.'

These speculations at least suggest that if we could find the correct (universal) concepts, all or much of the material we are examining could be included in one theoretical framework.

Modern religion might be described as the story of man's attempts to attain to higher and purer things. One is reminded here of the space astronaut in the earlier outline who on leaving Earth's gravitational field (the female, sexual area) found himself in full possession of the mental faculties mankind yearns for. But in fact, as a species, man never does seem to reach escape velocity – he aims for the light, for free space, but only too often like Icarus falls back to Earth. Often, moreover, setting out in search of enlightenment, he disconcertingly finds sex.

In Gnosticism this transpired as follows. While some Gnostics, despising the world, withdrew from it, others, arguing that the material world was irrelevant and unimportant, threw themselves into sexual orgies and every excess. Sometimes the Devil was actually worshipped, since, it was argued, he is ruler of this world and in any case hardly inferior to God. It was held too that the Devil had done mankind a service by releasing it from the tyranny of God. Whereas under God we are forbidden to do certain things, under the Devil everything is allowed. At this point the boundary of full-fledged Satanism is reached – which exists in any case independently of the Gnostics.

An important feature of Satanism, from our present point of view, is that when God is rejected, all his commands and rituals are reversed. The *opposite* is elevated to a guiding principle. Thus the cross is hung upside down, the altar is black instead of white; in place of chastity there are sexual orgies; marriage and procreation are rejected and sodomy encouraged.[11]

One does not need to consider outright Satanism to show the sexual side of the Devil, and incidentally of magic. Orgasm on the part of the magician, for instance, frequently formed part of the ritual of raising the Devil, and dried or fresh genitals are a magic ingredient. On the symbolic side, the slaughter of animals such as a cock (also, incidentally, a slang word for penis) and other animals noted for their virility featured prominently. So also did blood, a powerful sexual symbol and 'releaser', in the psychological sense. To have sexual intercourse with the Devil himself was the high point and the aim of many ceremonies.

One cannot attempt here a full review of the *methodology* of magic. What can be said is that magic, like dreams, is essentially symbolic and associative, and contains a large element of wish-fulfilment.[12] Sticking pins into the wax model of an adversary is *like* sticking pins into his real body.[13] Believing or wishing an enemy to be dead – that is, acting towards him or a picture of him *as if* he were dead, will have the result of making him dead. If the desire is to produce a fever in the victim, the ingredients of the ceremony must themselves contain heat, and so on. A black

11. One of the points one is trying to bring out in these early chapters is the constant preoccupation in all ages and many different fields with *oppositeness*. We are not therefore always over-concerned with the precise nature of the opposite involved, though this aspect is, of course, also of great interest.

12. In the fairy-tale the giving and granting of wishes is entirely explicit.

13. In more recent times a psychic devised a machine which he claimed would diagnose and cure disease. Later it was announced that a drawing of the machine worked equally well.

cock both represents virility and symbolizes the dark side of the mind, just as a lamb or a white cock or a new-born child symbolizes innocence. The slaughter of these last is, then, the symbolic destruction of the good with a view to releasing evil.[14]

The central message of magic is that wishing will make it so. Only one must wish really hard. Failures are often interpreted to mean that one has not wished (raged, commanded) hard enough. The process is, for some reason, aided if there are material props for the magic to work on – such as a lock of the victim's hair. Two white mice and a pumpkin can serve as the basis for a coach and two white horses. Why magic often cannot or does not work on thin air is a problem one may reflect on.

For the moment it is pertinent to recall again that the other phenomena found to be based on associative thinking, symbolism and wish fulfilment were dreams and legends.

Table I, which now completes this section, attempts to produce actual fairy-tale and legendary (merging into literary) figures for each view of the personality from the standpoint of every other part, as we discussed them earlier. The variables involved are male/female, old/young and sexual/a-sexual. The table is both incomplete and tentative.

A few final observations on the relationship of magic to religion and mysticism must be made.

While the magician seeks to dominate, the religious adherent seeks to *be dominated*. One is of course not dealing here with absolutes – though one *is* dealing with clear differences of emphasis. There have certainly existed those who sought to spread the word of God with their strong right arm, and who were not in-

14. Possibly deeper layers of symbolism are also involved. The magic circle or pentagram may represent the vagina. The magician himself is then a symbolic penis. The protective qualities of the circle however suggest also rather the other end of the vagina – the womb (as does for that matter, the comfort and ease of the cave of the Lady Venus in which Tannhäuser sojourned).

TABLE I

As Perceived By		MALE: Non-Sexual Male	MALE: Sexual Male	FEMALE: Non-Sexual Female	FEMALE: Sexual Female
Female Child	Pre-Adult	MALE FAIRY	DWARF	FAIRY	NYMPH
	Adult	FAIRY KING	GIANT	FAIRY QUEEN	WITCH
Male Child	Pre-Adult	YOUNG PRINCE	? URCHIN	(FAIRY) PRINCESS	NYMPH
	Adult	KING: HERO I*	ROBBER-CHIEF HERO II*	QUEEN	WITCH
Female Adult	Pre-Adult	CHERUB	IMP/URCHIN	VIRGIN: LITTLE GIRL	"LITTLE MINX"
	Adult	GOD: ANGEL	DEVIL	VIRGIN	WITCH
Male Adult	Pre-Adult	WONDERFUL BOY†	IMP/URCHIN	VIRGIN: LITTLE GIRL	'NYMPHET'
	Adult	GOD: WIZARD WISE MAN	DEVIL	GODDESS: VIRGIN WISE WOMAN	WITCH: SORCERESS

Note: In the pairings each item is the opposite in a certain sense of the one next to it.
The items immediately above/below each other are the young and old versions of the same archetype (stereotype).
* I and II meaning roughly good/bad, a-sexual/sexual.
† Jesus Christ, Hercules, etc.
For further elucidation of this table and these figures, see Chapter 12 and Tables 11 and 12.

clined to suffer oppression. These are, however, a minority compared with the passive, sufferant majority of, say, Christians – the martyrs and saints, and above all Christ himself. (One recalls, for instance, how at the last in Gethsemane Christ refused to allow his disciples to defend him.) The essence of at least the Christian message is submission – to the will of God – humility and turning the other cheek. What is demanded of the Christian is the merging and submerging of the individual identity with the cause (of which Thomas More and Thomas à Becket provide moving examples).

The magician or wizard, on the other hand, strives to control and dominate. His spells and ceremonies are not designed to destroy the devil, as is the case with exorcism, but to bring him under control – and at the same time to prevent him from becoming the controller, even though this is an aspect of the affair inclined to go very much awry, as we have seen.[15]

A further major difference between religion and magic is seen in their attitudes to sex. The magician invites, welcomes and uses sex. The priest denies sex and sexuality. In magic the devil is ordered to come:[16] in religion one commands him to go.

One needs, however, to exercise care in generalizing about religion, for the very ancient, and all pagan, religions were eminently sexual. Traces of this persist in surviving (e.g. Indian) religions. One cannot in any case distinguish between ancient and less ancient religions on a simple time-basis, since the time-scale varies considerably from region to region. Religion as we know

15. In apparent contradiction to the foregoing, one does frequently find the magician wittingly delivering himself into the hand of the devil – as Faust did. This would appear to parallel the Christian act of submission, but in fact that is not the case. The magician delivers himself to the Grand Master in order to be a master in his own right. By accepting the supreme overlordship of the devil (in the hereafter, perhaps), the magician becomes the lord (here and now). Similarly, a butler does not mind taking orders from his employer as long as *he* can give orders to the housekeeper and the maids. There is, in other words, a dominance hierarchy.

16. Interestingly, the verb 'to come' is a widespread slang expression meaning to have an orgasm.

it in the West begins some five thousand years ago (though not, of course, in Europe). And it is really this 'modern' tradition we are speaking of at the moment.

In the Western religious tradition the denial of sex (and hence of the body) has many forms. Some are: covering the body with extensive, shapeless, often deliberately coarse, clothing; shaving the hair of the head; congregating in single-sex monasteries and nunneries (the high walls of which are defence ramparts); fasting generally and avoiding meat (which, aside from its symbolic significance, reduces the sexual tone of the body); taking specific vows of sexual abstinence; worshipping the Virgin and virginity; stoning to death those taken in adultery; maintaining that sex is solely for procreation and not for enjoyment; circumcision (a symbolic castration, whatever else, reserved for example in ancient Egypt solely for the priesthood); insisting on the celibacy and bachelor status of ordained priests; and the exclusion of women from the priesthood. There is little need to labour these points, so openly and in so many words does religion condemn sex: what one is inclined to overlook is the *profound* nature of the denial, forgetting the essentially sexual nature of the Devil, the *earthly* opponent of God.

This denial for the majority of individuals is of course certainly both a hopeless and tragic farce. There seems very little doubt that the torturing of heretics, the burning of witches, the persecution of the Jews and all features of this kind are a major outcome of sexual urges denied expression. Symbolic sex persists in any case in a hundred ways. So nuns, for example, are termed the brides of Christ – an admission, perhaps, that like the witch, they need the male in order to become airborne.

One of the important *similarities* between magic and religion is seen in their employment of *physical representation* in the ceremonial. (They have different aims in so doing, but let us ignore this for the moment.) If the magician requires the victim's nail-parings, or two white mice and a pumpkin, Catholicism requires wine for the blood of Christ and the host for his body. 'Relics' (of the true cross, of Christ's shroud, and so on) are highly prized. The laying on of hands, the kissing of the hand or foot, the portable cross and the rosary – all these physical addenda we may choose to regard, in essence, and in point of origin, as the child's touching of its mother, its longing for physical reassurance (as discussed in the last chapter) and its trust in the touchable and holdable. Touch casts out fear, as Harlow's young primates so clearly show us. Simple physical touching and being touched renews our strength.

This intrusion of the physical in what purports to be a wholly spiritual undertaking, should and would be puzzling, except that, as we have already suggested, a continuum is involved here – that spiritual illumination or experience is (somehow, and somewhat paradoxically) the other end of earthiness and sexuality. We find and shall continue to find this expressed in many ways – for example, in the saying that the gods are often found to have *feet* of clay.[17] Here the total continuum is in effect described: Earth – sky, physical – spiritual, even perhaps mortal – immortal.

These observations lead us to the mystical, ecstatic experience itself.

Abstinence from food, and social and sensory deprivation (such as practised in religious contexts) are today known methods of getting 'high', that is, of having among other things mystical-religious experiences – or, in psychological terms, experiencing hallucinations and self-generated percepts. The mystic-ecstatic experience has, at any rate, emerged from the body of all religions – the 'whirling dervishes' of Sufism, the Hatha Yoga of Hinduism, the Cabala of Judaism, the Christian mysticism of Thomas à Kempis. And personal experiences which remarkably resemble the mystic's vision can now be obtained by anyone from minute doses of mescalin and LSD.[18]

In all these and other experiences of an allied nature to be

17. Suggesting their being rooted in or originating out of Earth. Similarly God fashioned Adam in clay.

18. See, for example, Timothy Leary's *Politics of Ecstasy* or William Braden's *The Private Sea*.

discussed later, emphasis can be laid on the 'opening-out' or 'mind-expanding' quality of the experience, the central role of light (brilliance, colour, vividness) and the symbolism of ascension, moving upwards, climbing, soaring, and so forth.

These experiences and their expression we may describe with the conveniently double-edged term of 'high' mysticism. There exists also what can be termed 'low' mysticism. It must be understood that no value judgement of any kind is implied by this usage. The two respective terms do, however, have a symbolic, perhaps even a literal, positional significance relative one to the other. The Greek Mysteries may be taken as one instance of low mysticism.

These Mysteries were ceremonies of religious purification practised in ancient, originally pre-classical Greece, although they survived into the classical period. The purpose and content of the ceremonies were, as their name suggests, closely guarded secrets. So closely guarded in fact that little information about them is at present available. It is known, however, that the ceremonies were very ancient, pre-dating the existence of the Greek people, and connected with nature-worship and agricultural life. Similar, though perhaps less evolved, rites existed among early peoples in other parts of the world. The clear association with the Earth and with ancient female deities such as Isis enable us to link the Mysteries essentially with the 'female principle'.[19] No explicit, first-hand account of the Mysteries survives, though Themistius, writing in the fourth century A.D., tells us the following:

> First come wanderings and wearisome runnings about, and journeyings through darkness, full of suspicion and without purpose, then before the rite itself all terrors, shudderings and tremblings, sweat and wonder; but after that a marvellous light meets him . . .[20]

19. The terms 'male principle' and 'female principle' will be familiar to anyone who has acquaintance with mystical and esoteric literature. They are convenient, if question-begging terms. They may be understood more prosaically as personifications (of a kind) of the typically male and female attributes of the personality.

20. An essay, *On the Soul.*

Contemporary in time and place with the Mysteries are the Labyrinths and the Pyramids. A Labyrinth – the word itself is of unknown origin – is an extensive building, wholly or partly underground, made up of intricate passageways and numerous chambers. As with a maze, which in fact it is, unaided exit was all but impossible once the structure was entered. The purpose of these structures is largely unknown. They may have been intended for religious ceremonies, for burials or as secret meeting places. The best known – though legendary – the Labyrinth of Knossos, housed at its centre the Minotaur which was eventually slain by Theseus.

Of great interest is the fact that drawings of labyrinths are frequently found on the floors of medieval cathedrals. No satisfactory explanation for this phenomenon exists.[21]

Pyramids as such need little introduction, except to point out that they are found in many other countries besides Egypt, including Greece. These have at least this much in common with labyrinths, that they have internal passageways, often running below ground.

Historians and archaeologists do not appear ready to argue any direct connection between the Labyrinths and the Mysteries. And yet Themistius' account of the rites strongly suggests a man running in an underground maze. That account is also not totally unlike a description of the trauma of labour and birth, the emergence from the womb into the light.

Whether there were any literal links between the Labyrinth, the Pyramid and the Mysteries, the psychological links seem obvious. Does not the account of Themistius remind one also, for instance, of a bad LSD trip – a trip, that is, with the horrors? It is at least possible that the Labyrinth and the Pyramid are the concrete, embodied expressions of some perceived psychological truth or experience. The journey of the Mysteries perhaps was in the mind only. The view gains from the fact that the Labyrinth of Knossos probably never existed. Moreover, the half animal,

21. One may venture to suggest that it has something in common with gods that have feet of clay.

mythical creature, who lived at its centre, had the two-horned head of a bull, for it was the offspring of a woman and a real bull. These are symbols we have met frequently enough before.

On a vaster scale the Mysteries and the Labyrinth are paralleled on the much larger canvas of the underworld, Hades, into which Odysseus, Perseus and other Greek heroes descended to perform their legendary deeds. (The journey *downwards* into a dark, enclosed region is incidentally the precise *opposite* of the trip procured by taking mescalin and in high mysticism.) Those stories also tell us that the journey to the underworld is fraught with dangers. But it appears to be in some sense a necessary, even desirable journey and the merit of one who returns successfully is great.

The drawings of Labyrinths on the floors of medieval cathedrals, and the mazes in the gardens of courts and palaces, seem, then, the dim memories of these once familiar regions – still at that time remembered as having had significance or having existed — but with the keys to the significance and the symbolism lost. Thus the Pyramids, the Labyrinths and other structures of ancient times may be the projection of an inwardly perceived psychological truth (or condition), one that perhaps was already old when Egypt herself was new, already at *that* time recalled rather than lived.

In ancient Egypt and later in classical Greece there is some evidence that an ancient Earth-religion and goddess-worship was in the process of being replaced by a sky-religion and god-worship.[22] In Egypt the prehistoric worship of the goddess Nuit – who then subsequently merges with Isis – is replaced by that of Ra (the sun-god and the son of Nuit), later again by that of Osiris (the grandson of Ra). In classical Greece Zeus is worshipped. But he is the grandson of Gaea – the 'deep-breasted Earth' whom Homer calls 'the universal mother, firmly founded, the oldest of divinities'. In order to become the ruler of the

22. Robert Graves, among others, appears in principle to support such a view – for instance in his *Introduction* to the *Larousse Encyclopedia of Mythology*.

universe Zeus has been obliged to defeat first the Titans (his own parents) and then the Giants – all of these being children again of Gaea. They, when defeated, were – significantly – cast deep underground.[23]

The worship of Isis (the female principle) does continue in Greece. But this worship is now the clandestine, underground religion of the Mysteries.[24]

The resemblance of a labyrinth to the intestines and organs of the human stomach – if we may now suggest this – may be merely coincidental, or at any rate unimportant. The broad-based, dumpy shape of the pyramid does, perhaps, resemble a female form, especially that of a pregnant woman. Certainly the Pharaohs and other dead (in being entombed within them) were being placed back into the womb to await re-birth. This was, it happens, one of the central tenets of Egyptian religion.

We have already proposed some kind of (psychological) continuity from low to high mysticism, and suggested a general shift through (evolutionary?) time from Earth-religion to sky-religion (at least in Europe and the Middle East). Can we additionally find any evidence of an *architectural* continuity, or shift, or whatever else?

The architectural forms of low mysticism (the Labyrinth and the Pyramid) are wholly or partly underground. The pyramid is large rather than high, when the ratio of height to base is considered.[25] The church, temple and cathedral of high mysticism are not only above ground, but strike up into the sky. The ancient structures extend laterally rather than vertically, and are ponderous and heavy. Those of high mysticism extend vertically rather

23. '. . . among impenetrable shadows and foul vapours, at the very end of the world, the Titans, by the will of the king of heaven, are buried'.

24. The otherwise myriad gods and legendary figures of ancient Egypt and Greece begin perhaps to assume a certain coherence and intelligibility when they are considered as multiple expressions of a protracted struggle for dominance between, what for want of other terms, we call for the moment the female and male, the unconscious and conscious, aspects of the personality.

25. cf. the Great Pyramid, base 765′ × 765′, height 481′.

than laterally, and can be miracles of lightness and grace – in for example the Gothic cathedral. The ancient structures are designed to exclude light, those of the later period to admit it. The former have no windows, in the latter these are their crowning magnificence. While the former are outside towns, often in secret places, the latter usually form the town centre.

So opposite, indeed, are these two classes of structure that one might be forgiven for seeing no connection between them. But while the spire of the cathedral reaches into the sky, down in the relative darkness on the floor, a Labyrinth is drawn.

Throughout this chapter we have struck examples of opposites which have been shown on examination to derive one from the other, or to be linked one to the other. Examination reveals the traces of the self-opposite. Thus the Prince of Darkness, Lucifer, is by name the Prince of Light – from Latin *lux*, *lucis*. Thus the spirituality of the Catholic Mass rests in part on a basic mechanism of physical survival, touch – to say nothing of that ceremony's equally basic 'oral' aspects.

A somewhat clearer understanding of this general phenomenon will perhaps be achieved in the course of the next two chapters.

To end now on the question, how does the womb of the labyrinth become the penis of the cathedral, would be perhaps a sufficiently arresting thought. We shall, however, suggest finally a far more disconcerting interpretation of those two structures.

Part Two:

Outlines of Consciousness

Polarity in Language: Connotation and Denotation 3

Conceptually at least it is possible to think of the universe as a gigantic Rohrschach ink-blot, a celestial Thematic Apperception Test, on to which we, mankind, project and thence perceive our own internal psychological processes.[1] This is especially true when we examine the universe for explanations: it is also true, in a somewhat different sense, when we search for so-called objective data.

Expressing the same concept in another way, it might be said that man inevitably, and largely without being aware of it, leaves his psychological fingerprints on everything that he touches. Regarding the past, that is, any kind at all of historical material, we should be able to find what are in effect the fossilized impressions of the mental fingerprints of long dead man.

In the earlier chapters passing reference has already been made to language and the history of particular words. We can now say that language is in a sense the richest of all the fossil-deposits available to us – for although it lives and changes, it

1. The Rohrschach 'ink-blots' are more or less just that – a series of bilaterally symmetrical but otherwise unstructured patterns or figures, such as are obtained when ink is spilled on the open pages of a book and the book then closed. A Thematic Apperception Test involves the use of pictures of individuals ambiguously engaged in non-stated tasks, or somehow dramatically but again ambiguously juxtaposed with other individuals. The subject is asked by a psychiatrist or a psychologist to say what he or she 'sees' in the blots and pictures. In fact they contain little or no actual information. All the subject sees is what he introduces to the situation from his own mind.

carries within it the history of its development. Certainly some features are now only vestigial, like the gills and tail of the developing embryo. But with suitably reinforced and directed vision, one still can detect them.

The above carries the implication that language is essentially an organic phenomenon, that is, a living and growing organism. This view, one should say, does not accord particularly well with the view of language taken by many psychologists at the present time – that it is essentially a stimulus-response, mechanistic phenomenon. I do not, incidentally, seek to deny the usefulness or the value of such conceptions – merely to indicate their limitations.[2]

Turning to less general considerations, the present availability to us of once existing forms and uses of language, now superseded, is on the one hand limited by the length of time *written* language has been in existence – a period of a few thousand years only at most, in many cases less. Fortunately, however, there does exist a method by which we can travel a good deal further back. Where two or more (modern) languages have had, or are suspected to have had, a common origin one can, by examining the present and as far as available past forms of these tongues, and by extending any convergent forms back to their hypothetical meeting point, discover the probable forms and characteristics of language in existence before the development of writing.

It is, for example, very widely accepted that the majority of languages in Europe, India and parts of the Middle East have a common origin. The re-constructed original parent language is usually termed Indo-European. It embraces, among many others, English, French, German, Latin, Russian, Greek, Persian, Sanskrit and Gaelic. It is this Indo-European family of languages which we shall be principally concerned with here, unless otherwise stated.

Of considerable interest is the undisputed fact of number in

2. An outline of stimulus-response and conditioning theories is attempted in Chapter 5.

the parent language.[3] While modern European languages show only two forms (singular and plural), their older counterpart showed three – singular, dual and plural. That is, Indo-European had a special form of the plural denoting two. Does this have any possible connection with the two-ness we have observed at so many points? The special dual form is also found in at least one non-Indo-European language – Hebrew.

Freud and other psychoanalysts, as well as many philologists, have proposed that in ancient languages, therefore also in Indo-European, the same word was often used to express both the thing and its opposite.[4] Freud terms this 'the antithetical sense of primal words'. A few examples follow.

Latin *altus* = high: deep
Latin *sacer* = sacred: accursed
Latin *hostis* = guest: enemy [As in English host/hostile.]
English *cleave* = to split: to adhere to
English *with* = together: against [As, respectively, in 'join with' and 'fight with'.]
Old Egyptian *ken* = strong: weak
German *Loch* = hole, English *lock* = to shut.

Objections are as follows. Latin *hostis* means originally an enemy or stranger; from stranger the meaning of guest derives. Similarly Latin *sacer* means initially only sacred, accursed being a secondary and later meaning. Finally, the most glaring 'error'. Cleave meaning to stick and cleave meaning to separate are from

3. The question of the use of gender in Indo-European is also interesting, but somewhat complex. It is, for example, in dispute whether there were three or only two genders. The oldest known Indo-European language, Hittite, had only two. For reasons outlined in Chapter 1 I would be inclined to opt for two. However, certain *African* languages, for instance, have a large number of gender-classes, so that one must beware of special pleading. What is perhaps of interest is that we still today apply the term 'she' to some neutral objects, such as ships and aeroplanes. Freud has pointed out that the majority of these have hollow centres – that is, a notional womb.

4. Sigmund Freud, *Introductory Lectures on Psycho-Analysis*, pp. 150–52.

two different roots, Anglo-Saxon *clifian* (or *cleofian*) and *cleofan*. W. W. Skeat (*An Etymological Dictionary of the English Language*) states flatly that all attempts to link these words in a common origin are 'fanciful'.

These and similar objections would appear to make serious holes in the basic proposition, but this is not the case. First, it remains true that the use of one word for two antithetical conditions or actions is more common in ancient languages than in modern languages. Second, there are no grounds for demanding that both parts of the antithesis should come into being at the same time, or even exist alongside each other. Some of the examples show this most interesting and unaccountable development of, or changeover to, their own opposite within relatively recent historical times – the word *with*, for example. That the process still occurs suggests that whatever principle or logic originally impelled or motivated the containing of opposite qualities within one expression, is still operative today – though perhaps for a variety of reasons (including the 'freezing' of language into standard, rationalized forms – an outcome of the emergence of the written word) less able to make itself felt. The falling together of the two words *clifian* and *cleofan* into one form is particularly surprising. Would one not expect two words of opposite meaning and different origin to diverge rather than converge, since clearly the use of the same word must produce more, not less, confusion?[5]

5. After writing this section I came across the following passage in Otto Jesperson's *Language: Its Nature, Development and Origin.*

'Very small children will often say *up* both when they want to be taken up and when they want to be put down on the floor. . . . In the same way a German child used *Hut auf* for having the hat taken off as well as put on. . . . But even with somewhat more advanced children there are curious confusions.

Hilary M. (2.0 years) is completely baffled by words of opposite meaning. She will say, "Daddy my pinny is too *hot*; I must warm it at the fire." She goes to the fire and comes back saying, "That's better; it's quite *cool* now." (The same confusion of *hot* and *cold* was also reported in the case of one

Aside from these considerations, the movement from a given position to a diametric opposite, with often the traces of the original position still visible, is an event we have already frequently observed elsewhere. What we found in the last chapter on the conceptual level appears to operate and to have operated on the purely verbal level also. May one at this stage accept that some kind of link is felt, apparently by a majority of people, between at least *some* opposites, and that this finds its expression in the circumstances we have been examining? This link might take the form of some actual (possibly physiological) 'oscillation'. Freud emphatically and frequently expressed the view that 'opposites *in the unconscious* [my italics] are close together'. It will be seen in the next section that an interesting beginning towards a structure of personality can be made on these very general premises. For the moment we leave this possibility, pointing out only that *with* meaning 'with' and *with* meaning 'against' can perhaps without too much effort of the imagination be converted into the emotions of love and hate.

It seems that there are a number of good reasons for believing that *origins* of language are connected with the female, or female *side*, of personality, not with the male. As we shall see in the further course of events, there are also good reasons for suggesting that language steadily *becomes* more the ultimate province of the 'male principle'. For the moment let us ponder the evidence from developmental psychology, that girls, compared with boys,

---

Danish and one German child.) . . . She confuses *good* and *naughty* completely. Tony F. (2.5 years) says, "Turn the dark out".'

After reading this I then approached married friends with young children on the issue. They, far from showing surprise, offered me further instances of their own.

This evidence appears to place the phenomenon under discussion beyond the realm of speculation. What is of the greatest interest is that we seem to have here the duplication, in the early stages of language learning among modern children, of an apparently once common feature of language among early man. Is this again a case of recapitulation, this time in language, of an earlier phase of *psychological* development? And if so must this not then have a biological – that is, an inherited – basis?

talk earlier; talk more; read earlier; read more; show fewer defects (all types) of speech and reading; and learn foreign languages better, both as children and adults – in short, excel in all forms of verbal fluency over males (but *not*, however, in comprehension).

What is certainly true is that as we proceed back through the development of our language we find a greater 'emotional' involvement with it.[6] This is seen, for example, in alchemy. In that branch of study gold and silver, for instance, were described as *noble* metals and lead, copper and so on as *base* metals. These are quite clearly what we term today value judgements, and demonstrate only too obviously the subjective frame of mind which the alchemist brought to his work. Under such a weight of prejudice, it was unlikely that any objective truth could or would emerge.[7]

As is often noted with at least some truth, the English speak Anglo-Saxon and write Latin. This has to do, unequally, with Roman civilization, the Norman invasion and the use of Latin as the language of the early and medieval church, as well as of

6. One should, however, disabuse oneself of any idea that, in particular, the parent Indo-European language was either rudimentary or unformed. It had at least seven grammatical cases (Latin has six), three numbers (singular, plural and dual), and either two or three genders. Each of these sixty-three possibilities could show a different form in noun, adjective and pronoun. Yet the language was not written down. Some idea of what is involved may be gathered from a consideration of the mere *eleven* forms of the adjective 'glad' in Old English: *glæd*, *glædre*, *glædne*, *glædra*, *gladu*, *glades*, *gladum*, *glade*, *gladena*, *glad*, *gladan*.

It is salutary to wonder what these people, with a highly sophisticated language, talked about through the millennia preceding not merely the advent of science or technology but even of a civilized way of life, as we understand that term.

7. I would suggest in fact that one major cause of the full emergence of science and the scientific method in the West – more specifically in England and America – and not in the East, where conditions were on more than one occasion favourable to its inception, was the circumstance of the English tongue being double-tiered with a scholarly, precise, almost ossified language superimposed on the living, everyday one.

education. The Normans who invaded England in 1066 became the aristocracy, the conquered Anglo-Saxons the peasantry. Between these groups there was little friendship and less intermarriage for several generations. The situation is commemorated in the fact that we today say sheep and cow (German *Schaf* and *Kuh*) because Anglo-Saxons tended the animals, but mutton (French *mouton*) and beef (French *bœuf*) because the Norman overlords ate them. The persistent enmity between the French-speaking, Latin-learning aristocracy and the body of the people (the peasantry) actually permitted Anglo-Saxon to survive.

While words like beef and mutton eventually became established in all classes, many words of Latin/French origin, particularly the former, remain even today 'meaningless' in the sense that they sprang from foreign soil. These, moreover, are not learned at the mother's breast, but in the course of formal education at a later age. For both these reasons they are 'non-emotional', or 'non-vital'.

A glance at modern scientific or 'formal' language in German, which does not have a Latin overlay, shows the words to be largely constructed from native roots – and hence meaningful, or at least familiar, to the average citizen. The literal root-meanings of the words which follow are given in brackets.

German	*English*
werfen	to throw (Anglo-Saxon)
Entwurf	a project (Lat.)
(literally, a 'throw off')	(literally, a 'throw forward')
stellen	to place, put
nachstellen	to postpone (Lat.)
(literally, to place after)	(literally, to place after)
rechnen	to reckon (Anglo-Saxon)
berechnen	to calculate (Lat.)
(literally, to 'be-reckon')	(literally, to count with small pebbles)

The distancing achieved by the use of Latin roots is self-evident. People habitually use them when discharging a difficult or unpleasant duty, or when 'putting someone in his place' – that is, reminding him that he is an Anglo-Saxon and you are a Norman. In moments of strong emotion, or when talking to children, we revert to Anglo-Saxon.

The literal rendering of German words into the nearest Anglo-Saxon roots has a curious effect – e.g. Wagenführer = car-driver = waggon-farer – as do 'country' words: scranlet, mang-fodder, girty-milk. Stella Gibbons has used such language to striking, though, of course, ironic effect in her novel *Cold Comfort Farm* ('The ice-cascade of the wind leapt over him as he guided the plough over the flinty runnels/His thoughts swirled like a beck in spate behind the sodden grey furrows of his face'). Anglo-Saxon perhaps affects us in this way because it is a language of the Earth. (Does language, then, have feet of clay?[8])

As a counter-coin to these observations, the sharp increase in the use of words of Latin origin in public life in recent years is, I suggest, of considerable significance, and a matter to be raised again.

That language can and does convey something other than direct information is shown by the fact that magic, as well as religion – Catholicism, Judaism – have until very recently used ancient, especially ancient foreign, languages in their rituals. There is more than one reason for this, but one of them is that they certainly have both a greater and a different effect on the listener than does everyday, prosaic speech. Magic even employs nonsense words and incantations, though these are usually in

8. Apart from more obvious factors, the evocative influence of ancient language may somehow arise from the fact that it is the product of an earlier evolutionary stage. It is likely that every product of a time in some way bears the stamp of that time. This, if true, is also true of the form, sound, structure and content of language. These may arouse in us the 'memory' of earlier or more primitive stages of development. It is, however, extremely difficult to see how this could work in *physiological* terms, which is what the view demands.

fact shown to be the distorted remnants of old languages. Clearly, however, the purpose of language in these areas is not to convey information. What then *does* it convey?

Earlier it was noted that a non-emotional vocabulary or language is a great help in the objective description of phenomena. The failures of the alchemists probably resulted partly from the absence of such a vocabulary. In his book *Straight and Crooked Thinking*, R. H. Thouless has described many features of emotional language and thought, and shown how subtly (let alone more obviously) these can distort logical deduction and produce invalid conclusions.

What Thouless in part discusses are the denotative and connotative aspects of language. The word 'swastika', for example, *de*notes nothing more than a pattern of a particular shape. It *con*notes – to us – cruelty, totalitarianism and Nazi Germany. A tribesman from central South America would experience none of these feelings on seeing this symbol nor, of course, on being told its name. As a further example, the word 'home' denotes where an individual lives more or less permanently. It connotes warmth, comfort and acceptance. There is, however, nothing in the word logically or etymologically speaking which supports those assumptions. When one says that home connotes acceptance, one is saying that one personally *associates* it with acceptance. Connotation is essentially associative. Denotation is disjunctive, and lays down precise delimitations.

The function of language in magical and religious ritual is not to convey information. Its function there appears two-fold: (1) to arouse emotion (2) to convey *experience*. The connotative aspects of language seem to enshrine both our personal experience and our emotional attitudes – while the conveyance of information is more properly the function of denotation. Connotative or 'emotional' language somehow seems to cause us to participate in or re-live an experience.

I shall not attempt to define 'experience' at present, but the distinction made will prove a useful one. It will be seen in due course (for example) that the psychotic patient converts experience into information, while the neurotic, conversely, converts information into experience.[9]

We note then that language appears to have *two* principal forms, the connotative (subjective or emotional) and the denotative (objective or rational). This statement leads us to a more precise forward step in the central argument of this book.

In ancient languages and the language of children we find traces of a practice which uses one and the same term to express both a quality and its opposite. In modern languages, and as adults, we use not one, but two terms to express such antitheses.

When pairs of opposite terms in modern English are examined, one finds that these tend to fall into relatively few conceptual groupings. In the list of paired opposites which follows (and which is in no sense to be considered exhaustive) I have tried to restrict myself to items likely to arouse little dispute. One says 'little dispute' and not 'no dispute' because in fact not all the inclusions are equally straightforward. For instance, while 'science' and 'art', or 'logic' and 'magic', are not uncomplicated opposites like objective and subjective, or thought and feeling, they do possess a generally antithetical sense – a sense, admittedly, not easy to define precisely. One would, therefore, enter a plea at this stage for a measure of licence on this issue, on the promise of further definition of the position adopted at a later stage, when we have considered other evidence.

A	*B*
objective	subjective
impersonal	personal
thought	emotion
thinking	feeling
rational	irrational

9. For definitions of the terms neurotic and psychotic, and a detailed consideration of the mental states associated with them, see Chapter 5.

A	*B*
fact	faith
proof	belief
logic	magic ('illogic')
science	{ religion { art
detachment	involvement
classical	romantic
denotative	connotative
discrete	associative

The terms within each sub-list are not, nor are they intended to be, synonyms – though, clearly, some are (objective and impersonal, for example). But on the other hand, is there not *some kind* of within-sub-list agreement or coherence? Is not objectivity somehow related to proof, just as subjectivity somehow has to do with feeling, and feeling in turn with irrationality?

Not to make a long story of this, the proposal is that what one is apprehending as one reads each sub-list are the vague or fragmentary outlines of two clusters of somehow related phenomena – perhaps two broad areas of the human personality or of human experience.

There is a diversion for children, which involves joining up in sequence an apparently random scatter of dots on a page. When this is done a face or some other picture emerges. If a similar exercise is performed here metaphorically with these pairs, and a large number of others which could be produced, what then emerges can be described as the outlines of two continents. One of these, continent B, is seen at once to have close associations with what Freud and the psychoanalysts term the unconscious – an obscure, shadowed continent (a continent indeed already extensively plundered by the analysts, which may yet, however, contain undiscovered treasures).

Does the material we produced in Chapters 1 and 2 in any way align with the pair-lists A and B? Let us attempt such alignment.

A	*B*
day	night
sky	earth
smooth	hairy
a-sexual	sexual
light	dark
angel	devil

These concepts or attributes should now henceforth be considered to have been added to the previous list of pairs.

Yet at first glance some of these make strange bed-fellows for some of the earlier terms. However, if one were obliged to make a choice (and I insist that we do), that is, were obliged to say whether we thought classical form was smooth or hairy, would we not, grudgingly, choose smooth? And if one had to make the same choice in respect of romantic, would one not (thinking perhaps of poets) settle for hairy? These connections will seem, perhaps, tenuous. They may become less so in due course.

A further list of paired opposites now follows which will be seen fairly readily to differ from the paired terms already given above – although some cause for confusion certainly exists. For reasons which will be discussed, the new sub-lists are headed B1 and B2.

B1	*B2*
good	bad
kind	cruel
white	black
love	hate / fear
passive	aggressive / active
God	Devil
relaxed	tense
warm	cold
affectionate	rejecting
optimistic	pessimistic

The general comments made about list A–B also apply here,

that is, the within-sub-list items are by no means synonymous, but yet have certain affinities with each other. Thus if one *had* to decide whether God was tense or relaxed, one would opt for relaxed – while no doubt protesting that the proposition was ridiculous.

Apparent and rather glaring contradictions in these lists will, however, have been noted. We have angel: devil in list A-B, and good: evil, God: Devil in list B1-B2. Surely these cover the same ground? Similarly we have light: dark under A-B, and white: black under B1-B2.

It is not easy to clear up these difficulties at this stage. The concepts we need to do so are precisely what the present book is attempting to produce. However, it will help first to recall the dual nature of the Devil discussed briefly in Chapter 2, and the dual nature of knowledge also. One may imagine that a similar duality exists in the concept of God – consider, for example, the two statements 'God is Love' and 'God is Truth'. To some extent purely semantic difficulties are involved here. As we shall see in due course, two terms are required for a number of what are actually dual concepts, but for which at the moment there exists only one term in each case. More importantly and more relevantly, it must be pointed out that when speaking of such matters as gods and devils we are dealing essentially with *symbols* and *symbolism*. As in dreams, there is nothing to prevent a symbol or content having more than one aspect, or a contradictory make-up of a kind which the rational mind could not and will not accommodate. While it is one thing to translate symbols into the language of common sense for the purposes of discussion (something which this book does continually), this does not cause the symbols either to disappear or lose their inalienable rights as symbols.[10]

10. Even so, these considerations do not provide the answer here. While the explanation of the contradictions in question emerges only later in the book, one might consider for the moment the following – which, let it however be clear, is not the explanation – that if two unimpeachable witnesses each swore that they had seen the same individual at the same time in two entirely different places, the solution might be identical twins.

As far as the light: dark, white: black enigma is specifically concerned, as was pointed out in the previous chapter, the light of list A is essentially that of the day and of sunlight. The white of B1 is variously the pallor of fear, the white of *non*-sexual arousal (for the colour of sexual arousal, like that of rage, is red), or the paradoxical whiteness of creatures that live in total darkness, the pale, silver-white of the moon, or the electrical glow of the firefly, of plankton and of the deep-sea fish.[11]

The headings of the later list, B1 and B2, will have revealed the relationship of list A-B, or rather of list B alone, to list B1-B2. The opposites of B1-B2 are contained within the terms of list B. For example, religion (list B) 'contains' both God and the Devil. Emotion contains both love and hate. One may express the position in diagram form, as follows.

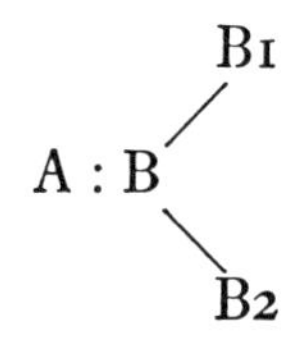

The diagram may be itemized:

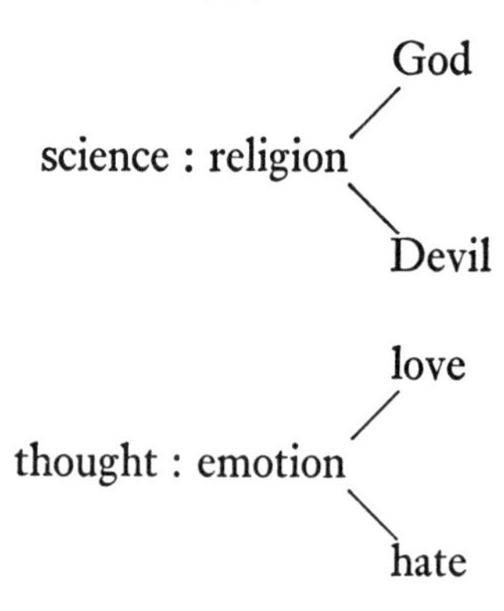

11. I feel occasionally as if I should apologize for becoming poetic – such a low estate has poetry reached in our culture. Nevertheless, it is *only* in 'poetic' terms that some of the phenomena we are examining may be fully understood.

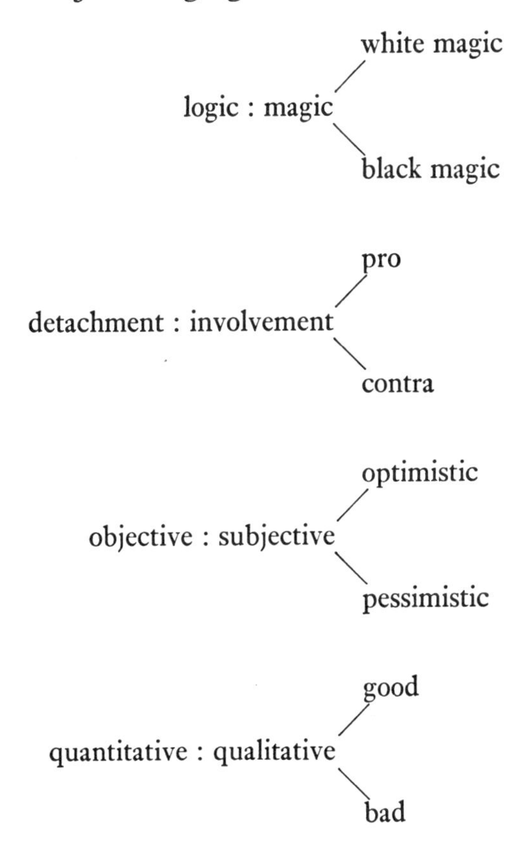

Naturally, one could spill seas of ink over the definitions of the terms involved. That, however, is precisely what we are not going to do. Nonetheless, one can agree that some of the terms in the earlier lists would not sit as easily in this form – art, for example. The fact of the matter is that a further dimension still is needed to accommodate certain aspects of these words and concepts, and this will be among the subjects of later chapters.

It may not have gone unnoticed that the terms male: female have not so far been placed. The fact is that these belong in both major divisions. The resultant 'paradox' is:

A	*B*	*B1*	*B2*
male	female	female	male

yielding:

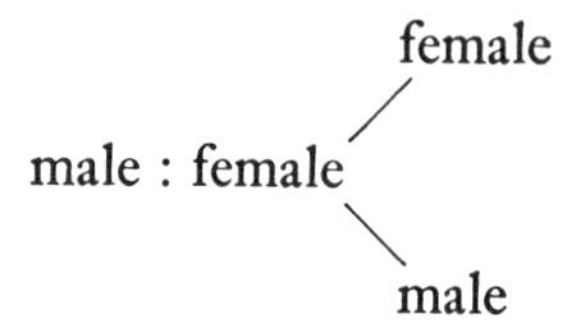

There is, however, no contradiction although, certainly, possibilities for confusion exist. Recalling some of the propositions in Chapter 2, the expanded paradigm becomes:

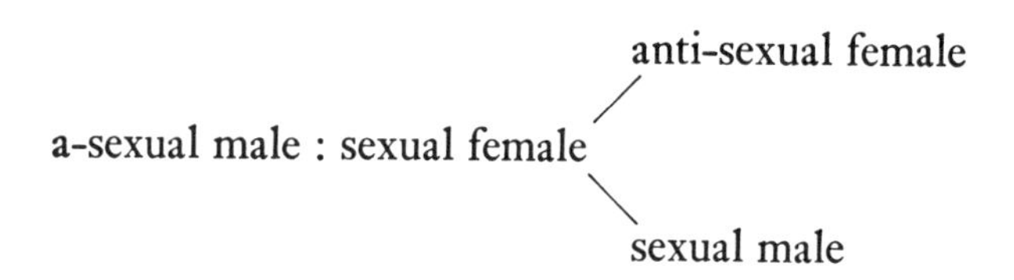

The terms *a*-sexual male and *anti*-sexual female are used deliberately. It is now suggested that the female (the 'female principle') can never be anything else but sexual. (The word 'sexual' is from some points of view an unfortunate term, but one which we are obliged to continue to use for the moment.) This principle can deny sexuality, but it cannot be *outside* it or *not part* of it. Only the male (the 'male principle') can be *a*-sexual.

Women readers may feel sardonically that the idea that men are a-sexual is news to them.[12] Part of the trouble here once again is that we are speaking at one and the same time of symbolic and of literal values, of the world of symbolism and the world of every day. This present area in particular does not allow of the easy separation of these components.

Let us attempt, nevertheless, a more purely symbolic pattern

12. Nonetheless it is a fact that the peak of sexual activity for males is in the late teens, and declines relatively thereafter. The consensus on women, although the data here is under various pressures, is that female sexual activity reaches its peak in the late twenties, *at which level it remains into the fifties.* This last statement is of considerable interest as regards the witch.

or paradigm. In so doing we bring out a further problem. *All* the personality elements under discussion here are present in the personality of every single human individual, whether male or female. The *emphasis* on particular elements tends of course – but only tends – to be rather differently placed in the human male and female. When therefore one speaks of the female aspects of personality one is not at all suggesting that these elements are absent in the male. But the expression (and experience) of those same elements in the male is not identical with the expression of those elements in the female. There is thus 'male femaleness' and 'female femaleness' – and of course 'female maleness', etc. There are no absolute values for these various components. Their value is moreover relative to, and a function of, the total matrix of inter-relating events and values which make up the complete personality of the individual.[13]

With this further addition to our problems the following alternative is proposed to the 'sex' paradigm above.

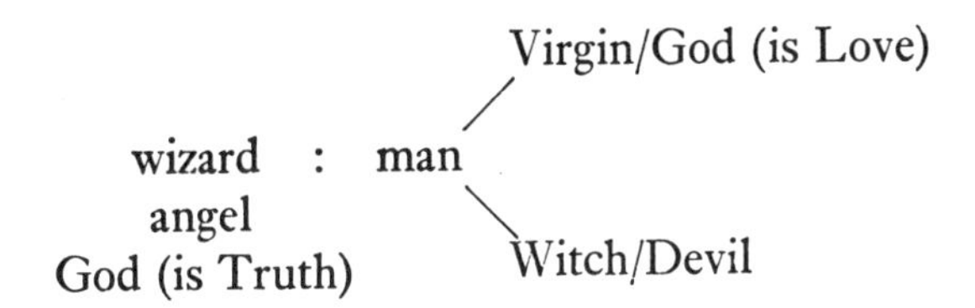

This version usefully underlines the a-sexuality of the figures on the far left. They are singletons only, not pairs, because they are outside sex and not therefore divided into 'male' and 'female'. *Man* is both (and separately) male and female. A single term is used in order not to over-complicate the diagram, and should be understood to stand for mortal male/mortal female. (The 'mortal' represents the human being living in the narrowest band of conventional experience. We all know what happens to

13. This, essentially, was the central tenet of Gestalt psychology, in which the value of the part may be known only from the role of that part in the whole. The whole, additionally, is then always greater and other than the sum of the parts.

'mortals' who wander outside this band – they become all manner of wonderful and terrible things.)

The paradigm also brings out the dual nature of God, and further the contrast between the Virgin/God (love without sex) and the Witch/Devil (sex without love).

Nevertheless it must be insisted that in the very broadest senses of these terms all the figures mentioned in the preceding paragraph – those on the right of the diagram – are still both sexual and of the female principle. And it must still be stressed that despite their a-sexuality the wizard and the angel, etc. – those on the left of the diagram – are nonetheless, in a sense we cannot yet adequately define, of the male principle.

List A, then, somehow refers to the sphere of influence of the 'male principle', while list B is concerned with the sphere of activity of the 'female principle'. These two principles, and their characteristics, are opposites. Within one of these, the B principle, exists a further set of linked opposites.

As we have seen Freud frequently stated that opposites are very close *in the unconscious*. May this be taken to imply that opposites outside the unconscious mind are not close? Of direct relevance here is the fact that the opposites described by the same word in older languages – the 'antithetical primal words', as Freud calls them – in general fit far better into list B1-B2 than into list A-B.

B1	*B2*
sacred	accursed
guest	enemy
weak	strong
adhere to	split off
with	against

In everyday life we speak of love-hate relationships, and of the dividing line between love and hate as a razor's edge. Other phrases in common use, such as manic-depressive and approach-avoidance, further support the idea of some kind of oscillatory movement between opposites.

The very existence of the term 'trust' implies its opposite – distrust, and fear. Love and hate are opposites but they are, as it were, two sides of the same coin. True (?) opposites are not part of each other.

We have arrived here at one of the central dialectics of the Christian faith in justifying the existence of evil in the universe – that if we did not know, did not have, evil we would be unable to recognize good. We cannot choose good, it is argued, unless there is evil to reject. So, for example, Christ was tempted by the Devil in order that he might declare his hand, one way or the other.

In contrast to this position, when we consider the opposites of list A-B we find that thought and emotion, for example, *do* conceptually exist by themselves in their own right. Similarly, neutrality or detachment is *not* involvement – it is neither *for* nor *against*. Proof does not require the existence of faith (neither does disbelief destroy proof in that sense, even if it refuses to recognize it). Faith is *irrelevant* to proof – and by the same token proof is irrelevant to faith. Thought is thought is thought in the way that a rose is a rose is a rose. It could itself exist in a 'universe' which had never heard of emotion (the overworked plot, actually, of many SF stories). Thought can say that good and evil are only concepts, mere semantic opposites that exist 'only in the mind'. Rephrase them in non-emotional terms and the 'problem' of these opposites disappears.[14]

It is clear from the foregoing that any embodied 'female principle' (list B) would have the greatest difficulty in genuinely comprehending the 'male principle' (list A) and vice versa. Strictly speaking, this is impossible. In so far as the attempt *is* made, in real life, the female side of the personality nevertheless perceives the male *subjectively*, while the latter perceives the former *objectively*.

14. Where, one wonders, does anything exist for us except in the mind – but that is not quite the point. The point here is that thought cannot destroy *itself* (that is, not by *taking thought*) in the way that it can destroy (negate, petrify) feeling and emotion – precisely, Wordsworth's 'we murder to dissect'.

Each 'principle' as it were projects or lends its own characteristics to what is perceived. Thus, for example, the female principle *must* perceive in terms of opposites – the opposites, naturally, of B1-B2. Of this a typical instance is the slogan 'Those who are not with us are against us'. The reverse – 'Those who are not against us are with us' – also holds, if not so thrillingly. But no third possibility exists. There can be no neutrality. (So runs the dogma of all types of 'commitment'.) The female principle cannot and will not understand the regard that neither praises nor condemns, neither accepts nor rejects, neither judges nor pardons. Detachment is anathema to the female principle.[15]

Returning to the notion that those who are not against us are with us, something further is accounted for. On first making the acquaintance of objective knowledge (that is, of Knowledge I) and finding it not hostile *per se* – because it is not involved with hostility (or any other emotion) – principle B concludes that that knowledge must therefore be 'with' and therefore good. Only after a time is it realized that knowledge, objective knowledge, is *outside* good and evil and has an equal place for both, i.e. no place for either. At that point principle A (Knowledge I) becomes an enemy, because of course 'those who are not with us are against us'.

Religion at this point turns against knowledge – just as the State turned against Socrates, and for precisely the same reasons. At that moment the Devil is he who *knows*. One need hardly recall that Adam was expelled from Eden on the day he acquired knowledge, and that this was specifically knowledge *of good and evil*. The emergence of objective understanding is the death knell of the rule (the one-ness?) of the female principle. This is one of the further layers of the Genesis legend.

In concluding this section it is worth emphasizing that the broad dichotomy we have here described in terms of lists A and

15. Thus, or so novelists say, a woman will accept the hatred of a former admirer, or the bad treatment she gets from a present one. What she cannot accept is indifference.

B is not itself some hypothetical arrangement. Only the terms A and B are that. The dichotomy – for instance between scientific or neutral words, and words carrying an emotional charge – exists. If it does not, then why do the terms themselves exist, and what on earth do we understand by them when we use them?

A brief selection not now of single words but of statements follows, each of which is based implicitly on this dichotomy:

'My heart wanted God, but my head got in the way' (Alfred de Vigny); the heart has its reasons; he's all head and no heart; business and pleasure don't mix; never do business with friends; don't bring feelings into this; the situation looked different in the cold light of day; I pinched myself to make sure I wasn't dreaming; one must allow for the human element; people are not machines; man does not live by bread alone.'

The position suggested appears to have some kind of reality. The assumption that there are two distinct ways of looking at life underlies each of the above statements. Of course, whether or not one then goes on to accept the model of personality which I in part base on this situation, is another issue.

From here we are in a position to proceed to some evidence provided by physiology, the observed physical properties of the human nervous system. This is not at all to say that speculation on the relevance of language, or other psychological phenomena, is exhausted. On the contrary, continuing stress will be placed upon language, in particular on the history – in most cases the forgotten history, of words. It is not, for example, an accident that *villain* (the devil's descendant) is in origin the same word as *villein*, meaning a peasant – as one might have predicted, a man of the earth.

Polarity in the Nervous System 4

In the previous chapter a tentative infra-structure for the human personality was proposed on the basis of certain linguistic and other evidence. This model divided the personality into two major parts, A and B, whose respective spheres of influence are both opposite and conceptually independent. Within B are two further sub-divisions, B1 and B2, whose functions, though again opposite or antagonistic, are this time conceptually inter-dependent. That is to say, B1 and B2 are the two sides of one coin, and not, as in the case of A and B, two different coins.

The question to be considered now is whether any *physiological* support can be found for this *psychological* model. Physiology, it should be said, is the study of the function of the structures and organs of the body – that is, of the brain, glands, nerves, muscles and so on. Psychologists in general look to physiology to provide the physical basis of personality – the physical hardware which underlies the psychological software.

Physiologists divide the body and its various organs into a number of systems. One such overall division – the one with which we shall be principally concerned in this chapter – is into the *central nervous system* and the *autonomic nervous system*. This division is one commonly accepted by psychologists and physiological psychologists. It should be emphasized, however, that it is not the only one possible.[1]

1. For the benefit of those familiar with physiology, I should point out

The central nervous system comprises essentially the brain and the spinal cord, and the large majority of structures found within these. The autonomic nervous system embraces groups of coordinating centres (in function not unlike parts of the brain) which lie outside the central system, as well as certain sections of the brain itself and, perhaps most typically, the various glands of the body.

The central system controls the skeletal (or 'striped') muscles and voluntary movements of these. It is *exteroceptive* – that is, it responds to stimuli outside the organism. One of its major functions is to make adjustments in the external environment. The autonomic system is *interoceptive* – that is, it responds principally to stimuli from within the organism (temperature change, hunger, fatigue and so on). It regulates the functioning of the glands and internal organs – that is, it makes adjustments to the *internal* environment. To this end it also has control over the so-called 'smooth' muscles (e.g. the muscles of the heart). Lastly, this system can when necessary produce involuntary or reflex movements of the striped muscles.

These two basic properties (1) the means by which organisms

---

that I am not concerned here with 'systems' which are really such in name only – systems, that is, which while definable in terms of common structural features and functions, do not actually possess, or comprise, any kind of control centre. The term 'peripheral nervous system', for example, exists, and refers to the nerve complexes carrying messages between the autonomic n.s. and/or the central n.s., and the various receptors and effectors. These complexes are therefore merely adjuncts to the a.n.s. and the c.n.s., not regulating systems, the sense in which I use the term above.

It is possible also to consider the brain itself from the standpoint of its evolutionary history – that is, in terms of its component parts as they are seen to emerge in the developing embryo. We shall be discussing these matters in some detail in Chapter 8, and they are postponed for the moment. However, it can be said here that the human embryo evidences what are termed a fore-brain, a mid-brain and a hind-brain. In the finished brain the mid-brain is a more or less vestigial – although by no means an unimportant – link or bridge between fore-brain and hind-brain: while the fore-brain and hind-brain have become the two major coordinating and directing centres of the total nervous system.

react to and move themselves about in the environment (contractility) and (2) the maintenance of internal equilibrium by means of secretion and excretion, are found even in the most primitive organisms.

CENTRAL NERVOUS SYSTEM

It will be helpful to refer to the figures between pp. 241 and 244 when reading these sections.

Under this heading are considered for the moment only the cerebral hemispheres – that is, the cerebral cortex and the hypothalamus. There are, of course, many other organs involved.

(1) *The Cerebral Hemispheres and Cerebral Cortex*

The leading role in the central nervous system is played by the outer covering of the two cerebral hemispheres – the cerebral cortex. The cortex is the heavily wrinkled external part of the brain immediately visible on removal of the skull. The term neocortex (new cortex) is frequently also used, since a large part of the cortex constitutes the most recently developed features of the nervous system in evolutionary terms. This development reaches its peak in man, and the neo-cortex is in fact the seat of the most typically human abilities. During the known evolution of man, average brain size has increased from 600–700 cc. to 1200–1300 cc., most of the increase being neo-cortical.

It is on the basis of considerable experiment and observation that the cortex is known to be the centre of volitional motor movement and the major terminus of sensory impressions. This part of the brain is likewise the seat of cognitive functions, thought and waking consciousness. Moreover, the learning of new, skilled activities is organized by the cortex. Language and speech are the concern principally of the temporal lobes at the lower rear sides of the surface brain, and here also is situated the memory store. Electrical stimulation of the temporal cortex produces, astonishingly enough, random scenes from one's past life *in toto*, as originally observed, together with the emotional feelings which accompanied them. The temporal lobes also house one of the several sleep centres of the nervous system.

Not only volitional control of *motor* movements but apparently volition generally (perhaps what used to be termed 'will') resides in the cortex, particularly, it seems, in the frontal lobes. The development of these lobes is the most distinctive feature of the brains of primates, particularly of man. Experimental lesions or accidental damage in this area produce a marked, sometimes a drastic drop, in general drive and the will to achieve.

Of great relevance here is the demonstration physiologically that stimuli originating in the cortex can block outright, or otherwise modify, impulses from centres lying below the hemispheres (perhaps emotional centres). 'Ideas' can frequently produce such controlling or modifying blocks. This fact is, of course, self-evident on the mental plane. Commands such as 'pull yourself together', 'for heaven's sake act like a man', 'think of the children', etc., may, though do not necessarily, have the effect of calming an hysterical individual, of overcoming panic, or rousing someone from a state of shock or torpor to a state of action. It has, additionally, been shown experimentally that, say, the name of a man's girlfriend or wife spoken while he is asleep is more likely to wake him than any other girl's name spoken at the same intensity. The cortex, therefore, can put an end to sleep when necessary. Alcohol, too, first depresses the higher centres, releasing 'lower' behaviour which is normally held in check. As we all know, this often constitutes undesirable or animal cavortings, but can comprise merely simple indiscretions, or the betrayal of secrets, or simply an increase in affection. These various kinds of evidence all demonstrate that, without any *effort* on the part of the individual, normal cortical consciousness is restraining or modifying a good deal of would-be behaviour. The cortex is, therefore, far from being a helpless victim of unconscious or emotional impulses.

What remains true is that as the autonomic system, to be more fully described below, is aroused, impulses from it to the cortex become progressively stronger – thus one speaks of rising excitement or mounting anger – and more difficult for the cortex to control. A relatively *small* amount of sub-cortical stimulation causes the cortex to function *more* efficiently. Arousal beyond that rather low level however causes a *deterioration* in performance. We all know that in states of even fairly moderate anger or anxiety it becomes difficult to perform such simple tasks as inserting a key in a lock or pouring a drink without spilling it. When stimulation reaches a point where the cortex is flooded we then experience states such as overwhelming panic, blinding rage or dazzling ecstasy. The 'loss of control' experience is a common enough sensation and to this we shall revert.

(2) *Hypothalamus*

The hypothalamus lies below the thalamus, an organ not of direct concern to the present discussion. Although connected with the thalamus, which is part of the central nervous system, and despite its name, the hypothalamus has a large measure of independence. It is in fact much more part of the *autonomic* system, constituting that second system's highest level of integration. Contained in this small structure are the most highly evolved centres for the regulation of sexual functions, aggression, general emotion, body temperature, hunger, thirst and so on. Like a number of other centres, it also controls sleep.

THE AUTONOMIC NERVOUS SYSTEM

As a generalization it can be said that whereas the central system is primarily a motor system, controlling the muscles and volitional action (and only in second place a sensory system), the autonomic is primarily a sensory system, controlling the glands and inner states of the organism (and only in second place a motor system). The autonomic system *does*, however, as already mentioned, control the 'smooth' (involuntary) muscles and reflex actions of skeletal muscles. The term autonomic means, effectively, automatic: the more precise meaning is self-governing.[2]

This is to say, the autonomic system is not only not under our volitional control – that is, we cannot in general consciously affect its operation – but it will frequently override or operate in advance of the rulings of the central system, particularly in emergencies of all kinds. A person will find, for instance, that he has removed his fingers from a hot object before consciously appreciating that it *is* hot. More complexly, we may find ourselves running away 'against our will' in the face of danger, even despite a conscious effort to stay and be brave. A person, too, who tries to stay awake indefinitely succeeds only in the short term. Similarly, it is not possible to hold one's breath until one becomes unconscious.

The autonomic system is itself divided into two sub-systems – the *sympathetic* and the *para-sympathetic*. This division is based on a variety of objective considerations – for example, the sympathetic nerves act by liberating noradrenalin into the space between their endings and the structures on which they work, while the para-sympathetic nerves liberate acetylcholine. Structurally, the synapses of the para-sympathetic system (a synapse is a junction in the nerve) are found in isolated groups of ganglia (masses of nerve bodies) adjacent to the gland or organ controlled, while the synapses of the sympathetic system are organized mainly in chains of ganglia lying along the spinal column. As a corollary to this, the action of the para-sympathetic system is piecemeal, while the sympathetic system tends to function as a whole.[3]

2. One needs to be alert to the idea of the very different levels and forms of behaviour mediated by the autonomic system – ranging from the simple reflex action or function, to extremely subtle shiftings of chemical balance, leading ultimately, at the conscious level, to changes of mood.

3. The labels sympathetic and para-sympathetic are more than a little misleading. They mean in fact the opposite of what they at first sight imply. Thus it is the sympathetic system which is concerned with aggression (or mobilization), and the para-sympathetic with passivity (conservation and regulation).

Describing the general functioning of these two sub-systems, a standard text[4] has the following to say:

> The two divisions of the autonomic system are largely, although not completely, antagonistic in their effects. In general the sympathetic system *mobilizes* the resources of the body for use in work and special emergencies, while the para-sympathetic *conserves* and stores bodily resources. In other words, the first helps *spend* bodily resources and the second helps *save* them. This statement, although true, does not hold in certain specific instances. Naturally, too, these systems never act independently of each other but are brought into correlated activity in varying degrees, depending on the demands made on the organism.

Of particular interest perhaps is the phrase 'never act independently of each other'.

(1) *The Sympathetic System and the Posterior Hypothalamus*

As experimental physiology has effectively demonstrated, the posterior, or rear-end, of the hypothalamus is the chief regulator of the sympathetic system.

This system prepares the organism for the emergency reactions of fight and flight, rage and fear. Its action therefore is, variously, to augment the blood supply to the brain and muscles while denying it, for example, to the stomach, to inhibit saliva ('mouth dry with fear'); to increase the heart rate ('his heart raced'); to inhibit the secretion of gastric juices (hence bad temper and indigestion as well as fear and indigestion go together); and to *in*activate the alimentary and uro-genital canals. The phenomenon of defecating and urinating in extreme fear is due *not* to sympathetic action, but to a prior burst of parasympathetic activity.

4. C. T. Morgan and E. Stellar, *Physiological Psychology*.

(2) *The Para-Sympathetic System and the Anterior Hypothalamus*

Experimental stimulation of the anterior, or frontal, hypothalamus sets the effects of the *para*-sympathetic system in operation.

This system encourages the secretion of gastric juices (aids digestion), facilitates salivation, urination, defecation, and aspects of sexual readiness, such as erection of the penis. Here is also located one of the so-termed pleasure centres. When an electrode is implanted in such a centre, an animal will endlessly press the lever which activates the electric current. A rat or monkey will stimulate itself for as long as forty-eight hours at a stretch, oblivious of other needs, till it drops with exhaustion.[5]

The anterior hypothalamus is also concerned with the regulation of body temperature. In this connection it is interesting how many terms relating to temperature we apply to sex. A sexually active female is called a 'hot' woman; 'hot-lips' is a vulgar term of sexual endearment. A bitch ready for copulation is said to be on heat. We speak of being aflame with desire; of burning passion. Affectionate people are called warm. Reserved people are cold.

The anterior hypothalamus regulates also hunger and thirst. One speaks, interestingly, of hunger for love and affection, and one is consumed with desire. Life without love is called empty.[6]

5. Does this not remind one just a little of the cave where Tannhäuser lingered oblivious of his knightly duties and heedless of passing time?

6. It might be objected that such terms as these are found in a variety of other contexts – 'hungry for facts', 'thirsty for knowledge', 'hot news', and so on. There are several points. First, many of these usages are not so 'non-physiological' as they at first appear. When we are *really* eager for a piece of information (or a decision) we are in fact in a (mildly) aroused state. One's heart is beating slightly faster, one's eyes are slightly wider open than usual, and so on. We experience then also the slight drying of the mouth and throat which the sympathetic system mediates – so that we are, indeed, in a sense actually 'thirsting for information'. However, such expressions are really more by way of *transferred* metaphors. When on the other hand we are,

What has been said thus far is now summarized and its implications considered.

The central system is primarily a 'doing' or motor system, while the autonomic is chiefly a 'feeling' or sensual system. The former is also a 'thinking' system and active 'by day' – i.e. during waking hours – at which times, under normal conditions, it controls or modifies impulses of autonomic origin. In emergencies, or in states of drunkenness and similar conditions, the autonomic system can, wholly or partly, override this central control. The autonomic system, therefore, *also* functions by day – but it is additionally a 'night' system, at least in the sense that it controls sleep, and perhaps too in the sense that dreams contain generous helpings of sex, aggression, fear and other autonomic responses.[7] It is also true that vital functions such as breathing and digestion continue uninterrupted throughout the night. The *central* system, however, is then largely inactive. In some senses, therefore, the two systems are alternatives. And certainly it is impossible, for instance, to think clearly and be in a state of panic at one and the same time.

The autonomic system is itself divided into two closely linked, though in general antagonistic, sub-systems. The characteristic mode of the sympathetic system is arousal, tension and activity. The para-sympathetic system is characterized by quiescence, relaxation, passivity and vegetation.

It will by now be clear where this consideration of the nervous system has led us. For the sake of the record, however, this is

---

say, in the presence of our lover, the sensation of heat is very literally and directly related both to our passion and our body temperature.

It seems, in short, that phrases like 'hungry for facts' and 'hot news' (an expression deriving in any case rather from the overloading or too frequent use of an electrical wire system) may be fairly regarded as simple metaphors – while those quoted above and in other parts of this book are supported by a physiological infra-structure which is far from metaphorical.

7. Male animals actually have erections during some 80% of the time they are dreaming.

now spelled out. First, a round-up and juxtaposition of the terms applied by physiologists to the central and autonomic divisions of the total nervous system.

Central Nervous System	*Autonomic Nervous System*
(primarily) motor	(primarily) sensual
doing / thinking	feeling
voluntary	involuntary
'day' (waking)	'night' (sleeping)
exteroceptive	interoceptive
'looks out'	'looks in'
neo- (new)	paleo- (old)
cortical	sub-cortical
human	animal / sex

Are these not the *kinds* of alternative already seen in the previous chapter in lists A and B?

The terms applied to the para-sympathetic and sympathetic sub-systems provide us with some further close correspondences to our earlier groupings, this time in terms of lists B1 and B2:

Para-sympathetic (B1)	*Sympathetic (B2)*
quiescence/passivity	activity / aggression / fear / arousal
conservation	mobilization
save	expend
digestion	interrupted or disturbed digestion

Thus the psychological structures earlier proposed appear, in very broad terms, to be paralleled by structures independently described by physiologists, on an objective and experimental basis.

Paradigms, here, are as follows.

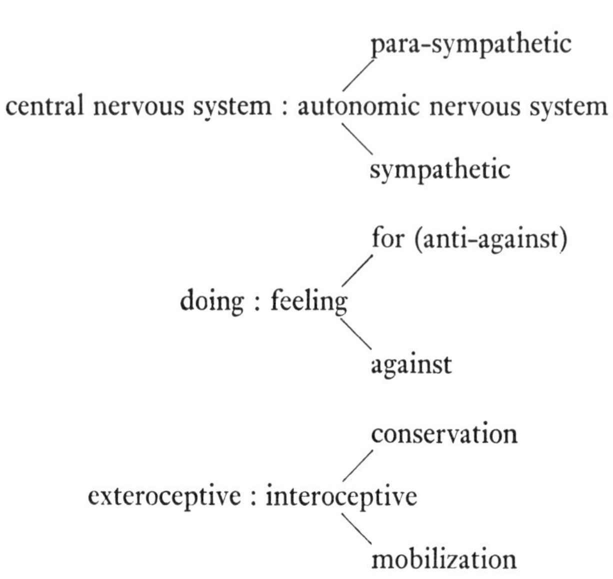

If paradigms are a little thin on the ground here, this is partly because one is considering the relatively simple (and physiological) points of *origin* of complex psychological phenomena. Despite what some physiologists appear to think, physiology, from the behavioural or experiential standpoint, is an extremely gross and coarse description of the human organism.[8] Thus if one asks a physiologist what the physiological correlates are of pathos, loyalty, sentimentality, longing, admiration, cynicism, whimsicality, fortitude, remorse, firmness, insolence, etc., he has (for the time being at any rate) no answer.

Nonetheless, the parallels described between our two levels of explanation, the psychological and the physiological, appear undeniable. Man's psychological experience and the physical basis of that experience are not merely linked, but are perhaps so on a scale and in ways that are not currently imagined. The

8. The objection here is not to physiology but to the reductionist approach of some physiologists – that is, the attempt to *reduce* psychological events to physiological events. The reverse direction should apply. Properly, physiology is the firm surface from which the exploratory balloon rises into a more tenuous element. A better analogy still is that of physiology as the sea and psychology the myriad life forms swimming in it.

'shape of conscious' may prove to have something very much to do with the physical 'shape' of the nervous system.

The central system as a whole is a 'do-er' (i.e. it is volitional and controls the normal means of body movement). The autonomic system as a whole is a 'feeler' (i.e. it generates and transmits emotion). The autonomic experiences of anger and fear, however, lead to action – fight or flight. Thus we have:

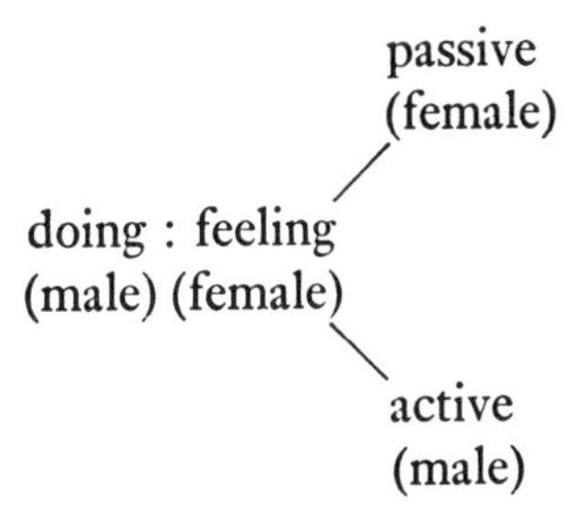

A point of additional interest is that in both parts of this physiologically based paradigm, it is the male element which is active and the female element which is passive.

A number of obvious difficulties for the proposed basic structure of personality must be considered.

Although it cannot be disputed that 'emotional' states both originate, and are to a great extent integrated in the lower brain centres (e.g. the hypothalamus), the conscious *experience* of pain, pleasure and so forth takes place in the cortex – in what are termed projection areas. As already indicated, the memory store, complete with the emotional feelings associated with each experience, is likewise contained in the cortex, apparently in the temporal lobes.[9]

The cortex is however closely connected with what we shall from now on refer to as System A (a term deriving from list A). Our present argument has placed also consciousness and cogni-

9. However, it should be pointed out that associative thinking is something a good deal more dynamic and creative than a straightforward memory bank in which emotional experiences are routinely associated with given events, and these with each other – cf. both Freud and Noam Chomsky, *inter alia*.

tive thought in System A – while emotion and associative thinking, on the other hand, were assigned to System B. Would it not be at least desirable that System B – if it really exists – should have somehow a different location from System A? Have we not at least been led to expect this? Yet, undeniably, the conscious experience of emotion *is* in the cortex, as is a comprehensive (?) memory store, complete with emotional ties. This would seem a fairly serious difficulty.

It would in fact appear that some relatively straightforward model as now proposed would adequately cover the position so far outlined, removing the need for the more complex picture we have already hinted at. Would it not be sufficient to say that the cortex receives stimuli from two different sources, i.e. from the internal and external receptors, which either differ in themselves, or in the sense that they arrive by different routes at possibly separated areas of the cortex, and which are therefore processed, even perceived, rather differently by this organ? Let us now admit at once that some such account *could* accommodate at least some of the phenomena so far reviewed – and is one that would have the additional advantage of causing little disturbance to existing concepts of the nervous system.

But instead of pursuing this and other possibilities, we will turn briefly to some additional linguistic and highly speculative considerations. In examining these one should keep generally in mind comments already made on the nature of metaphor.

When speaking of memory one commonly uses phrases which contain the idea of going back. Is it merely coincidental that the memory store happens to be at the rear of the cortex? And does 'back' refer in any case rather to time, not space? But, then, what of such a phrase as 'to be of high birth'? Does the phrase mean that one was born on the third floor? No, it does not. Similarly, to be of low birth does not mean that one was born on the ground floor, except in a metaphorical sense. (*Does* it, perhaps, mean closer to the Earth?) Why do we say that a king is set *over* us, or that a government rules *over* us? Is it because the king sits on a raised platform? Or is the fact of it rather that we place him on a platform to symbolize some state of affairs already agreed – say, inwardly. Why for that matter is God above us and Zeus on Mount Olympus? Are these positional metaphors entirely unrelated to the fact that the cortex is *on top* of the rest of the brain? And what of *base* emotions, *low* cunning, *under*hand, as opposed to *above* board, *above* suspicion? Let us, of course, not forget that these turns of phrase long preceded any detailed knowledge of the anatomy of the brain.[10]

A point worth mentioning in the present connection is that as a joint function of the increase in size of the cortex and the adoption by man of an upright posture, the cortex now completely covers and encloses parts of the brain formerly 'free-standing' and at the surface – as they still are in many animals.[11]

Figure 1 shows some of the changes in the relative size and positioning of a few parts of the brain as one ascends the evolutionary scale. Although not mentioned in the present chapter, because reserved for detailed discussion in Chapter 8, the reader's attention is directed here to the *cerebellum*, in particular to its final position and size relative to the *cerebrum* (that is, the cerebral hemispheres) in the evolved primate brain.

Reverting to the question of impulses from the lower centres, it was observed earlier that too strong a barrage of stimuli from that source effectively floods the cerebral hemispheres, temporarily destroys their fine tuning, and totally removes the controlling influences of past experience, socialization, original

10. I do not deny here that some at least of the phrases cited have well-known reality-antecedents. The point is rather, that when deriving our metaphors from the wide variety of situations available – be they gaming, sea-faring, commercial practice or whatever – whether certain possibilities appeal more to us than others: and if so, why?

11. A point related to the general issues we are opening up here is that the organization of the sympathetic system into *one* functioning chain of ganglia, as contrasted with the scattered organization and piecemeal functioning of the para-sympathetic ganglia, already perhaps gives some grounds for the supposition that one's subjective *experience* of, for instance, aggression and passivity (the Devil and God?), might differ, and moreover along lines generally determined by these physical facts.

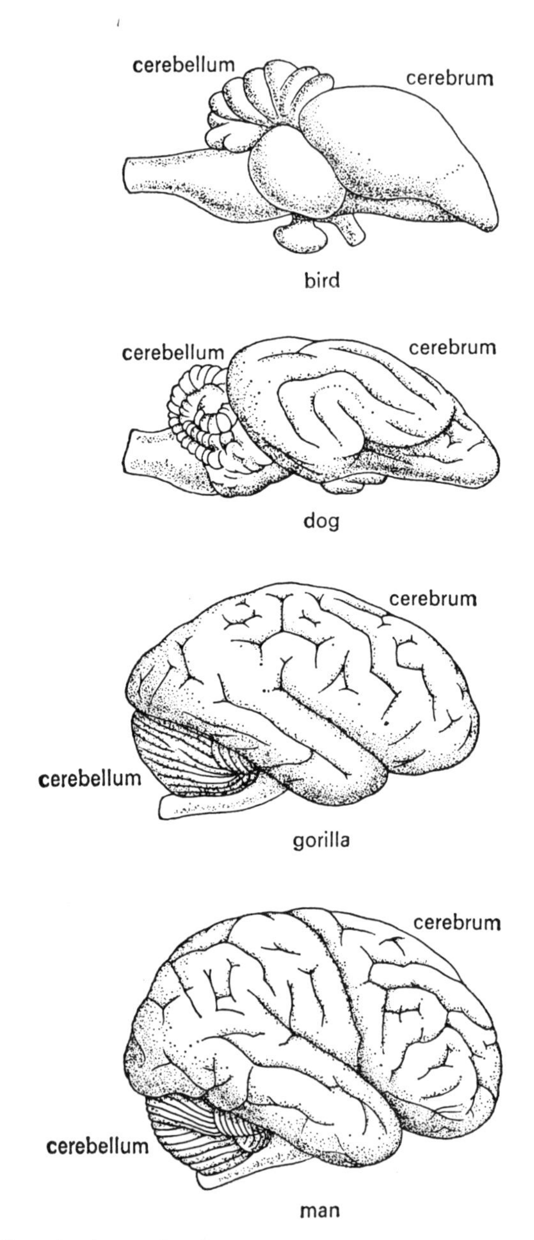

Figure 1. The relative reduction of the cerebellum and its 'masking' by the cerebrum in the course of evolutionary time.

conscious intention, and so on. At such times one may run from danger, leaving friends or relatives to die, may strike or kill someone one loves deeply.[12]

This 'loss of control' experience proves more than usually disturbing for some individuals – even where produced by basically pleasurable excitement. So much so that they avoid not only all forms of excess but even merely unpredictable situations. Such a person might abstain from alcohol, not drive cars (or drive unnecessarily cautiously), tend to subscribe to a rigid moral, or religious, code and in all probability deny having dreams at night. In respect of these individuals it would not be too fanciful to say that they are afraid of being overwhelmed or perhaps even 'possessed'. In everyday speech, after all, we use expressions such as 'he worked/drove like a man possessed'.

Would it be sufficient, however, having led the discussion round to this point, to suggest that the widespread fear of possession described in detail in Chapter 1 could be ascribed wholly to the fear individuals have of losing their tempers, being seen to be cowards, or whatever? In all probability not. It might be enough of an explanation in respect of the kind of person cited, and is perhaps indeed *part* of the explanation of the phenomena of possession generally. But were the magicians of the Middle Ages then simply trying to work themselves into states of anger or lust? It seems unlikely. In particular, the 'possession' we have just discussed is essentially *dis*integrative, essentially merely a loss of control. It has nothing to tell us, for instance, of the *alternative organization* which is a prominent feature at least of so many stories of possession – the possession which can produce complex acts performed over a long period, and which allegedly confers special powers.

An incidental piece of linguistic parallelism. The posterior hypothalamus controls rage (hate) *and* fear, while the anterior hypothalamus controls the opposite of *both*. Linguistically, and confirmationally, it had seemed necessary in the previous

12. France and some other countries, as if in recognition of this, pass only moderate sentences for these 'crimes of passion'.

chapter to give *one* opposite – love (or trust) – for both hate and fear.

A final linguistic point. It is interesting to note that passion is from the same root as passive. (The Passion of Christ is, of course, the suffering of Christ.) The words for these two opposite emotions or states are once more from the same root.[13] The use of the same root for these opposite (though linked) states looks again like a comparatively recent example of the already noted practice widespread in older languages.

13. I have recently come to suspect a further pair of words of opposite meaning – 'raze' and 'raise'. Why do we prefer the former to perfectly good words like 'flatten' and 'level'?

Part Three:
Trance and Anti-Trance

Aspects of Western Psychology 5

The further development of the central argument is slowed at this point, in order that we may consider how the position so far tentatively adopted squares with the current positions of various schools and persuasions of orthodox modern psychology. These considerations are grouped under four headings:

Pavlovian and Skinnerian conditioning (also termed classical and operant conditioning respectively).
Freud and the psychoanalytic movement.
The relation of conditioning and learning theories to psychoanalysis.
Psychosis, with particular reference to the views of R. D. Laing and Bruno Bettelheim, and the relation of these to psychoanalysis.

It will appear that far from conflicting with the general position so far adopted, these various approaches are readily accommodated by it, and each in fact offers fresh support.

(1) *Pavlovian and Skinnerian Conditioning*

The attention of Ivan Pavlov, a Russian physiologist active in the early part of this century, was drawn to the psychological accompaniments of physiological reactions, in the course of studying digestive reflexes under laboratory conditions. He noted that dogs salivated not only when food was placed in the mouth, which was natural enough, but also at the sight of various items of laboratory equipment. Experimental investigation in due course established the following – that *any* stimulus, be this a sight, a sound, a smell or whatever, that regularly precedes the giving of food to an animal will, in time, itself produce salivation and other preparatory digestive reactions in the animal, even if it is no longer followed by food. The novel stimulus (say, the sound of a bell) which, prior to its experimental coupling with food produces no salivation, is called, after the successful coupling, a *conditioned stimulus.* The salivatory response to it is called a *conditioned response.*

The process of linking new stimuli with existing responses, so that these come to elicit those responses, can be achieved in respect not merely of salivation, but of any natural, reflex response – that is, any *autonomic* response. These include anxiety, rage, heart-rate, eye-blink, contraction-expansion of the pupil, reflex withdrawal or flexion of the hand or foot, and many other reactions.

The process of conditioning an autonomic response (of producing such a response at will by the introduction of a conditioned stimulus) is termed classical conditioning. This title distinguishes it from another form of conditioning, discussed later. There are qualifications and riders to the simple description of classical conditioning given above, not all of which are central to the purposes of this inquiry. One or two, however, must be mentioned. First, any other stimulus which *resembles* the conditioned stimulus will also produce the conditioned response, though not to the same extent. The strength of the response actually varies regularly with the degree of similarity or dissimilarity of this further stimulus to the original conditioned stimulus. This feature of the process is termed *stimulus generalization.*

The likelihood of producing the original response with a conditioned stimulus, a bell or whatever else, increases as a function of the number of times the bell has been paired with the unconditioned stimulus, in this case food. Every time the bell is presented with the food the connection is said to be reinforced, and a given programme of reinforcement is called a *reinforcement schedule*.

The terms conditioned and unconditioned are actually an inept translation of the original Russian terms used by Pavlov. Quite aside from that fact, it is more meaningful to speak of learned and unlearned responses, and learned and unlearned stimuli – instead of conditioned and unconditioned respectively. Salivation is an unlearned reflex response. Food-in-the-mouth is an unlearned stimulus, which automatically produces salivation in the dog. The sound of the bell is a learned stimulus. In the course of the experiment the animal 'learns' that the bell indicates food. This may all seem rather self-evident. It is less self-evident when one considers, for example, that expansion of the pupil of the eye can be conditioned to the sound of a bell. Does one then say that the eye has learned the sound of the bell?

Finally, let it be emphasized that *any* stimulus associated with an original stimulus which naturally evokes a particular response, itself comes to produce that response, whether the association is meaningful or not, whether the association is deliberate or not, and even whether the association is consciously noted or not.

Many parallels to the experimental situation are found outside the laboratory. The manager of a business perhaps finds himself inexplicably uneasy in the presence of one of his clerks, possibly allowing him a degree of licence not accorded to other members of staff. In the course of counselling or therapy the manager might realize that the clerk had a moustache like his (the manager's) father once had, or that he was tall or walked like the father: this would then be a case of stimulus generalization.

A psychologist, perhaps not very wisely, caused a loud noise to occur in the presence of a young infant whenever the cat of the house came into the room. Human beings have an instinctive (unlearned) fear of loud noises. Very soon the child showed anxiety whenever the cat appeared, although the noise was now discontinued. The child's anxiety spread also to such objects as the mother's fur wrap. The experiment was then discontinued, and presumably the conditioning gradually weakened through non-reinforcement. (It is not known, however, what 'little Albert's' feelings were towards animals – or psychologists – in later life.)

Of the various points of interest in classical conditioning for the present argument, the first is that it is essentially passive. The events happen *to* the individual concerned: he or she can do little about preventing their occurrence, and may indeed be unaware that they have occurred. That is to say, the process is often unconscious. An individual may consciously attempt to do something about the *re*currence of a classically conditioned response. He may, for instance, force himself to eat in restaurants, even though this causes him acute (and quite irrational) anxiety or embarrassment. Such direct action, for reasons which need not be gone into here, may or may not have much effect on the underlying structures. A second point of interest, already stressed, is that classical conditioning heavily involves the autonomic system. A third is that classical conditioning is basically associative. One thing, idea or situation becomes associated with another, often for no other reason than that the two occur together.

A second type of conditioning, operant conditioning, is usually associated with the name of B. F. Skinner, an American psychologist. The work of Skinner and others began as an attempt to bridge the gap between classical conditioning and more complex, particularly human, learning.

Operant differs from classical conditioning in a number of important aspects. In order for it to occur, the organism must first make some volitional movement. Without this first motor movement the events described as operant conditioning do not take place. From this fact alone it is clear that this form of conditioning has an *active*, as opposed to the *passive* basis of classical conditioning.[1]

1. One needs to be clear about the meaning of the term passive as applied

A rat is placed in a small box or cage, into which projects a small lever. Unknown to the rat, the pressing of this lever will produce a pellet of food or a drop of milk. In the course of general exploration the rat will eventually press the lever. He may do this with his body, snout or paw. At some subsequent point he will, either accidentally or perhaps already half deliberately, again activate the mechanism. After only one or two 'trials' it is not uncommon to find the rat confidently pressing the lever for the purpose of obtaining pellets. Sometimes the original movement which activated the lever – leaning on it, pressing it with the snout – is preserved for a time, the rat 'superstitiously' assuming that this is the only way to achieve the desired (magical?) effect.

At later stages, by the staggered delivery of pellets, the rat can be trained to press only at fixed intervals of time (other responses go unrewarded and hence gradually disappear) or at a fixed rate per minute, or in sequences of slow and quick presses, and so on. Operant conditioning is the basic process underlying all animal training.

An important point is that the *first* stage of the operant process differs from that of classical conditioning. In the operant instance it is a *motor* response that is rewarded or reinforced by food (the unlearned stimulus), and which in time becomes a learned or conditioned response.

Attempts have certainly been made by theorists – in particular the Russians, who have ideological reasons of their own for wishing it to be so – to reduce the two conditioning situations to one and the same, basically Pavlovian, model. The most generally held view, however, is that *two* processes are operating and that two models are required, which have some points in common.

---

to associative conditioning. The individual *suffers* the association passively – it is made *for* him – but association itself in its effect is an active process. A man walking down the street finds himself, for instance, whistling a Maurice Chevalier tune. Thinking about this, he realizes that the name of the girl he is taking out for the first time at the weekend is called Louise. The situation has *actively* affected his behaviour.

As will be imagined, the present book subscribes to the two-process view.

The differences between the processes are summarized as follows. Classical conditioning begins with a stimulus affecting a passive or receptive organism. Operant conditioning begins with a response made by an active organism. In the former case *the environment acts on the organism*, in the latter *the organism acts on the environment.*

The *similarities* between the two processes reside in the end-product, where in both cases a now known or learned stimulus elicits what is now a conditioned response.

Evidence of the role of operant conditioning in human affairs is as impressive as that for classical conditioning. It has been shown, for instance, that in the course of a conversation with a subject, an experimenter can cause the incidence of given words or phrases in the subject's speech to rise significantly merely by nodding or grunting assent whenever the chosen phrase occurs.

In another experiment[2] subjects learned the names of examples of twelve classes of Chinese characters, allegedly as a test of memory. Embedded in each example of each class was a component peculiar to that class only. The subjects were not aware of this. At the end of the learning session the subjects were unexpectedly asked to guess the names (classes) of a further selection of characters. Although above-chance scores were obtained in allocating the correct names to the new characters, those involved had no conscious idea why they had got them right. Unconscious operant conditioning had occurred during the first part of the experiment.

The statement that this second, operant, class of conditioning is frequently an unconscious process contains an apparent difficulty. The intention of the present book is to propose that *classical* conditioning is a function of System B, while *operant* conditioning is to be ascribed to System A. If the latter is frequently unconscious, however, does this not rather link it with

2. C. L. Hull, *Quantitive Aspects of the Evolution of Concepts*, Psychological Monographs, No. 123.

System B? The difficulty here is actually one only of semantics. In respect of operant conditioning, unconscious means really 'at present out of conscious awareness'. The human subject can readily be made aware of what is taking place, at which point he can (largely) assume active, conscious control. While it is true that in classical conditioning lack of awareness of what is happening may also, and usually does, exist, *making the subject aware in general gives him no control*, certainly no direct control, over the process. The actual *mechanics* of classical conditioning tend to remain unconscious. The classically conditioned response *continues* to be involuntary. One has, for instance, no direct volitional control over the expansion and contraction of the pupil of the eye, a classical response. One *can*, however, stop oneself pressing a lever, even though when operant conditioning has taken place the urge to do so may be surprisingly strong.[3]

Operant conditioning is, then, proposed as a function of System A. This is *a priori* reasonable in view of the fact that it involves volitional, motor movements, which are a function of the cortex: the cortex has already been ear-marked for System A. Yet further functions of the cortex are cognition and the higher thought processes. It would be of value, then, if a link could be shown between operant conditioning and these higher processes. Fortunately, such a link has already been demonstrated by H. F. Harlow, of the Wisconsin Primate Laboratory.

Before proceeding to describe it, something more must be said of conditioning and learning in general. If a cumulative graph is plotted of the strength (amplitude) of the conditioned response following given programmes of reinforcement, an S-shaped improvement curve is produced, as shown in Figure 2.

3. Compare, for instance, grinding one's foot on the floor of a car (i.e. on a non-existent brake) when someone else is driving. This reaction is probably, however, also supported by classical processes involving anxiety. A purer example of operant conditioning is experienced when driving a car with a column gear change when one has been used to a floor change. The left hand is continually reaching for the non-existent gear lever. The conditioned movement can be readily corrected or halted half-way through *whenever one becomes aware of making it.*

This indicates a period of gradual improvement, followed by a time of rapid gains, with a levelling out as the ceiling of maximal response is approached.

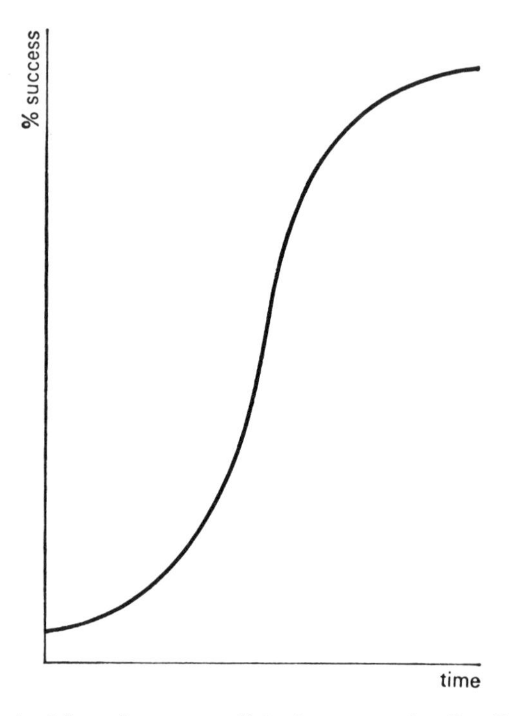

Figure 2. Typical learning or conditioning curve (stylized).

It has been demonstrated that this same basic curve underlies learning, both simple and complex, of many different kinds, in a wide variety of real-life and experimental situations. It is frequently referred to as the normal learning curve.

Critics of the alleged over-application of conditioning principles have, however, pointed to insightful learning – the sudden flashes of understanding, be these the answer to a calculus problem or the correct solution to a crossword clue – which contrast with the process of gradual learning and which would seem to be among the more typically human, higher forms of learning.

Harlow noted that apes solving problems seemed sometimes able to extract the principle of a series of learning experiments, which they then applied successfully to new examples of the problem – solving these on the first trial or after one exploratory mistake.

A monkey is so placed in an apparatus that he sits at a table across from the experimenter. A solid screen can be lowered and raised between them. While the screen is down the experimenter hides a food reward, say, a peanut, under one of two inverted, differently coloured cups. The monkey is allowed to lift one cup. If he finds the nut, he can eat it. For the purposes of the experiment the experimenter always places the nut under, say, the red, and never the blue cup – left and right positions of the cups being randomly alternated. After a number of trials the monkey begins always to choose the correct, red cup. What has taken place so far is fully explicable in operant conditioning terms. The correct response has been reinforced, the incorrect extinguished.

If the monkey is now presented with a new example of the problem – e.g. two cups of the same colour, but one in wood, one in metal – the process begins all over again. On this occasion, however, the number of trials taken to reach success level may be slightly lower. (Success is arbitrarily designated by some criterion such as six correct choices in a row.) A third example of the problem may be solved with still fewer initial errors.

Eventually the point is reached where the monkey has 'got it'. Whenever presented with a new variation – e.g. one involving a square cup and a round cup – he either makes one exploratory mistake on the first choice and none thereafter, or his first choice is right by chance and he stays with it, ignoring the other cup throughout. The learning principle involved here Harlow has called *Learning Set*.

What makes this demonstration so impressive is that Learning Set *cannot* be formed by lower organisms, such as lizards or fish, even after very large numbers of experiments. They lack the capacity to develop Learning Set. Although like other animals they can be successfully operantly conditioned within any single experiment, in the case of these lower organisms the number of trials required to reach success does *not* diminish from experiment to experiment. These facts suggest that Learning Set is associated with the neo-cortex, which the lower organisms do not possess, or possess in reduced measure.

Of crucial interest for present purposes is the fact that if the proportion of correct/incorrect choices made by the monkey within each successive example of the problem prior to solution is examined and plotted, the plotted points show up as a normal learning curve. The final moment of 'sudden' insight or understanding has apparently in fact been arrived at by just such a gradual process as takes place within the individual operant conditioning experiment.

Learning Set thus offers one bridge between operant conditioning and higher forms of learning.[4] This demonstration opens up an interesting theoretical possibility. If Learning Set evolves from operant conditioning, and still higher thought processes are perhaps also in some sense linked to Learning Set, might there not exist some equivalent both to Learning Set and to 'higher thought' *deriving from classical conditioning*?

Table 2 (a) shows the position as it exists in current psychological thinking.

What, however, might one put on the left-hand, System-B side of the table where only blanks exist, to match what exists for System A?

The second part of the table, (b), offers a theoretical possibility. It is suggested that dreaming has something of the same relationship to classical conditioning that Learning Set has to operant conditioning. Dreaming is then seen as a kind of abstraction of the principles of classical conditioning. In third place we have a theoretical concept already proposed in Chapter 2 – Knowledge II.

Whether or not the model proposed in Table 2 (b) is acceptable as it stands, the theoretical possibility that overriding principles can be extracted from classical conditioning in the same way that they can from an operant series, still remains.

4. Human beings, of course, can go on to deduce that there are 'principles of principles', and so on.

TABLE 2

(a) *The current 'orthodox' position*

	(SYSTEM B)	(SYSTEM A)
(1)	classical conditioning	operant conditioning
(2)	———————	Learning Set
(3)	———————	higher thought processes (Knowledge I)

(b) *A theoretical extension of the above position*

(1)	classical conditioning	operant conditioning
(2)	dreams	Learning Set
(3)	Knowledge II (higher dream processes)	Knowledge I

(2) *Freud and the Psychoanalytic Movement*

Freud has often been accused of (deliberately) describing his theories in such a way as to make them unverifiable by experiment. What seems rarely considered, especially by critics, is that the phenomena described may be *by their very nature* not susceptible to this kind of proof.[5]

5. With this statement I am not attempting to begin a defence of some mystical half-world where all is flux and nothing certain, and where arguments need never be lost because at the moment of defeat one merely shifts one's ground or claims to have been talking of something else altogether. This view of psychoanalysis is by no means uncommon among experimental and academic psychologists. Let us indeed admit at once that at least some psychoanalysts are fully deserving of the suspicions they arouse – and perhaps all deserve them sometimes. With this, however, I certainly do not wish to excuse the equal and opposite follies of experimental psychology.

We thus have the 'mists of obscurantism' on the one hand, and the obtuseness of the narrowly academic view on the other – both of which it seems to myself must be avoided.

What I am in brief suggesting is that there may exist more than one kind of proof, and more than one kind of evidence: that the word proof will in fact bear a wider interpretation than that accorded it, in particular, by the physical sciences. If we could demonstrate, for example, that some other principle besides cause and effect existed in the universe, we might thereby at the same

Be this as it may, while it is possible to make relatively brief statements about the essential nature of theories of conditioning, without either being unfair or suppressing necessary information, the same does not hold for psychoanalysis. The insistence on brief, standardized descriptions of the processes involved – principally by opponents of psychoanalysis – is almost certainly inimical to genuine understanding of those processes. One can, however, be less circumspect about the *overall* significance of psychoanalysis.

It is not wholly incorrect to say that the position of psychoanalysts is strong in general and weak in particular, while conditioning and learning theories are strong in particular and weak in general. This is to say, the latter provide precise, acceptable accounts of what occurs in simple learning situations – but are much less satisfying when applied to more subtle and complex life situations. Psychoanalytic views seem to guide us far more meaningfully through the labyrinth of real-life situations in all their perversity and subtlety, but are of much less use in approaching – and psychoanalysts are in general contemptuous of – the standardized and artificially simplified laboratory situation.

While in the popular mind psychoanalysis is a form of treatment for some kinds of mental disturbance, principally neurosis, it has in fact a much wider scope. It is a means of arriving at a better understanding of oneself in relation to the life-experience and, in some sense, to the universe itself. It embraces both normal and abnormal function. Its outstanding contribution, at least from the standpoint of the present argument, is the demonstration not so much that the unconscious mind exists, but that it possesses a *coherence* and a *dynamism* that are other than those of consciousness.

Although it is the insistence on the power and autonomy of the

---

time have removed the need for, for instance, either quantification or repeatability. Some other criterion, or criteria, would apply instead.

It is said frequently that seeing is necessary for believing. Perhaps, conversely, believing is necessary for some kinds of seeing.

unconscious which perhaps all of us immediately associate with psychoanalytic theory, it is no less true that Freudian dynamics depend equally on the power and autonomy of *consciousness*. This is something which Freud himself insisted on over and over again – as if aware that what might emerge from his work would be a one-sided view of personality. His own words, quoted later in this chapter, offer ample evidence. Nonetheless he had no choice but to champion the unconscious and focus all eyes on it – precisely because his own era, the late nineteenth century, evidenced so completely the supremacy – that is, the apparent supremacy – of the conscious personality. It is possible in fact, as obliquely proposed in Chapter 2, to view the relatively recent history of *homo sapiens* as the story of the rise of consciousness and the fall of the unconscious. Fall here does not mean destruction – while implying some limitation of activity – so much as a driving underground. Retaining this imagery for the moment, it is not by any means misleading to think of the functioning of the unconscious in modern Western society as a kind of guerilla action.

The Freudian view of personality is based on two opposing forces, the id and the ego. The ego (among other things, consciousness), 'the most superficial layer of the id', is in touch with reality and is modified by it. The ego is said in fact to be ruled by the *reality principle*. It is considered to have evolved from the id. The id is wholly unconscious and is the source of libido – that is, of sexual drive. Another outgrowth of the id, the super-ego – a kind of conscience or censor – sets boundaries to the ego by limiting, sometimes severely limiting, the amount of gratification, dereliction of duty or whatever, which the ego may allow itself.

The general picture derived from the Freudian model is of a beleagured ego (consciousness) in imminent danger of engulfment by the libido, limited on all sides by reality and from above by the super-ego. Yet the importance which Freud actually attached to the ego is seen from his own words, in extracts (1), (2) and (4) below. As already suggested, one probable reason why Freud chose to emphasize the role of the unconscious was, paradoxically, because the unconscious was in fact getting such an extremely bad deal in the Western culture of the day – a view he himself expresses in extract (4). His emphasis on the unconscious at the expense of consciousness was demanded by the lop-sidedness of his, and our, time. However, the central notion of two forces – the ego an as it were equal and opposite force, pitted against the like strength of the libido, with the outcome favouring now one, now the other, is well brought out in the following extracts from the *Introductory Lectures on Psycho-Analysis*, particularly the last.

(1) It is undeniable, however, that the course of the development of the libido may be influenced from the direction of the ego. (p. 295)

(2) Psychoanalysis has no conceivable motive in denying the existence or significance of the ego-instincts while it investigates the part played by the sexual instincts . . . (p. 294)

(3) A person only falls ill of a neurosis when the ego loses its capacity to deal in some way or other with the libido. (p. 323)

(4) The way in which [the neurotic protects himself against neurosis] is certainly a very interesting subject, well worth studying. . . . *The danger is that the unconscious will be overlooked* . . . and that everything will be judged as it appears to the patient's own ego. . . . The ego is after all the force which denies the existence of the unconscious and has subjected it to repressions. . . . As soon as the nature of repression begins to dawn on us we are advised not to allow *one of the two contending parties, and certainly not the victorious one*, to be judge in the dispute. We are forewarned in being misled by what the ego tells us. (p. 318, my italics)

The antithesis of id and ego in Freudian psychology as two fairly matched and highly evolved systems – two alternative organizations – is of very great interest to our theme. Here at last we have a theoretical psychological framework which offers some hope of accommodating the phenomena of possession discussed in detail in Chapter 1.

This book equates the Freudian id (along with its energy, the libido) with System B, and the ego (drawing its energy from the ego-instincts) with System A. Freud himself has described both of these in terms very acceptable to the present argument.

First, you will notice how closely connected the libido and the unconscious on the one hand, and the ego, consciousness and reality on the other show themselves to be . . .[6] (p. 302)

There remain, however, apparently significant differences between what Freud assigns to the ego and what is assigned here to System A – differences which it will prove nevertheless possible to accommodate in due course. For example, Freud places some aspects of ego in the unconscious. At all levels, and not least at the instinctive, however, he maintains the antithesis, the fundamental division between id and ego.

As brought out in the extracts above, Freud ascribed the origin of neurosis to conflict between the two components of personality. He was able to substantiate this, at least theoretically, in respect of what he termed the *transference* neuroses. He was, on his own admission, far less able to account for the source and nature of the *non-transference* neuroses – that is, of what are normally referred to as the psychoses and schizophrenia.

These patients, paranoiacs, melancholics, and those suffering from dementia praecox, remain on the whole unaffected, proof against psychoanalytic treatment. (p. 366)

The present book is, however, more than happy to ascribe the psychoses *also* to the conflict of interests of libido and ego. Freud's unwillingness or inability to draw the inevitable conclusions from the evidence lay perhaps in his relative neglect of the latter ('We certainly understand the development of the ego much less well than the development of the libido . . .' p. 295). Yet, over and over again, he describes the basic situation which can serve as a ground model for the aetiology of psychosis:

6. The completion of this sentence, however, reads: '. . . although there were no such connections between them originally.'

> The connection between neurotic anxiety and objective anxiety may be found with the help of the antithesis, so often put forward, between the ego and the libido. As we know, the development of anxiety is the reaction of the ego to danger and the signal preparatory to flight: it is then not a great step to imagine that in neurotic anxiety also the ego is attempting a flight from the demands of its libido, and is treating the internal danger as if it were an external one. (p. 338)

Let us change this picture a little. Is the *neurotic* really in flight from the demands of his libido? Is he not rather its *prisoner*?[7] Is not the ego in neurosis, in however circuitous a way, the slave of the libido? Surely the neurotic, however unwillingly, *takes orders* from his unconscious?

Freud maintains at many points that an indispensable prerequisite for the cure of mental illness is the ability to form an affective relationship with the analyst; this is what transference means. It makes little difference, incidentally, whether that relationship is loving or hostile, as we shall further see later.

It is the *psychotic* who cannot *or will not* form a relationship with the analyst. Is it not then rather *he*, the psychotic, who could be said to be 'in flight from the demands of the libido'?

How would it be if the libido's demands were felt to be so great or so bizarre that they seemed to herald the total engulfment or destruction of the ego, that is, of the conscious personality? What if in self-defence the ego erected *ultimate* barriers against the libido in order to preserve itself, that is, some part or semblance of self-government?

> In the traumatic neuroses, especially in those arising from the terrors of war, we are particularly impressed by a self-seeking, egoistic motive, a straining towards protection and self-interest. . . . This tendency aims at protecting the ego from the dangers which led by their imminence to the outbreak of illness; nor does it permit of recovery until a repetition of the dangers appears to be no longer possible . . . (p. 139)

7. Here again one thinks of the enslaved Samson, the swine of Circe, and the many other such symbolic figures.

May not the 'flight' described also be what is termed madness? And would one be surprised – if events *are* as we have described them – that any affective, emotional relationship, perhaps the very thing which might breach the barriers, would be avoided at all costs? Might not the catatonic schizophrenic who sits for years on end in motionless silence have withdrawn his ego to the only real place of safety?

As will appear in a later section, R. D. Laing's many insights into the nature of schizophrenia have established a very good case for such a view.

The mechanisms of Freudian psychology considered in detail further lend themselves to the position outlined. It is a commonplace both in and outside psychoanalysis to speak of 'the defence mechanisms' – a term, however, deriving initially from Freud's formulations. This next seems not to be generally acknowledged, but does not the existence of defence mechanisms imply an attacker? Let us therefore, at least for the moment, consider that there might also be identifiable 'attack mechanisms' (by forces unknown). Let it be clear, however, that in discussing attack we are here still speaking purely from the standpoint of Freudian psychoanalysis, which (even though it indulges in something like double-think on the issue) essentially regards the unconscious as the aggressor, and the conscious mind as the beleagured defender. It is not without significance that Freud contemptuously stigmatized the theories of Jung and Adler as 'conscious psychologies'. It would in fact be more reasonable to regard *neither* consciousness *nor* the unconscious as an aggressor (in quite that sense) – but to think of these rather as two unfortunately opposed systems each engaged in the business of furthering its own ends.[8]

However, continuing to maintain the essentially Freudian position, there follows in Table 3 a sample list only of some

8. A view incidentally also supported by Freud: '. . . from the outset [the id and the ego] each have a different relation to the task-mistress Necessity, so that their developments are different, and they acquire different attitudes to the reality-principle.' (p. 344)

typical 'defence' mechanisms, together with some of the now proposed 'attack' mechanisms, which are then discussed in detail. As to how one decides which is which, it is simply a matter of asking which part of the personality is calling the shots – consciousness or the unconscious?

TABLE 3

(*Attack mechanisms*) The influence of the libido on ego (system *B* on system *A*)	(*Defence mechanisms*) The influence of the ego on libido (system *A* on system *B*)
(1) Freudian slips; motivated errors	(6) repression; denial
(2) neurotic tiredness; psycho-somatic illness	(7) projection
(3) rationalization	(8) reaction formation (*see text*)
(4) prejudice	(9) intellectualization
(5) major life decisions based on unconscious wishes	(10) displacement
etc.	etc.

The comments following apply as numbered to the items of Table 3.

Attack Mechanisms

(1) The 'Freudian slip' or motivated error. These slips are considered by Freud to be not really mistakes but to have a meaning, to be give-aways, as it were the tips of icebergs projecting from the unconscious. Opponents of this view argue that these slips are simply nothing more than retrieval errors of memory. The frequent external resemblance of the mistake to the correct form is stressed by the conditioning theorist.

A man writing a cheque for forty-two pounds might find that he has written *fourty*-two. The motivated error (or Freudian) view proposes that the man writing the cheque e.g. resented the expense ('I wish it had been four pounds'). To regard this as 'just a mistake' seems, perhaps, in this instance to be more reasonable. But what if a man, in 1970, enters the date on a cheque as 1932?

Alternatively, a shy person who has been more or less dragooned into giving a public lecture says in the course of it 'The point I want to strain is that . . .' Is it the attitude of the speaker which has substituted 'strain' for 'stress' ('stress and strain' being of course a common expression)? The slip appears to be saying 'I find this lecture a strain' – which agrees rather well with the known facts.

If this evidence still does not convince, what are we to make of the man who mistakes the date of his wedding? What of the woman, who, noting her husband in the street, referred to him as a friend, having completely forgotten that she had been married to him for some weeks? (This marriage, Freud tells us, came to a very unhappy end some years later.[9])

(2) Neurotic tiredness. This is something which most people experience occasionally. In fairly extreme cases the individual is always 'too tired (or unwell) to go to work', 'too tired to entertain friends', and so on. The main observable (though not absolute) difference between neurotic and genuine fatigue is that the former *can* disappear or modify from one moment to the next with a change of circumstance.

Psycho-somatic illnesses are allied here in the sense that they are, expressed simply, partly or wholly produced by an unconscious attitude of mind, or by a situation, rather than by a disease. Thus moving away from one's mother's neighbourhood might have the effect of causing a bad skin condition to disappear.

(3) and (4) Rationalization and prejudice. Psychologists normally include rationalization (and prejudice) under defence mechanisms. It is evident that this is an error. All one needs to ask is – who wins, the unconscious or the ego? In both these cases it is fairly clear that the unconscious has effectively won.

This is illustrated by simple experiment. A subject is hypnotized in the course of a social gathering and given the post-

9. *Introductory Lectures on Psycho-Analysis*, p. 45.

hypnotic command to fetch his umbrella into the drawing room and open it, fifteen minutes after the session, and to forget that he has been instructed to do so. The subject duly does as commanded. When those present ask why he has done it he replies matter of factly that he needed to see whether the umbrella had any holes. The subject is rationalizing, that is, giving an apparently rational explanation for an (unconscious) irrational act.

(5) Major life decisions based on unconscious wishes. This phenomenon is harder to illustrate to the satisfaction of the sceptic. (In all fairness, it is difficult to see how one could produce experimental evidence in its support.) Examples are provided, however, by those individuals who go somewhere for a holiday for a few weeks – and then, astonishingly, never return, but spend the rest of their lives in that hitherto strange place. Or by the businessman who wanders into a lunch-time lecture or evangelist meeting – and is 'saved' or converted on the spot. While this is frequently only a transitory state, it can happen that a man's or woman's life is totally, irrevocably altered from then on by this chance (?) event.

It is, incidentally, extremely difficult even to begin to account for *these* particular phenomena in terms of conditioning and reinforcement schedules.

Defence Mechanisms

Here one sees the reverse side of the coin – the conscious mind dealing more or less successfully with libido or unconscious impulses.

(6) Repression; denial. According to Freudian theory any mental content in consciousness which is painful or otherwise disturbing to the conscious mind or ego may be repressed, that is, pushed down into the unconscious. The memory of this act of repression is also repressed at the same time, so that the whole incident from start to finish is 'forgotten'. This act produces at least a temporary solution to the difficulty or conflict, and one

that is satisfactory to consciousness. That, however, is not the end of the matter. The repressed content is said to gather energy to itself, and attempts to re-enter consciousness. As it grows more powerful, more and more energy is expended by the ego in maintaining the state of repression (hence, possibly, neurotic tiredness). The more powerful the repressed content becomes, the more it can counter-influence conscious behaviour, by such means as Freudian errors.

Denial is allied, but is perhaps more the refusal to consider that something (external) is so. A land-owner might deny that his tenants were suffering, when it was obvious to an outsider that there was actually very real suffering. Denial perhaps is the blocking of disturbing ideas seeking to enter awareness through conscious perception.

(7) Projection. Appropriately placed among the defence mechanisms, this is a warding-off, a *defence-against* System B activity. In Freud's terms it is an attempted flight from the demands of the unconscious.

The existence of the mechanism is readily demonstrated. A large number of people who know each other, say, students following the same course at a given college, are asked to rate each other and themselves for meanness with money. The individuals voted by the large majority as possessing this trait not only attribute a significantly high level of meanness to all other individuals, but a low level to themselves. That is to say, the person agreed by the common consent of others to be mean does not only not see himself as mean, but instead, by projection, finds this trait in everybody around him.

(8) Reaction formation. This (as its name implies) is yet another form of denial of an inner state, by means this time of the outward display of the *reverse* of the internal state. Thus an outward bully may be an internal coward. A person noted for giving his opinions loudly and at length may secretly have strong feelings of inadequacy.

(9) Intellectualization. The discussion of various personal or human elements (pain, loneliness, love) in dry, scientific,

objective terms, or within academic, legal or business frameworks is a way of defusing potential emotional explosives. It is a common device for reducing one's emotional involvement, feelings of guilt or whatever, and of denying the valid emotional rights of others. As we shall see in due course, this mechanism is rife in Western culture, as are in fact all the defence mechanisms discussed here. The matter was already touched on in Chapter 3 in connection with the latinization of public language.

An example of this all too powerful weapon in the hands of the intellect is that of naming the unnameable, thereby trapping or destroying it. The process is described in many fairy-tales, such as *Rumpelstiltskin* and *Ali-Baba*. (In early Judaism, conversely, the name of God was not to be spoken or written down.) Discovering the name, the magic phrase, the 'open Sesame', both tames the phenomenon and places the power within conscious grasp. (Unfortunately, these contents frequently arrive there dead. So fairy gold examined in the light of day becomes dross.)

A yet worse form is the replacement of a name by a number. This is employed in concentration camps and prisons as part of a process of dehumanization, and partly to protect the guards against their own feelings.

(10) Displacement. Emotion deriving from or appropriate to a particular situation felt for some reason as too threatening, may be displaced to another, less threatening one. A youngster feeling aggression against his parents may go and instead hit his smaller brother or sister, or his pet dog. Experimentally, a rat can be induced to attack, say, a rubber doll, in preference to another bigger and more dangerous rat.

This necessarily all too brief review of un-free behaviour was undertaken partly to emphasize the existence of two basically opposed relationships with the unconscious. These are, to repeat them, an *involvement-with* and a *defence-against* (or flight from). An extension of this basic position, incidentally, could constitute a definition of mental illness. Both too much involvement and too strenuous a defence constitute a hazard to mental health, an imbalance of the psychic economy, leading ultimately to serious abnormality.

Involvement-with may, however, be on a willing or unwilling basis – and in the latter case it is therefore better described as *domination-by*.

Willing involvement of various kinds with System B is displayed by the enthusiastically superstitious, the conventionally religious, the uncritical believer in psychoanalysis, many pacifists, many do-gooders, and hordes of others. Unwilling or only partially willing involvement is displayed, for instance, by the dominated and ineffectually resentful husband, the unmarried son (or daughter) unable to escape from the mother, the severe neurotic nevertheless prepared to undergo a long and painful psychoanalysis in order to become free.

Conversely, *defence-against* (so clearly displayed in psychosis, as we shall see later) differs from unwilling involvement-with, in the sense that the defence-against situation, though demonstrably un-free, has been established *in some sense* on the individual's own terms – as opposed to someone else's. He has, so to speak, chosen his own poison. He cannot be said to be free of the influence of his unconscious, but he is fighting a grim and at least partially successful rear-guard action. This individual is, as it were, prepared to cut away the rigging in order to remain afloat; or risk bursting the sails in order to continue to run before the wind. Or we could say that he has retreated to his castle, within which, despite the barbarian hordes without, he enjoys a measure of security and exercises a measure of self-government and autonomy. It is only (!) that he has cut himself off from the surrounding, natural countryside – that is, from warm, affectional and spontaneous relationships, from the experience of meaning as opposed to significance – which are his birthright. His sole remaining contact with it must then take the form of brief, armed forays, protected by his intelligent contempt and his sterilized scalpels, under the banner of his university degree.

These analogies are possibly over-emotive, or perhaps not emotive enough. A fuller discussion of the further ramifications of the defence-against position is best kept, however, for the detailed consideration of psychosis.

Before closing the discussion of Freudian psychoanalysis a word must be said of the mechanics of dreaming, since this is such an important part of Freud's theory.

The dream, in Freud's view, is essentially a fantasy wish-fulfilment expressed in fantasy or symbolic form. And this holds true, he maintains, also for unpleasant, even horrifying dreams. Here the wish is disguised. The dream is also in a sense a message from the unconscious to consciousness about the general state of the nation and in particular the needs of the libido. Most of the requirements of the libido are unacceptable to the ego, however, especially in later life after many years of social training. It is for this reason, according to Freud, that the libido dresses up and disguises the dream. What we recall on waking is the disguised, so-called *manifest* dream. Dream-analysis (associative and symbolic interpretation) breaks the code and reveals the true, so-called *latent* dream – the dream's real meaning.

Much of the manifest dream can be shown to be made up from waking experiences of the previous twenty-four hours – people one has met, remarks one has heard, and so on. Moreover, pin-pricks experimentally administered to a sleeping individual, or drops of water on the forehead, may be immediately incorporated into an ongoing dream. Most people occasionally experience the fortuitous incorporation of knocks on doors, alarm clocks ringing and so forth.

The fundamental error of the experimental psychologist engaged in attacking the Freudian view of dreams has been to base the attack solely on the manifest dream – e.g., by causing the kinds of incorporations we have just mentioned to occur under experimental conditions. The Freudian view is based quite explicitly on the *latent* dream – on its associational and symbolic values. Freud does not deny that water experimentally dropped on the forehead may be incorporated into the dream. He asks instead why every individual incorporates it differently. One dreams he is drowning in a river, another that he has found an oasis in the desert, a third that he is being tortured by Indians.

Freud's view that the manifest dream is an evasive smoke-screen or a sop to the prudery of the conscious mind seems to me somewhat over-complicated, however. Is it not simpler, and more in line with other observations made of the unconscious (of System B), to propose that the 'disguised' nature of the dream is an accident – that is, that dreams are that way because it is in their nature to be so? On this view, the products of the unconscious are symbolic, obliquely expressed and associative, because these happen to be the characteristics of that system.

In the same way, the 'mysterious' utterances of the Delphic Oracle and other diviners and soothsayers would not be intentionally so (in the way that the conscious mind understands intention), any more than it is a Russian's intention to be mysterious when he speaks Russian.[10] As already pointed out, Freud himself, however, supported the view of the deliberate, concealing encodement of unconscious material by the unconscious.

Freud has stated the ego to be under the control of the reality principle. This view is acceptable to the present argument.

10. While the possible nature of unconscious truth is a matter for later chapters, some comments are appropriate here. As suggested earlier, a function of the unconscious may be to summarize and collectivize experience (perhaps only certain kinds of experience) along lines *other* than those practised by the conscious mind. These lines, as we know, are likely to be associational rather than logical or rational. The 'language' in which any 'conclusions' might be expressed would be that of symbols, not words. These, incidentally, would probably contain a universal component as well as a learned component based on the experience of the individual. A white street-lamp might be the best summary possible for 'cowardice + erection + night-time + playing in the street as a child'. A bear might be the most appropriate symbolic rendering of 'father + embrace + terror + masculinity + a visit to the circus'.

It may also be the case that in contrast to the permanent connections established by the intellect, those of the unconscious are – at least to an extent – reformed to suit the demands of the moment. This suggestion, as we shall see, has much to recommend it.

Freud's further declaration that the libido is ruled by the pleasure principle – hence dreams are wish-fulfilments, and indeed many Freudian errors contain a hidden wish – is not, however, acceptable, unless it is clearly understood to apply *only* to the *sexual* instincts, and not to the whole of the unconscious.

In this connection nevertheless it is interesting and apposite to recall the behaviour of the rat with the electrode embedded in one of the pleasure centres, who spent all his waking hours activating the electric current. Would one consider that the experimental psychologist had liberated the rat's libido from the restraining influence of his ego?

However, the pleasure principle cannot serve the argument of this book as an adequate description of the principle governing System B. Some other principle will have to be sought; though this will need to incorporate the pleasure principle at least as a sub-case.

At this point we end our consideration of Freudian psychology. The analytic theories of Jung, too, are postponed for more appropriate consideration elsewhere.

(3) *The Relation of Conditioning and Learning Theories to Psychoanalysis*

There exists at present no answer to the question, which of the two main persuasions of modern psychology, the psychoanalytic or the experimental, has produced or will produce in the long run the better description of the human psyche. At present it appears to the outsider – to the person, that is, not committed *a priori* and therefore probably for not wholly reasonable reasons, to the position that either of these approaches is the only approach possible – a case of match drawn.

To myself, as one such outsider, it seems obvious that the evidence offered by each faction will never wholly be explained away, least of all by the opposing faction. Nor does the reason for this view lie merely in the simple differences of methodology and terminology of the two factions. It is my view that each has hold of a different and equally important approach to the investigation of personality *in each case deriving from different aspects of the personality itself*, since where else could they derive from – the nature and modes of functioning of which differ along the kinds of lines continuously suggested by this book. The very existence of the two major persuasions of psychology under discussion itself gives support to this view.

However, avoiding for the moment the ultimate confrontation with this central proposal, let it instead be asked whether there exists an area of behaviour where the two approaches in question have met as it were head-on – and where the outcome appears to be a partial victory for both. There exists such an area. It is that concerned with the identification of stimuli and the processes of stimulus association – the stimuli being in this case words, presented briefly under experimental conditions such that both straightforward, i.e. retrieval, errors, and 'Freudian errors', can occur.

Regarding association, experimental psychology has unfortunately demonstrated its usual penchant for gathering large numbers of facts, without having much idea what to do with them once gathered.[11]

Table 4 provides an example of this. The data in question was gathered by a number of psychologists at the beginning of this century.[12] So far no real explanations have been forthcoming either in respect of this or of large amounts of similar material.

The precise make-up of the sample groups involved is unfortunately not known. Considering these results in passing, however, it is clear that children very seldom give what one

11. It has been said that psychologists collect facts for which they have no explanations, while philosophers collect explanations for which they have no facts. In fairness, then, let it be agreed that many 'wholistic' or 'integrational' theorists are for their part content to devise universal theories on the basis of small quantities of fact (e.g. Freud, Piaget), skating cheerfully across gaps and around objections. Possibly, however, this is the only way to make any real progress.

12. See R. S. Woodworth and H. Schlosberg, *Experimental Psychology*, 1955, Chapter 3.

TABLE 4

Associational response made to a verbal stimulus by three groups

stimulus word	*associated response*	*1,000 children*	*1,000 men and women*	*1,000 men in industry*
TABLE	eat	358	63	40
	chair	24	274	333
DARK	night	421	221	162
	light	38	427	626
MAN	work	168	17	8
	woman	8	394	561
DEEP	hole	257	32	20
	shallow	6	180	296
SOFT	pillow	138	53	42
	hard	27	365	548

might call the logical opposite of the stimulus word as a response, whereas adults frequently do.[13] Men in industry (in a certain sense, masculine men) give this *opposite* noticeably more often than a group of less masculine (?) men, or at any rate than a group of women. The average group frequency (of 1,000 individuals) for responding with an opposite over the whole list, only part of which is reproduced in Table 4, is: for children, 43; for men and women mixed, 298; and for men in industry, 473.

Thus children, both male and female, are inclined *not* to see the world in terms of direct opposites. A mixed group of men and women – and therefore *presumably* women in particular – are

13. Strictly speaking, 'table' is not the *opposite* of 'chair'. Nevertheless, they do seem to have much of the force of opposites.

much less inclined than a group consisting solely of men to see the world in such terms. There is thus some case for considering children as more related in this aspect of behaviour to women than to men (to the female, rather than the male principle?). Experimental psychology, as noted, has no incisive comment to make on the data. Regarding the view of personality proposed by this book, however, the data seem supportive of the general position we have taken up, and further reference will be made to it in due course.

Reverting to the proposed discussion of stimulus recognition and stimulus association, psychoanalysis has always maintained that 'taboo' and similar words (i.e. swearing, colloquial names of parts of the body, and so on) are avoided because of their relationship to repressed and/or forbidden mental contents and wishes. Jung in particular held that attitudes or reactions to certain words provided clues to unconscious conflicts or complexes.

One of the prime difficulties which has always faced the psychoanalyst is that of gaining access to the unconscious. Dreams provide one avenue, free-association, as it is termed, another. The latter (in favour of which Freud abandoned hypnosis) consists of persuading a patient to say whatever comes into his or her head (however irrelevant, irreverent or inconsequential it may seem), in connection with topics suggested by the analyst. Jung narrowed this technique down to the practice of giving single words to which the patient had to make a single word response (as in the data of Table 4).

While all the features of the response which Jung considered significant need not concern us here, some of these are: an unusually long pause in making an association; blushing, speaking softly or mumbling, and other displays of embarrassment; a mis-hearing or misunderstanding of the stimulus word; failure to recall the original response in a second run-through. When any of these indicators are present they are taken as evidence of an emotional disturbance or conflict.

In the general context of the experimental study of reaction times, a device called a tachistoscope – simply a modified slide projector – projects images or printed words on to a screen for pre-determined fractions of a second. There is an exposure time, which varies from individual to individual, termed a threshold, below which the stimulus cannot be identified by the subject. Among the many factors which can raise or lower a subject's threshold, one is the nature of the word itself. That is to say, some words are identified, and not others, at one and the same exposure time.

With the exposure time set just above the subject's previously determined threshold, such tachistoscopic projection demonstrates clearly that many individuals fail to 'see' taboo words (like 'fuck', etc.) and anxiety-arousing words (e.g. vagina, menstrual), while identifying ones like house, gramophone and so forth. But the changes in skin conductivity registered when these other words appear show that they have indeed been recognized at some level. There is no suggestion, however, that the subject is consciously lying.

Critics of such demonstrations argued that taboo words were far less often seen and heard than most words; that, in effect, they had undergone a very different (that is, less intensive) *schedule of reinforcement* during the subject's lifetime. Experiment showed that neutral words which occurred with roughly the same frequency as the impolite words in the spoken and written media – such as diaton, prole, etc. – earned the same lack of recognition when given the same exposure time as taboo words. In association tests these rare words also produced delayed responses, forms of uneasiness, etc.

There are a number of points here. First, it is likely that many of these unfamiliar words arouse similar reactions to taboo words because the person suspects that that is what they are. If this seems far-fetched, let us remember the use of such words in comedy shows and farce *precisely* to obtain this very effect (she asked to see my testimonials, and so on). Possibly the subject is also put out by his failure to understand the word, and so on.

Third, and more importantly, it is hard not to be impressed by the clear associational evidence from the lie-detector of knowledge of a crime on the part of an individual who denies having any such knowledge, but who is subsequently by some independent means proved to be the criminal, and who meantime reacts appropriately on the detector to words associated with the misdeed.

In respect of the findings discussed here the typical situation has arisen in which (a) the psychoanalysts and their supporters deny that the alternative findings of the experimentalists in any way invalidate their views on the significance of taboo and other index words *vis-à-vis* unconscious conflict and (b) the experimental psychologist claims to have demolished the psychoanalytic case.

The overall picture is, naturally, rather more complex and confused than this brief review suggests. It is perfectly possible to retain a single-minded psychoanalytic or a single-minded 'experimental' view of the results in word recognition and word association, without losing the respect of one's colleagues – depending, of course, on who those colleagues are. In other words, many psychologists on both sides of the fence steadfastly refuse to look through the Galilean telescopes of their opponents – for fear, possibly, of what they might see.

(4) *Psychosis*

In this section we shall look in more detail at the mentally abnormal states of psychosis and schizophrenia. Both these terms refer to the same broad area of deviant behaviour, but the first is the more general – that is, while all schizophrenics are psychotic, not all psychotics are schizophrenic. Each is a blanket term covering wide ranges of 'mad' behaviour. There is some tentative agreement on more sharply defined categories within these very broad divisions, but, on the other hand, the patients themselves rarely oblige by conforming to, or remaining within, the category assigned to them. Many psychotics display a wide diversity of symptoms.

I do not seek to deny the real or theoretical differences

between forms of psychosis, but this is not my primary concern; nor were such differences our concern earlier when neurosis was considered. What basically concerns us here is the distinction between neurosis and psychosis as a whole. That *these* broad divisions represent two fundamentally differing conditions is the view of many, though not all, psychologists. As far as the present book is concerned, the neurotic is to be seen as the *prisoner* or *victim* of his autonomic nervous system (of System B as a whole, and in particular of the kinds of classical conditioning to which he was subjected as a child). The psychotic, however, is viewed as one sort or another of *fugitive* from his autonomic system.

These and other possible aspects of psychosis are described here with particular reference to the writings of R. D. Laing, and with some detailed reference also to a case study by Bruno Bettelheim. On first encountering this material the general reader may be surprised to hear 'mad' patients talked about as if they were in some sense sane, and exercising in their astonishing world at least on occasion a kind of volition. And it remains true that the views expressed by the two writers named are not the views of every psychiatrist, particularly those of the older schools. The views of Laing and Bettelheim are, nevertheless, among the most advanced of our times, and currently gaining wide acceptance.

Before proceeding, one point of method adopted here and maintained now through subsequent chapters requires clarification. In the literature on personality and mental health, two terms occur frequently, which refer to the overriding unity, or to the centrally integrative force(s), of the personality. These terms are 'ego' and 'self'. There is, unfortunately, no generally recognized or consistent difference of usage respecting them. Even where one author is himself consistent, this does not carry over to other authors. The convention observed from now on is as follows. The term 'ego' is used to refer to the experience of oneself that is associated with waking consciousness and the conscious personality. The term 'self' we shall avoid defining too closely at this stage, but fairly obviously it will have links with and relate to the unconscious side of the mind. Where necessary the usage of other authors will be amended in the sense that the appropriate 'translation' will be inserted in square brackets after the term used by the other author. This is not, of course, designed to alter what that writer is saying, but to enable an orientation to be maintained to the present book.

Laing's basic premise is that psychosis is the pathological form, the exaggerated version, of the condition experienced by most human beings occasionally, difficult to describe, but which gives the feeling that one has become somehow detached from one's body, or from external events (for instance, in the course of serious accidents, or during some other great stress or shock), namely *dissociation*. He suggests, among other things, that some individuals are more predisposed to this experience than others, undergoing it more often and more strongly. In Laing's own words:

> Quite apart from these 'ordinary' people who feel in moments of great stress partially dissociated from their bodies, there are individuals who do not go through life absorbed in their bodies, but rather find themselves to be, as they have always been, somewhat detached from their bodies. It is with certain of the consequences of this basic way in which one's own being can become organized within itself that . . . this book will be principally concerned. This split will be seen as an attempt to deal with a basic underlying insecurity. In some cases it may be a means of effectively living with it or even an attempt to transcend it; but it is also liable to perpetuate the anxieties it is in some measure a defence against, and it may provide the starting position for a line of development that ends in psychosis. This last possibility is always present if the individual begins to identify himself too exclusively with that part of him which feels *unembodied*.[14]

The process of becoming more disembodied is, then, an attempted solution to an (unspecified) problem or problems – problems which, by definition, involve the body, or some aspect of it, since it is from the body that flight is taken. As one sees from the next extracts, body turns out to mean, really, feelings.

14. R. D. Laing, *The Divided Self*, pp. 65–6.

The quotations are taken from the case study by Bruno Bettelheim of a nine-year-old schizophrenic notable for believing himself to be a machine, which depended for its functioning on a number of other machines.

> Joey was convinced that machines were better than people. Once when he bumped into one of the pipes in our jungle gym he kicked it so violently that his teacher had to restrain him to keep him from injuring himself. When she explained that the pipe was much harder than his foot, Joey replied: 'That proves it. Machines are better than the body. They don't break: they're much harder and stronger.'
>
> Joey had created these machines to run his body and mind because it was too painful to be human. . . . Reared by his parents in an utterly impersonal manner, he denied his own emotions because they were unbearably painful. . . . He wanted to be rid of his unbearable humanity, to become completely automatic.[15]

Needless to say, this solution to the problem, the divorce of mind and body, or feeling and body, does not turn out as planned. In Laing's words:

> The self [ego] then seeks by being unembodied to transcend the world and hence be safe. But a self [ego] is liable to develop which feels it is outside all experience and activity. It becomes a vacuum. Everything is there, outside; nothing is here, inside. Moreover, the central dread of all that is there, of being overwhelmed, is potentiated rather than mitigated by the need to keep the world at bay. . . . This in fact defines the essential dilemma. The self [ego] wishes to be wedded to and embedded in the body, yet is constantly afraid to lodge in the body for fear of there being subject to attacks and dangers from which it cannot escape. Yet the self [ego] finds that though it is outside the body it cannot sustain the advantages it might hope for in this position. (R. D. Laing, *The Divided Self*, pp. 80 and 161)

As Laing and Bettelheim point out, the dissociation of the psychotic is not simply a temporary reaction to a specific situation of great stress or danger, reversible when the danger has passed. It has become a fundamental reaction to *all* life. More-

15. Bruno Bettelheim, 'Joey, A "Mechanical Boy" ', *Scientific American*, March 1959.

over, it is one that is often observed already in the first months of existence. 'But in the patients here considered . . . they seem in fact to have emerged from the early months of infancy with the split already under way.' (Laing, p. 79)

> How had Joey become a human machine? From intensive interviews with his parents we learned that the process had begun even before birth. Schizophrenia often results from parental rejection, sometimes ambivalently combined with love. Joey, on the other hand, had been completely ignored. 'I never knew I was pregnant', his mother said, meaning she had already excluded Joey from her consciousness. His birth, she said, 'did not make any difference'. (Bettelheim)

Precisely why this split occurs in some children and not others, even within the same family, is due probably to the interaction of any of a number of predisposing factors – genetic make-up, chemical imbalance, perhaps brain damage, etc. – with any of a number of precipitating factors – a particular relationship to parents, severe isolation, too many changes of environment, and so on. Some of these factors possibly have both roles, and perhaps no one of them by itself is either necessary or sufficient cause.

Of the types of family situation frequently numbered among the possible precipitating factors one, as we just saw, is the *indifferent mother.* Laing likewise stresses this view. The extract refers to Peter, a male patient.

> His parents were never openly unkind to him . . . *they simply treated him as though he wasn't there* . . . he was never cuddled or played with . . . . His mother hardly noticed him at all . . . she had eyes only for herself. (pp. 120–21, 126)

The question is how such indifference – which appears again and again in the case histories presented by Laing – leads to psychosis and schizophrenia. What mental processes are actually involved?

Somewhat oversimplifying – though not falsifying – the personality of the new-born infant, it can be said that the cruder and more obvious emotions are fully developed at birth, while the ego is present only in the sketchiest of forms. All will have observed babies in the grip of the most extreme rage (the whole body red with anger) as well as in similarly extreme states of fear and pain. Freud said that if a baby had the strength proportionate to his rage, he would destroy the whole world, and regarded the rise of civilization as following on man's ability to repress these extremes of infantile rage.

The ego – which as distinct from self-awareness mediates awareness of other individuals, as well as of a physical world that is distinct from the infant and not under his direct control – is a central aspect of personality which only *develops*, along with motor ability, intelligence, and so on, as the child grows up. And to an extent which is difficult to specify precisely, and which no doubt varies from one individual to another, the ego seems also made or marred by the conditions under which it develops.

Let us now recall a number of points – first, Freud's view that the ego reacts to *internal* danger in the same way and with the same anxiety as it does to *external* danger – the former being as real as the latter.

Let us recall, too, Harlow's demonstration that a baby monkey will face danger – e.g. the drumming teddy-bear – bravely, and indeed *will go forward to explore it*, providing that it can touch the wire and cloth 'mother' and be reassured by her physical presence. When the 'mother' is absent from the compartment the infant monkey crouches frozen in abject terror.

Let us note also that many psychotic patients as children were, typically, never cuddled, never played with, often bottle-fed, frequently ignored or left alone. Let us note, finally, that in work with maladjusted and disturbed children, hugging or holding them is a known method of quieting an emotional outburst. One frequently suspects indeed that one of the functions of the child's outburst is to force one to touch him.

Even aside from disturbed children, or babies who may be somehow predisposed to anxiety, normal children often show fear of being alone, or of being in the dark, and of their dreams. Are these fears not clearly *of the children's own manufacture*

– since what *real* harm would be likely to come to a child tucked up in its own bed in its own house in the next room to its parents? All children crave routine (bed-time stories and so on), structure, reassurance. These are props to the faltering, developing ego.[16]

In view of all the foregoing can one hesitate for a moment in suspecting grave developmental damage in a baby who is not merely denied even a minimal amount of normal reassurance and physical contact from its parents, but who in addition has some in-built, exaggerated sense of insecurity? Is it so unlikely that the desperate ego of such a child, as soon as it is able, will take whatever emergency measures it can in regard to its fears? Is it not likely to attempt to seal them off permanently, in the way that a dam is built to keep back flood waters, or a cage built for a wild animal? How at any stage, one asks, may such a deprived child go forward to explore its own emotions, even begin to understand them, learn to control them without denying them, and at last be able to accept them as a part of himself?

The second kind of mother that emerges from the case histories of psychotics is what I shall refer to as the *operant* mother. As contrasted with the indifferent mother, this parent is notably businesslike, efficient, training-oriented. The child in the care of such a mother is unlikely to experience loneliness and anxiety to the extent just discussed. (His contact with his mother, however, resembles in many ways that of an animal with its trainer.) He obtains ego-stimulation, but without emotion, without warmth – or possibly with a certain amount, though not necessarily a marked amount, of covert hostility and rejection. An example of such an approach to a child is provided by the mother of Julie in connection with the 'throwing-away' game which is often played with babies by adults. In the usual form of

16. 'Julie had a spell at about the age of ten when she had to be told everything that was going to happen in the course of the day and what she was to do. *Every day had to begin with such a catalogue.* If her mother refused to comply with this ritual she would start to whimper. *Nothing could stop this whimpering, according to her mother, but a sound thrashing.*' (Laing, my italics.)

this, of course, the baby throws the object (the rattle, or whatever) and the adult retrieves it. Julie's mother, however, relates: 'I made sure that *she* was not going to play that game with me. *I* threw things away and she brought them back to *me*.' (Laing, p. 185)[17]

The rearing process in the hands of these mothers has a very strong resemblance to the *operant reinforcement schedule* of the experimental laboratory. The child *does* what he or she is told to *do*, and learns what it is required to learn. Spontaneous and self-initiated behaviour is at the same time rigorously discouraged. This is reflected in the reports of many of these psychotic patients that they were 'model children', 'always good', 'never any trouble'. Of Julie her mother said precisely 'she always did what she was told'. Laing comments that he shudders with foreboding whenever he reads such statements in the case reports of patients referred to him. He goes on (p. 186):

> One may note . . . that it is not unusual to find in schizophrenics a precocious development of bodily control. . . . One is certainly often told by parents of schizophrenics of how proud they were of their children because of their precocious crawling, and walking, bowel and bladder functioning, talking, giving up crying, and so on. One has to ask, however, in considering the conjunction between what the parent is proud to tell about and what the child has achieved, *how much of the infant's behaviour is an expression of its own will . . . whether the child develops a sense of being the origin of his own actions*, of being the source from which his actions arise. [My italics]

A circus dog does not jump through a hoop because he wants to, but because you want him to.

With these statements have we perhaps not now also found the basis for man's intense interest in and involvement with *puppets*, over so many thousands of years and in all civilizations – discussed in detail in Chapter 1 as one aspect of possession?[18]

17. As, of course, with a dog.

18. cf. also this remark of Laing's: 'The dissociation of the self [ego] from the body and the close link between the body and others is conceived not

That the behaviour under discussion is really operantly conditioned, and does not arise simply from a reluctance or not wishing to act in any other way, is shown by the remarks of James, an adult patient of twenty-eight. What is called good behaviour in the child is seen in its almost nightmarish quality when compulsively displayed by the adult:

> For instance, his wife would give him a cup of milk at night. Without thinking, he would smile and say, 'Thank you'. Immediately he would be overcome with revulsion at himself. His wife had simply acted mechanically and he had responded in terms of the same social mechanisms. Did *he* want the milk, did *he* feel like smiling, did *he* want to say 'Thank you'? No. Yet he did all these things. (Laing, p. 144)

Both the indifferent and the 'operant' mother have one thing in common – they fail to 'feed' the child's emotions or to put him in normal touch with them. In the first case the development of the ego-personality is also neglected. In the second, a kind of ego-development is encouraged, but it is a controlled, subservient, conditioned ego-development, from which volition and self-determination are excluded.

A brief word can be said here of probable differences between the neurosis-producing and the psychosis-producing mother (or, of course, family). The neurosis-producing mother is likely, for instance, to be an extremely emotional, hysterical or histrionic personality – in contrast to the marked coolness or objectivity of the psychosis-inducing parent. The former may well love her child too much. She may spoil him, smother him with affection, indulge in *too much* physical contact of a not particularly healthy kind – for instance, by allowing him to share her bed up to adolescence. Or she may hate the child – not the covert rejection of the psychotic mother, but an obvious, open dislike or worse. The hatred, like the love, will be physical. The mother may often strike the child, feed it roughly, even tie it up or lock it in a

---

only as operating to comply with others, *but as being in the actual possession of others.*' (Laing, p. 144, my italics)

room. Unlike the indifferent mother, this parent from *her* side has a clear neurotic involvement with the child.

In all the cases cited, the child of the neurosis-inducing mother 'enjoys' two great advantages: he observes emotion in someone else; and he has his own emotions stimulated. These, of course, are likely to go unsatisfied in any real sense. They will be incorrectly associated, and their expression may earn him a good deal of pain and punishment. But *at least they are there.* Hate is almost as good as love in the prevention of psychosis.[19] This surprising statement is borne out by the patients themselves. The following are the words of a twenty-six-year-old psychotic, Joan. (Laing, pp. 166 and 172)

> It is wonderful to be beaten up or killed because no one ever does that to you unless they really care and can be made very upset . . .
>
> If you hate you don't get hurt so much as if you love, but still you can be alive again, not just cold and dead . . .
>
> I kept asking you to beat me because I was sure you could never like my bottom but, if you could beat it, at least you would be accepting it in a sort of way. Then I could accept it and make it part of me.

The child who is later to become psychotic, it seems, has erected temporary or permanent barriers of one kind or another against his feelings (also represented by the body and body functions), against the unconscious. These, however, prove to be inadequate at times of great emotional or other stress and, in the long run, altogether. Interestingly, we use the term 'breakdown' for this time of failure. In a sense the pre-psychotic walks in fear of his own and other people's feelings. This suggests another reason why this child cannot risk being disobedient. The consequences are too unpredictable. Unlike the normally – abnormally – peaceable and diligent pre-psychotic, the *neurotic* child is the source of unending inconvenience to those around him – presenting a continuous pageant of bed-wetting, soiling, crying, oversleeping, insomnia, constipation, stealing, tantrums,

19. In Chapter 4 we suggested, on other grounds, that the female principle (the unconscious) could accept love or hate, but not indifference.

aggression, timidity, jealousy, poor health, loss of appetite, over-eating and whatever.

What is the evidence, apart from the strong theoretical position we have established, that the future psychotic is terrified of his feelings and emotions? The confirmation emerges in the course of the psychosis and during its cure.

Going to the toilet, like everything else in Joey's life, was surrounded by elaborate preventions. We had to accompany him. . . . He could only squat, not sit, on the toilet seat. He had to touch the wall with one hand, in which he also clutched frantically the vacuum tubes which powered his elimination. He was terrified lest his whole body be sucked down. (Bettelheim)

'No one seemed to realize that if I went back to my family I would be sucked back and lose myself.' (Laing, p. 173)

Mrs D., a woman of forty, presented the initial complaint of vague but intense fear. . . . Her fear was 'as though somebody was trying to rise up inside and was trying to get out of me'. (Laing, p. 59)[20]

'My interviews [with the psychiatrist] were the only place I felt safe to be myself, to let out all my feelings and see what they were really like without fear that you would get upset and leave me. I needed you to be a great rock that I could push and push and still you would never roll away and leave me.'[21] (Laing, p. 166)

In the following extract, concerning Joey when he was getting better, we understand why going to the toilet was for him such a terrifying experience.

20. It will be clear from this and other examples in this section that many reports of 'possession' during the Middle Ages were simply of individuals undergoing a psychotic breakdown. This is widely agreed, and is in no sense a discovery of this book.

Our closer concern, however, is with the theme of possession *in respect of normal human beings*. The extremes of the psychotic and neurotic personalities will furnish us with clues to the nature of this. The legend of Charybdis, for instance, the whirlpool which dragged mariners to their doom, an everyday item of classical Greek consciousness, is clearly paralleled in the previous two cases.

21. This is what the patient should have experienced as a child with her mother.

> [Drawing his fantasies] was the first step in a year-long process of externalizing his anal preoccupations. As result Joey began seeing faeces everywhere: the whole world became to him a mire of excrement. At the same time he began to eliminate freely wherever he happened to be. (Bettelheim)

Can we doubt that Joey had secretly *always* thought of the world as potentially a sea of excrement, into which he would be sucked down were it not for his machines – i.e. his ego in projected and artificially enlarged form? Can one fail to suspect that all these psychotic patients see the emotional (and autonomic) side of their personalities as in some way totally devouring, overwhelming or destructive? As Joey's (true) ego-strength grew with protracted therapy the controls and defences could be gradually dismantled and the risk could be undergone of the world's becoming – being? – a mire of excrement.

It may well be that all babies have experiences of the world similar to those described by adult psychotics. But in the normal course of events – that is, with the implicit and explicit support of the mother and other adults who are patently not swallowed up, or destroyed, or driven away by the child's fantasies, but are reassuringly unharmed by them – these experiences may not be terrifying.[22] They may in some ways even be pleasurable (cf. the thrill of terror). It may be that only when the working through and experiencing of the fantasies and accompanying sensations is delayed beyond the appropriate time that the prospect of them becomes, paradoxically, increasingly more terrifying to the developing ego – *precisely* perhaps because of its increased sense of reality.

In some psychotics the 'life-saving' (actually, life-destroying) initial stratagem of denial has been only too successful. The feeling side of the personality is almost completely beyond the reach of the ego. This is frequently perceived by the patient, who may be desperately aware that some precious part of life – that is, real life – is being or has been lost, perhaps for ever.

22. Psychoanalysts report that children whose parents die often believe unconsciously that they have killed them.

> One schizophrenic woman who was in the habit of stubbing out her cigarettes on the back of her hand, pressing her thumbs hard against her eyeballs, slowly tearing out her hair, etc., explained that she did such things in order to experience something 'real'. It is most important to understand that this woman was not courting masochistic gratification; nor was she anaesthetic. Minkowski reports that one of his patients set fire to her clothing for similar reasons. (Laing, p. 145)

There is, finally, the 'spy in the enemy camp' situation, where the psychotic consciously or deliberately takes on the aspect or character of what he most fears – a step that somehow prevents the final annihilation. It is illustrated here by David, an eighteen-year-old schizophrenic. Like so many of the others, he had always been a very good child, who did everything he was told and 'never caused any trouble'. Until the death of his mother he had simply been whatever his mother had wanted him to be.

> But then he found he could not stop playing the part of a woman. He caught himself compulsively walking like a woman, talking like a woman, even seeing and thinking as a woman might see and think. . . . For, he said, he found that he was driven to dress up and act in his present manner as the only way to arrest the womanish part that threatened to engulf not only his actions, but even his 'own' self [ego] as well, and *to rob him of his much cherished control and mastery of his being*. (Laing, p. 72, my italics)

This was essentially the theme of Hitchcock's film *Psycho*.

Joey also, in another sense, is an example of this type of response. Machines were better and harder than people, and ran things. Therefore the best course of action was to become a machine.

A more general instance of the behaviour being discussed, however, is found in the speech of schizophrenics. So typical is the evolved language or 'word-salad' of these individuals that it is sometimes termed 'schizophrenese', and regarded as a diagnostic symptom of the illness. As the examples show, it is unquestionably associative in character.

Julie in her psychosis called herself Mrs Taylor. What does this mean? It means 'I'm tailor-made. I'm a tailored maid: I was made, fed, clothed and tailored.'[23]

'She was born under a black sun.
She's the occidental sun.'

. . . She was an 'occidental sun', i.e. an accidental son whom her mother out of hate had turned into a girl. (Laing, p. 204)

Nor could he Joey name his anxieties except through neologisms or word contaminations. For a long time he spoke about 'master paintings' and 'master painting room' (i.e. masturbating and masturbating room). (Bettelheim)

It is as if the patient has decided or at least is able *to take over the process of association that is normally unconscious.* What one sees here is essentially a creative process. It is not *mindless*, as it at first sight appears, but extremely *mindful* – frequently of so high a quality that one does not exaggerate the position to regard it as art.[24] The general relationship of the unconscious to the creative process will be more fully discussed later.

In other cases, however, one has the impression that the unconscious has somehow got into waking consciousness. It is not easy to think of a suitable analogy here – perhaps something as bizarre as an army of octopus rising from the depths of the ocean to invade a tourist beach. So:

I read newspapers because they gave me newspapers and things to read, but I couldn't read them because everything that I read had a large number of associations with it. I mean I'd just read a headline and the headline of this item of news would have – have quite sort of– very much wider associations in my mind. It seemed to start off everything I read and everything that sort of caught my attention seemed to start off everything I read and everything that caught my attention seemed to start off, bang-bang-bang, like that with an enormous number of associations moving off into things so that it became so difficult for me to deal with that I couldn't read.[25]

23. A clear parallel to the 'operantly shaped' (or tailored) behaviour of the experimental rat or circus animal.

24. It differs from art, perhaps, in that the patient cannot really help doing it.

25. R. D. Laing, *The Politics of Experience*, p. 124.

In discussing the origins and aetiology of psychosis Laing, Bettelheim and others are explicit in the view that the phenomena observed are either not explicable, or are much less adequately explicable, in psychoanalytic terms than in some other terms.

During his first year with us Joey's most trying problem was toilet behaviour. This surprised us, for Joey's personality was not anal in the Freudian sense; his original personality damage had ante-dated the period of his toilet training. (Bettelheim)

An insensitive application of what is often supposed to be the classical psychoanalytic theory of hysteria to Mrs R. might attempt to show this woman as unconsciously libidinally bound to her father; with, consequently, unconscious guilt and unconscious need and/or fear of punishment . . .

However, the central or pivotal issue in this patient's life is not to be discovered in her 'unconscious'; it is lying quite open for her to see, as well as for us (although this is not to say that there are not many things about herself that this patient does not realize). (Laing, p. 56)

One proposes at this point that the terms in which psychosis, as opposed to neurosis, is best accounted for are those of an *ego-psychology.*

In the extracts from Freud's writings it was clear that the influence of the libido (the unconscious) on the personality required the existence and resistance of juxtaposed ego forces. Conversely, it seems equally clear from the multiple evidence of the present section, that explanations of the actions of psychotics and schizophrenics require, not simply an ego-psychology in isolation, but the existence of juxtaposed libido-forces if they are to make adequate sense.

In the interests of maintaining the distinctions proposed, from here on we shall refer not so much to psychoanalysis – since this term in the hands of its proponents has a history of being able to account for the totality of human behaviour – but rather to libido-analysis or *libido-psychology.*

There will be more to say of both psychosis and schizophrenia at another stage. For the moment, however, we turn to a brief consideration of the *experimental* position regarding mental abnormality, to see what evidence, if any, is forthcoming to support the views of this last section in general, and in particular the proposed divisions into ego-psychology and libido-psychology.

Whereas psychoanalysis regards neurosis and psychosis as having essentially the same origins – as being in most senses the *same* illness (so that schizophrenia is seen as a very bad case of neurosis) – experimental psychology views these as essentially different. This latter view is that also of the present book, and appears, as we have seen, to have the additional support of at least some leading psychiatrists.

The kinds of pointer which the experimental psychologist looks to, and his methods, are described by H. J. Eysenck in several of his publications – the one directly concerned here being *The Dynamics of Anxiety and Hysteria.* Eysenck reports a study in which a number of psychiatrists were asked to indicate in respect of a large number of patients those traits or symptoms, from a long list of these, which applied to each particular patient.

Some of the items frequently indicated by these psychiatrists in respect of *neurotic* patients were: lifelong or episodic anxiety; neurotic traits in childhood; unsatisfactory early life; lifelong or episodic hysterical symptoms; anxiousness; symptoms of over twelve months' duration before admission; unsatisfactory adolescent adjustment; family history of neurosis; lifelong or episodic obsessional symptoms; low energy output; and bad work record.

In respect of psychotics, frequently mentioned symptoms were: delusions; hallucinations; ideas of reference[26]; motor disturbances; mood disturbances; impairment of thought or memory; retardation; suicidal impulses; social withdrawal; suspiciousness; and family history of psychosis.[27]

26. The patient imagines that everything, e.g. the Queen's Speech, refers to him.

27. The present book is not concerned to criticize either the method or these

One can broaden one's impression of these two groups of patients further by reversing the statements about the opposite group. If the neurotic child has an 'unsatisfactory early life', the pre-psychotic child presumably has a satisfactory early life. From the mother's point of view at least, as we have seen, this is true.

From the theoretical side, experimental psychology sees the neuroses and psychoses as the outcome of bad, inadequate or inappropriate learning (conditioning) during the early years of childhood. They are, as it were, agglomerations or hierarchies of bad habits. The experimental treatment of these two conditions is usually termed behaviour therapy. It involves the application of the principles drawn from conditioning experiments – first to undo or dismantle the faulty habits, and second to establish new, correct ones. The more precise details of the procedures involved do not concern us here.[28]

However, what *is* of extreme significance for our present argument is that *classical* procedures are used in the treatment of *neurosis* and *operant* procedures in the case of *psychosis*.

As far as the writer is aware, while it has been frequently suggested that some sort of classical conditioning situation underlies neurosis, the proposal put forward in the previous section

findings, as such. One notes a certain amount of overlap, however, between items like 'unsatisfactory adolescent adjustment' in the first group and 'social withdrawal' in the second.

28. Quite aside from what may and may not eventually prove to be the facts and the final truth about neurosis and psychosis, I suggest that the tendency to perceive situations as (single) wholes is in any case – and quite independently of the facts – a function of the type of personality that is or tends to be attracted to psychoanalysis; while conversely the opposite tendency to break situations down into parts – again regardless of the 'facts' – is a feature or function of the personality-type which is drawn into experimental psychology. The former psychologist will be inclined to try to keep the whole person in view. The latter will be more concerned with discrete aspects of behaviour, from which when required he builds up a 'whole' person. In the two cases we detect the separate influence of System B on the one hand, and System A on the other.

that some, or perhaps all, psychoses have a similar *operant* basis, has not so far been made elsewhere. Yet does not the fact that it is operant methods which are successfully used in the behavioural treatment of psychosis, quite aside from any other considerations, urge this conclusion?

One might argue further that operant responses established in the early years of life, like classical responses in the same period, are possibly unusually resistant to extinction. The generally greater 'plasticity' of the young organism, compared with the greater resistance to impressions and stimuli of the older organism, is in any case a fairly well-agreed fact of psychological life. Actual examples of the persistence of early operant conditioning may be seen in the 'imprinting' of newly hatched ducklings on moving objects. (A topic to be fully discussed in Chapter 11. As has been established experimentally, the amount of effort, that is, of physical movement, which the duckling expends in following the moving object is the major variable in determining the strength of the resultant imprinted learning.) A more everyday instance is seen in the persistence of regional and class accents in adults, often despite attempts to change or disguise the accent. Even where the cover-up is apparently successful, such a person almost invariably reverts to his original accent under conditions of stress.

A comparison of the relative merits and demerits, and in particular the respective aims, of behaviour therapy and psychoanalysis would not be wholly relevant to the central argument in hand. It is suggested, however, that they be regarded as complementary, rather than as alternatives.

This text in no way seeks to deny that the mechanisms underlying the classical and operant processes are an important, even a major, part of personality structure. In terms of evolution and development they certainly pre-date the emergence of the higher personality functions. Probably the latter could not have evolved, and in fact did not evolve without them. And yet there is no justification for what one can only describe as the astonishing assumption made by experimental psychologists that a complete explanation of personality and the facts of psychological existence can be constructed in such terms.

In concluding this chapter, several points may be emphasized.

There are *two* basic mechanisms of human conditioning. This statement still holds even if one prefers the alternative minority view that they are two forms of one process.

There are *two* broad groups of mental illnesses – neurosis and psychosis. This statement still holds even if one prefers the less extreme view that they are two forms of one process, e.g. transference and non-transference neurosis.

There are *two non-experimental* forms of treating these: (1) psychoanalysis – which admits that it can do little if anything for psychotics and (2) what this text has termed ego-analysis – the type of approach advocated by Laing, Bettelheim and others, who in addition explicitly reject the psychoanalytic interpretation of these conditions.

There are *two* forms of behaviour or conditioning therapy, one based on classical conditioning and used with neurotics, the other based on operant procedures and used in respect of psychosis.

This book has earlier postulated *two* major areas of personality – on the basis of course of quite other evidence – which are referred to as System A and System B.

Without wishing to open up any detailed discussion of these next points at this juncture, the argument begins to perceive, among other things, links between System A, psychosis and experimental psychology; and between System B, neurosis and psychoanalysis. Conceptually, conditioning principles – *all* conditioning principles – are a System A product. System A products are by nature reductionist and dissociative in character. The analytic view deals with, effectively, the same phenomena as does conditioning, but in System B terms. System B theories are always by nature wholistic and associative in character. At this stage no further explanation of this tentative position can be attempted.

Some Socio-Behavioural Phenomena of Western Society 6

A point has perhaps been reached where the ground has been sufficiently prepared for the central theory of this book to risk advancing to be recognized. From now on the theory itself, rather than discussion about the need or the justification for the theory, takes the centre of the stage. Evidence will, naturally, continue to be offered.

There appears to me to be no way whatsoever of avoiding the conclusion that the human personality – the personality of every member of our species – is essentially of a dual character. Not simply this. The two parts each have a very large measure of autonomy. They subserve quite different ends. They are, indeed, centred in two different 'universes'. The first of these is the inner, existential universe of man. The second is that of the physical, material universe around us, which exists in our presence and our absence, both before we are born and after we are dead. This second universe is, certainly, governed by laws which we as human beings can come in a sense to understand and manipulate. But in its actual creation we had no part whatsoever.

Thirdly, and perhaps most importantly, the manifest attributes of the two divisions or systems of the personality *are in every conceivable sense*, and every single aspect, opposite and antithetical. One cannot stress too highly the central importance of this last statement, which will be justified.

These two main systems of the personality we have termed

System A and System B. We have spoken, respectively, also of the male principle and female principle. For principles, however, we would be better served by two less sexually oriented expressions – which is not, however, to deny that the male and female of the species have close respective links with those principles. In respect of System A we can appropriate an existing term from Freud's psychoanalytic theory – the *Reality* Principle. For the second a suitable term also already exists, but this is not widely known. In order to have a term of reference to hand until the other can be further defined, System B will be said to be governed by the *Alternative* Principle.

Some further refinements to the tentative basis of personality outlined in earlier chapters are now in order. Previously it was stated on numerous occasions that System B was female and System A male; similarly, that the former was unconscious, and the latter conscious. As generalizations these statements continue to hold good. They are, nevertheless, incomplete. For part of 'maleness' (System A) is also unconscious; and part of System B is not female. In this connection, incidentally, it is worth recalling Freud's insistence on the separation of the libido and ego instincts in the unconscious.

The stylized form of the 'personality triangle' shown in Chapters 3 and 4 is this:

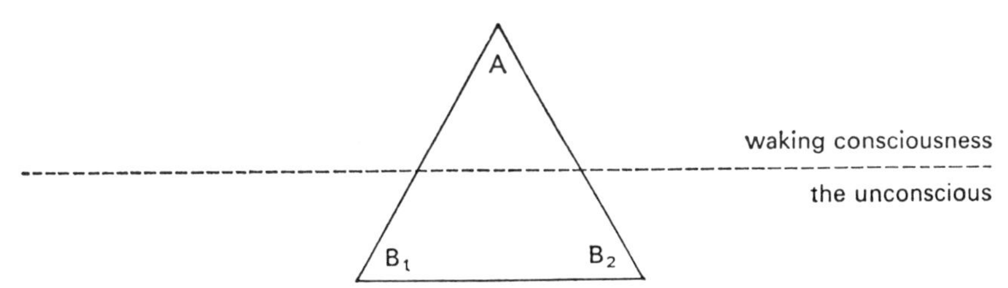

Figure 3

We know already many of the attributes of these sub-elements. Sub-System B2 (derived from the sympathetic division of the autonomic nervous system) is concerned primarily with aggression and flight; with states of rage and states of fear. Sub-System

B1 (the parasympathetic division of the autonomic nervous system) is concerned with states of rest, of quiescence, of simple *being*; and, importantly, with *sexual* arousal.

Women, one observes, are not generally noted for their aggression. They are, at least in men's eyes, noted for their sexuality. Interestingly, it is difficult, if not actually totally inappropriate, to refer to a woman as a coward. Certainly women experience fear, and certainly they sometimes run away. More usually they hide (perhaps because men are better runners!) or submit in some way. But, still, they are not as a rule termed *cowards* – certainly not in the physical sense. No one really expects them to fight, or at any rate, to fight and win. Men *are* expected to fight, and to refuse or to run away is cowardice.

Thus, for these and the many other reasons already discussed in previous chapters, sub-System B2 is more logically assigned to the male, while sub-System B1 may be more appropriately considered female.

System A, on the other hand, remains essentially a male province. It is concerned primarily with (1) volitional motor and muscular movement – and the male is clearly more muscular – and (2) with *logical*, objective thought. Already the implication behind the last statement will seem to cast a slur on the intellectual that is, the logical, ability of women. There are, unfortunately, worse 'slurs' to come. Yet these line up very well with observed fact, as we will show.

The typical male and the typical female personality profiles are exhibited below in Figure 4. The use of capital letters indicates

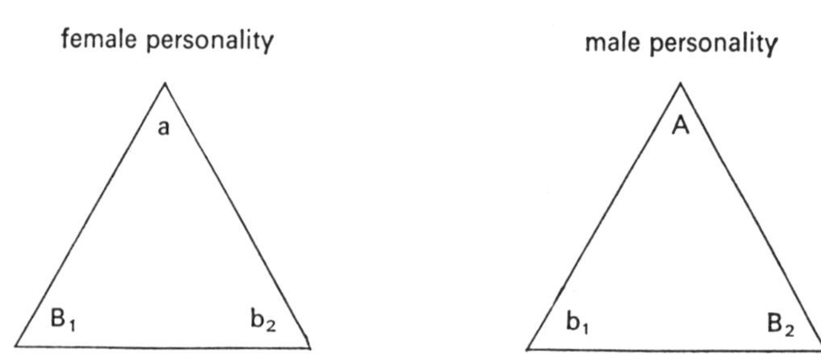

Figure 4

the dominance or preponderance of the sub-system concerned; small letters indicate that the sub-system concerned is not dominant.

Studying Figure 4 one may be tempted to make quite unnecessary value judgements and say that the male personality is 'better' endowed than the female. It is preferable to say instead simply that the human male and female personalities are differently endowed, with particular strengths and weaknesses in opposed areas.[1] In any case it should be constantly borne in mind that *all* human beings, male and female, possess *all* the attributes in question in *some* measure. *Of course* women are capable of aggression and logical thinking. Of course men can be gentle and considerate. The point one is making is that certain of these attributes tend to be typical of most men, and others to be typical of most women.

If a horizontal dividing line is inserted between the conscious and the unconscious, as in the two triangles of Figure 4, the acceptable result is produced that the female personality is predominantly unconscious – i.e. therein lies its major endowment – while the male personality is predominantly conscious.[2]

Regarding the use of capital and small letters in these models, it will be appreciated that this is a drastic oversimplification of the actual position, for demonstration purposes. There are clearly many degrees of aggression possible between B2 and b2, just as the precise level of logical thinking (A to a) will likewise differ from individual to individual. Moreover – to say this once

1. We are considering in this section allegedly observable facts (or, at minimum, tenable theories). Praising or lamenting those facts does not alter them. In this area at least, wishing does not make anything so.

2. With this the Jungian view of personality is in part rejected. Jung held the female side of the male and the male side of the female to be unconscious. In the present model the male side of the male personality *and* the male side of the female personality is always conscious; while the female side of the male, *and* of the female, is always *unconscious*. Women, by construction and relative endowment, are, however, less directed by their conscious minds than are men. Their particular gifts lie mainly in the unconscious. The reverse holds for men.

again – it is of course clear that not *every* woman is a wholly typical or average woman, just as not every man is representative of the male norm. With this clearly understood, we can continue to handle the component parts of personality not unmeaningfully in these gross terms.

Various permutations of the basic model are possible which also yield acceptable results – results, that is, which produce some reflection of what can be observed in everyday reality. If the figures now considered are, however, stereotypes rather than characters in the round, that is actually no weakness. More will be said of the nature and genesis of stereotypes in a later chapter.

A certain type of dedicated or singleminded academic, the pedant or scholar – lacking in aggression, but in no way considered feminine – could well have the first personality profile shown below in Figure 5. The model here brings out, too, that the personality of this individual is basically a conscious (volitionally directed) one. The unconscious, or libidinal, side of personality has *relatively* much less influence. The second profile of Figure 5 could variously be that of certain homosexuals; or of a certain type of artist or poet. Here the unconscious as a whole is rather stronger than the conscious – which, however, is not weak – and the unconscious itself is predominantly female.

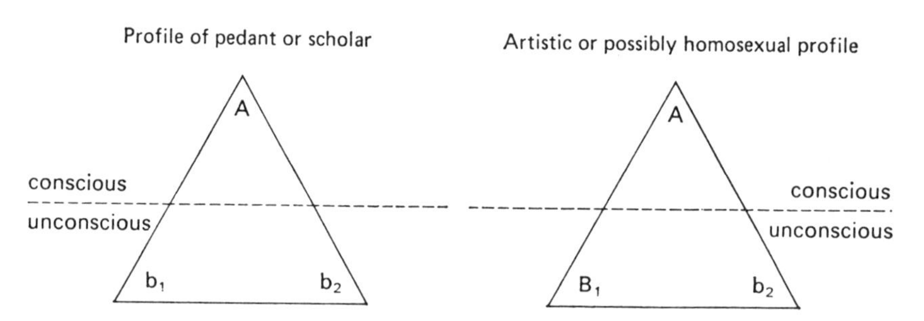

Figure 5

One needs to remind oneself when producing these models that this is not intended as a new parlour game, or even an academic 'game' having no relevance or real significance for actuality. It is true that one here generates what one might call a certain sense of astrology or fortune-telling. This is partly because typology, unfortunately, *is* associated with such dubious subjects as palmistry, phrenology and various somatic psychologies of this kind. Typology as such, moreover, is very out of favour in the West, perhaps the only respected psychologist advocating any kind of typology at the present time being H. J. Eysenck.

Nonetheless, we proceed. From now on actual diagrams can be dispensed with, and instead solely the alphabetical shorthand used. The models can be made slightly more sensitive, and less narrowly typological, by the application of some of the normal conventions of algebraic notation.

Some lesbian tendencies presumably arise because the woman concerned has a larger than usual share of (physiological/psychological) male characteristics. Such a lesbian, nevertheless retaining all or much of her feminity, a sort of 'cognitive' woman, might have the profile: A B1 b2. A butch lesbian, on the other hand, endowed perhaps with male aggression rather than intellect, might show: a B1 B2, or in extreme cases: A b1 B2.

It will be noticed that the 'feminine' lesbian has been given the same profile as the artist/homosexual of Figure 5. Indeed, one would not be looking for marked differences between these various types of individuals. The position can be somewhat retrieved by the resort to 'algebra'. Thus:

$$\frac{\mathrm{A}}{\mathrm{B1}}\ \mathrm{b2} = \text{artist/homosexual}$$

and

$$\frac{\mathrm{B1}}{\mathrm{A}}\ \mathrm{b2} = \text{intelligent, 'feminine' lesbian}$$

The implication is that the maleness of this homosexual and the artist in general is somewhat more dominant than in the case of the lesbian.

A similar distinction can be drawn between the extremely masculine or butch lesbian and the average male:

$$\frac{B2}{A}\ b1 = \text{butch lesbian}$$

and

$$\frac{A}{B2}\ b1 = \text{'normal' male}$$

While there is no intention, and perhaps in fact no possibility, of refining the notation proposed beyond this level, the device is retained as a useful shorthand. It has the advantage that one can apply it not only to real people, but also to mythological and hypothetical figures, and usefully even to situations, as will be seen.

For the moment, however, we consider two additional figures only, God and the Devil. The Christian God is strong on love, strong on anger but weak on (i.e. anti-) knowledge – that is, Knowledge I. His profile is therefore B1 B2 a. The Devil is strong on aggression, strong on Knowledge (I or II), but a non-starter in love, although he enjoys ego-centred sex. The profile is therefore B2 A b1.

Symbolically now rather than algebraically these last are perhaps best expressed as follows:

$$\frac{B1\ B2}{a}\ \text{(God)} \qquad \frac{B2\ A}{b1}\ \text{(Devil)}$$

indicating that here the major features of the personality seek to destroy or at least dominate that in the personality which is a threat, or anathema, to the dominant personality organization. This offers us possible pointers again towards neurosis and psychons respectively. Here, too, the 'feminine' basis of God and the 'male' basis of the Devil, as discussed in Chapter 1, are very much in evidence.

These considerations have taken us a little aside from the normal male and normal female profiles of Figure 3, to which attention should now be redirected. In this connection it would seem appropriate at this point to make a more searching examination of men and women in a general setting.

The view of this book is unequivocally that if the choice *had* to be made, more profit would emerge from regarding woman as a different species from man, than from regarding both as members of the same species.

This attitude is in no way shared by the experimental psychologist. Such data as actual experiment have produced nevertheless supports the views of the present book.

There are in psychology three approaches open to the experimentally inclined: (1) actual experiment (2) failing this (or prior to this) controlled observation (3) the gathering, under rather less controlled and controllable conditions, of very large amounts of information or data, which by their very size overcome the chance fluctuations that the, usually small-scale, experiment seeks to eliminate by the controlled manipulation of variables. It is the third of these approaches which best suits the very broad orientation of the present inquiry, and it is here in fact that the most supportive evidence is found.

As an initial step, nevertheless, the findings of small-scale controlled experiments are considered. It will be seen that these tell us little that we did not know already.

It appears first that girls in general excel somewhat over boys in general in all forms of verbal fluency – but not in word definition. Second, that boys in general excel over girls in general in all mathematical operations involving reasoning and problem-solving – but not in straightforward computation. It would rather seem, then, putting these two findings together, that girls are stronger in the more concrete aspects of verbal and number ability, boys in the more abstract properties of these. This observation appears to be borne out by the fact that on tests of vocational aptitude, girls score significantly higher than boys on tests of clerical aptitude and dexterity (skilled movements).

There is further a large and significant difference in favour of males on tests of spatial visualization. This links to the allied demonstration, for instance, that boys are significantly better at ignoring the visual field in which a pattern being studied is buried, than are girls. Girls have therefore been described as *field-dependent*.

In her article on sex differences, under the general entry 'Individual Differences' in the *International Encyclopedia of the Social Sciences*[3], Leona Tyler reports in the context of a summary study by E. Bennet and L. Cohen[4] some of the general conclusions of the literature – these being: (a) masculine thinking is orientated more in terms of the self [ego], whereas feminine thinking is orientated more in terms of the environment; (b) masculine thinking is associated more with desire for personal achievement, feminine thinking more with desire for social love or friendship; (c) masculine thinking finds value more in malevolent and hostile actions against a competitive society, whereas feminine thinking finds value more in freedom from restraint.

One notes here the tendency for ego to be associated with masculinity, which agrees well with the personality profiles of the previous section. The link between masculinity and aggression is also clear.

This is just about all that experimental psychology has to report on general personality variation between the sexes – except for large amounts of pure description, as contrasted to explanation, in highly technical language, of differences even more self-evident than those offered above.

A further word may be added, however, in respect of intelligence. The following extracts are taken from the article by Leona Tyler, already cited.

3. Macmillan & Free Press, 1968.

4. *Men and Women: Personality Patterns and Contrasts*, Genetic Psychology Monographs, no. 59.

One persistent question has stimulated research interest throughout all these periods: why have women's achievements failed to match those of men?[5] Why are there so few outstanding female artists, scientists or statesmen? ... From many sources evidence was accumulating that intelligence tests of all varieties do not and cannot measure pure native *capacity*. Although genetic differences in intellectual potential undoubtedly exist, the only way we have to *test* a person is to ask him questions or give him problems to solve. His performance in such a situation always reflects the experience he has undergone as well as his native capacity. It is, therefore, only natural that males show superiority in dealing with some kinds of questions and tasks and that females show superiority in dealing with others. In order to produce intelligence tests that will be equally fair to members of both sexes, psychologists try to include approximately the same number of both kinds of item. . . . When this kind of test is given to samples of the population . . . sex differences in over-all score or I.Q. turn out to be negligible.

There are two main points here. First, it seems agreed that native, that is, inborn or inherited differences in level (and in type?) of intelligence exist. Second, that the different upbringing of boys and girls tends to give them, respectively, a different kind of training.

If the article is merely saying in the statement, 'in order to be equally fair to members of both sexes', that an attempt must be made to counteract or offset the influence of the differing opportunities for learning which boys and girls have even in the early years of childhood, then well enough.

To the present writer, however, it seems that what is implied here is that differences in types of ability *dependent on sex* should be 'compensated for' in tests of ability. It is not then being suggested that boys and girls should not be tested or treated differently, even though they *are* different?

Should one have tests which are fair to *both*? Should one not rather have tests which are fair to *each*?

If one devised a joint test, say, for both elephants and croco-

5. The question may have stimulated research. One can only comment that it has brought little in the way of answers.

diles ('in whom genetic differences of ability undoubtedly exist'), it would be necessary to leave out many items precisely at which the two creatures differentially excel. For the crocodile scores zero on squirting water through his nose, and the elephant does badly swimming a measured course under water. But what on earth would be the *point* of giving the two animals the same test? And let us not say, to achieve a fair comparison. There never *can* be such a thing between two differing organisms.

Even then an essential point is somehow missed. The point is that (these) girls are being tested on devices produced in a male-oriented Western society of the twentieth century – one could say male-dominated – the expressed aim of which, meaning the test but as much the society which produced it, is *the maximum standardization both of question and response.* The 'psychotic' overtones of this last statement will, incidentally, be re-examined in full.

Let us consider some further aspects.

On the basis purely of the evidence supplied by experimental psychology, what would be one's prediction, for instance, of the relative proportions of men and women at present in psychiatric hospitals in this country? The only sensible reply to this question is to say that experimental psychology has provided us with nothing (so far) which would be of any real help in formulating an answer. Still, if one were earnestly required to make an estimate on the basis of the information that *is* provided, would one not be obliged to assume roughly equal proportions? After all, both boys and girls have some aptitudes in their favour, and as far as intelligence goes are (allegedly) more or less equal.

These are the actual figures for patients occupying psychiatric beds in the British Isles in 1967.[6]

Mental Illness		*Psychopathic Disorder*	
males	*females*	*males*	*females*
65,269	99,826	2,601	1,363

6. *Annual Report of the Department of Health and Social Security*, H.M.S.O., 1968.

The sharply reversed trends of the two categories show 35% fewer men than women in the first category, and 50% more men than women in the second.[7] Summating the two categories still produces an excess of women over men of some 34%. Is it allowable to deduce from this data that our society is more inimical to the female than the male organism? Or is this merely a differential response to stress – since the psychopathic category contains twice as many males, and prisons show a similar and still stronger bias in this direction.

The statistics of the literature are in fact littered with instances urging explanations based on sex-differences rather than any other variable or set of variables. For example, maximum admissions to hospitals of men suffering from schizophrenia are between the ages of twenty to twenty-four. For women the high point of admissions occurs between thirty-five to thirty-nine. As we noted in Chapter 1, the peak of sexual activity for men happens to be in the years fourteen to seventeen. For women the peak is reached around age twenty-eight. The 'coincidence' is interesting. Of course, one should beware of simple explanations. After all, the early twenties for a man is also a time in which he is called upon to take his place in an adult, highly competitive society. For a woman the late thirties are a time when her sexual attractiveness may diminish rapidly, her chances of having children if she has not already done so are almost past, and the menopause looms ahead. This time would prove perhaps especially difficult for women who have not married. However, one's faith in a more 'sexual' (that is, libidinal) explanation is continually revived. Kinsey[8], for instance, notes that twice as many boys who reach adolescence early have homosexual

7. The psychopath, incidentally, is described as egocentric, impulsive and asocial, and as having no conscience. H. J. Eysenck reports that psychopaths are difficult to condition classically, and that any conditioning achieved decays rapidly. These statements support the picture of a male personality severely lacking in a balancing female, unconscious, or 'classical' component. It would be of very great interest to know whether the psychopath dreams less than the normal male.

8. A. C. Kinsey, *Sexual Behaviour in the Human Male*, Saunders, 1949.

experiences as boys reaching adolescence at the normal time and reaching it late, and that this difference persists until the age of twenty-five. Kinsey suggests:

> As a factor in the development of the homosexual, age of onset of adolescence (which probably means the metabolic drive of the individual) may prove to be more significant than the much discussed Oedipus relation of Freudian philosophy.

As far as this book is concerned what is of most interest here is that early adolescence is actually characteristic of the female as opposed to the male, and that females are normally attracted to males. (Yet, here again, one must perhaps be watchful for alternative explanations. The boy who reaches adolescence early is still a schoolchild. He may, after all, be drawn to other males by lack of other outlet at the time.)

It is when one sets out to examine the incidence of assorted phenomena in terms of sex differences that one discovers what amounts to a conspiracy of silence (whether intentional, or not, is hard to say). It is rare to find breakdowns by sex in tables even where they would seem to be relevant, e.g. in records of types of mental illness and social deviance. A rather surprising majority of the readily available reference works both of a general and a psychological nature are at fault in this. In a twenty-page article on suicide by three contributors, again in the *International Encyclopedia of the Social Sciences*, for example, sex differences are not even mentioned.

In order not to leave this on a purely impressionistic basis, I examined current numbers of the five major psychological journals published in this country.[9] The first considered was the *British Journal of Psychology*. In the fourteen papers which this contained no mention occurred of sex differences at any point. In the seventeen papers which the next publication, the *British Journal of Education Psychology*, contained, sex differences were reported in one table of one paper. No discussion of these differences occurred in the text. (In both of the journals named men-

9. All published by The British Psychological Society.

tion was made in one or two papers that the sample considered was entirely male. But no instance occurred where an all-female population was studied.) In the *British Journal of Social and Clinical Psychology*, four out of thirteen papers showed sex in tables and discussed differences between the sexes. A fifth paper stated that its sample (which happened to consist of Australian aborigines) was composed of 50% males and 50% females. But no further mention of sex was made in the text. In the *British Journal of Medical Psychology*, however, *all* eight papers dealt in one way or another with sex differences, and this in many cases was the central purpose of the paper in question.

In proceeding from the first to the last of these journals one had the feeling of returning from an area of non-sanity to one of sanity – despite the bizarre nature of the subject-matter of the last-mentioned journal – to a world in which men and women do different things for different reasons. In professional circles associated with psychology it is often queried, in semi-humorous vein, whether the contributors to the psychological journals have ever seen a human being. In respect of the fifth journal, the *British Journal of Mathematical and Statistical Psychology*, one can only assume that it was intended for a readership on some other planet. It does not, at any rate, appear to be about people.

Instead of men and women, psychology talks of *subjects*. The dictionary defines subject as someone who is under the domination of another. Speaking of the psychotic individual Laing wrote: '*The body is felt more as one object among other objects in the world than as the core of the individual's own being.*' Would it be unfair to say of the experimental psychologist: 'A man or woman is seen more as one object among other objects in the world than as the core of an individual's own being'?

As stated earlier, large-scale surveys, preferably repeated over time, are an approved method of demonstrating permanent trends. The extension geographically and over time tends to ensure that the distorting effects of local or chance conditions are smoothed out. Of nature, too, it has been said in a general scientific context that she reveals all her secrets to us all of the time. Certainly it is true that she conducts at every moment an infinity of experiments, and presents us with the results of a billion surveys. Does it not show a lack in ourselves rather than any deviousness on the part of nature if we are unable or refuse to make judicious use of the information given us? The data which, say, Freud or Darwin gathered and presented had after all existed on open view since mankind began. The question now is whether there are other kinds of such evidence available which may be of use to us in the present inquiry.

I suggest that at least some of the persistent behavioural differences between men and women on show throughout history and at the present time are not due to 'local' conditions, or to a lack or surfeit of opportunity in respect of either sex, but are a reflection of permanent, genetically based differential endowments.

The topic is an inflammatory one. Quite aside from its emotional charge, however, the purely factual investigation of the interaction of environment (opportunity) and potential, which is what is being discussed, is among the most complex issues facing modern psychology. We need, therefore, to phrase what is to be said very carefully indeed.

It must first be understood that we are dealing in generalizations. A generalization does not merely admit of, but actually implies, occasional exceptions. These do not necessarily of themselves destroy the generalization or render it invalid.

With this provision we may now consider one or two instances. The reader might like to take a piece of paper and write on it the names of all women composers of music, past and present, that he can think of. On a second (or the same) piece he should then write down the names known to him of all female writers of comedies or comedy material. This exercise should sufficiently illustrate the point that women do not write symphonies nor make jokes. At least, they have not *yet* done so.

The observations in question are of interest because there *are* large numbers of female singers, instrumentalists and accompanists: and many outstanding women writers. In other words, it is not easy to see how women could be said in any immediately obvious sense to have been debarred from the general provinces, music and literature, from which our examples are taken. The environmentalist position, however, in general argues a lack of opportunity in cases of such non-achievement.[10] In the present instance at least, it would seem insufficient to employ the

10. An itself not unconnected area – that of education – can also function here analogously. It is frequently suggested in the current literature that the school failure fails academically not because, or not primarily because, he is not bright, but because his background is lacking in such items as parental encouragement, books in the home, a tradition of learning, and so on. Or, more subtly still, it is proposed that 'brightness' or intelligence is very much a product of the very early environment of the child – that is, of his early interaction with his mother, and a stimulating or non-stimulating relationship.

Let us without any further argument agree that the present wide gap between the academic failure and the academically successful would undoubtedly be narrowed, given genuinely equal opportunity for all in every sense at all stages of development. However, what I personally cannot accept is the more extreme view that under these conditions the attainment differential would disappear, or even become marginal.

In the kinds of reasoning just described the tacit assumption seems to be that the academically successful child *is* operating and has always operated under optimal conditions. The view is certainly erroneous. In the 'feminist' position – to revert now to our subject matter – it is similarly assumed that the male in our society likewise operates under optimal conditions of stimulus and opportunity.

Even a cursory inspection of the lives and background of, for example, many outstanding male artists nevertheless demonstrates extremes of both material and spiritual deprivation, of discouragement, of repeated failure, of the endurance of the contempt not merely of society but often of fellow-artists, which, one imagines, would have proved sufficient to crush the individuals concerned many times over. In such cases I at least am compelled to look for explanations in terms of a 'push from within' and not a 'pull from without'.

There is of course much more that should be said on both sides of this argument. That, unfortunately would take us beyond our present brief. In summary, one questions simply whether the achievements and non-achievements of woman are best to be understood in terms of lack of opportunity – or whether they are not better explained by genuine differences between the sexes of inner motivation and potential.

undifferentiated, blanket term 'environment'. The environmentalist must be prepared to say much more precisely how it is that our society prevents women from writing music, without preventing them from performing and interpreting it.

One cannot hope to open up here in any adequate sense the complexities of the question of the differences in male and female achievement in terms of opportunities offered or withheld. One would be obliged, for instance, to attempt a definition of what constitutes a major contribution (in a wide variety of fields) – or more importantly, to define the difference (never in any case an absolute difference) between a truly innovative and a merely (?) developmental or applied contribution.

Two comments can nevertheless be made. The first is that when the achievements of women over the last few hundred years are examined, one finds these to be contributions to the Arts rather than the Sciences. As it happens, we shall later be arguing a strong System B component (more precisely, a B1 component) in artistic creativity – so that an observation that women excel more readily in the Arts is quite acceptable. The second point is a purely personal conclusion. This is that, as a generalization, the difference between what does and does not lie within a woman's grasp is the difference between 'here is an existing situation – do something with it' and 'here is a vacuum – fill it with something entirely new'. That, in other words, the question of the achievement potential of woman, and its precise nature, devolves in the last analysis on the *field-dependency* of woman,[11] noted earlier, on their attachment to the existing, the actual and the concrete.[12]

11. The German dramatist F. Hebbel remarked: 'What man discovers in the stars, woman finds a use for in the kitchen.'

12. One has already pointed out that generalizations admit of exceptions – so that I do not have the feeling of being unfair to the genuinely innovative woman.

The statement that the majority of women are field-dependent, not merely in the microcosm of the standardized psychological test, but also in their macrocosmic activities, is one on which the reader is, of course, free to make up his or her own mind. In considering it, however, I would suggest that

Unlikely as it must by now seem, the purpose of this section is not to denigrate women. One is attempting to *define* woman – and to show, certainly, that her attempts to function as a male are probably seriously misguided. One of the aims of this book is to arrive at a Charter for Women that is not based on 'generously' granting them opportunities under the Charter for Men. This perhaps begins to sound like a defence of apartheid. Hopefully, it is not.

In magical thinking, but also in much mysticism, religion and philosophy, sex differences – including stages of sexual development and sexual practice – play an extremely important part, both figuratively and literally, as was indicated briefly in Chapter 2. Psychoanalysis (together with medical psychology) also supports the view that men and women should be regarded as essentially different, at least to the extent to which they are regarded as similar – i.e. the differences are as real and as great as any similarities.

Experimental psychology, on the other hand, and the twentieth century in general, do not support this view. Socialism has been one of the major forces in the shaping of this thinking. *However*,

---

attention be directed to the social sciences – a group of disciplines which are both new and which attract large numbers of women.

The field of psychoanalysis in particular is found to be one in which women are making notable contributions – one thinks of Melanie Klein, Karen Horney, Anna Freud and others. One asks, nevertheless, whether these contributions would have been made in the absence of the work of Freud, Jung, Adler and Reich. (The contribution, then, would be developmental rather than innovatory.) Further, if we turn to the fields of academic and experimental psychology we find the names of Pavlov, Skinner, Hull, and so on, unaccompanied by those of women. It appears, then, from such a very cursory glance at the issue, that the more purely *cognitive* psychologies have not so far aroused an especially responsive echo in the female psychologist.

These observations are moreover well in line with our general theory. It was suggested in a previous chapter that psychoanalysis is a System B phenomenon – and one would therefore expect, or at least not be surprised to find, women involved in it. Conditioning and learning theories, on the other hand, were proposed as System A phenomena – and consequently could be expected to show a lack of female participation.

it is extremely important to note that while experimental psychology says that *women* are the same as *men*, Socialism says that *men* are the same as *women*. This rather oracular statement will be further elucidated in Section 4.

There is good reason to suspect that much of the congruency between males and females demonstrated in the controlled laboratory setting is a product, that is, an artifact, of that frequently trivial and in any case transient situation. The (deeper) non-congruency is perhaps seen in the complex, enduring and relatively unstructured real-life situation.

In cafés and restaurants, if one chooses to listen, groups of men and women discuss (*choose* to discuss) quite different matters from quite different viewpoints. Certainly, of course, some of these differences can be ascribed to social and environmental influences.

However, a friend of the author's, a Scot, was living with his girlfriend who was English. A mutual friend called on this couple one evening to find them playing Scrabble. The visitor pointed out that 'Jo' was a proper name and could not be used. To which *the girl* said immediately: 'Yes it can. In Scotland it's a term of endearment.'[13]

This small but significant incident illustrates the notable fact that wives and girlfriends almost invariably take over the opinions of their husbands and boyfriends. They become 'experts' in whatever the husband is an expert (or thinks he is an expert, an even greater nonsense). The woman does not check out, and has little interest in checking out, the facts. This is not the purpose of her exercise. At a lower social level, the wife 'sticks by' the husband – that is, sees little wrong with his behaviour, even when the husband is a thief, a counterfeiter or a murderer. In these examples one sees yet further instances of field-dependence.[14]

13. The *Shorter Oxford English Dictionary* gives: *jo* (Scot.) a sweetheart, darling, beloved one.

14. The reactions of men and women to suggestion, and its stronger counterpart hypnosis, are considered in detail in Section 4.

There are other pointers also which suggest that the wife supplements or replaces her own System a with the husband's System A. She becomes more self-assured. Some women after marriage report that for the first time they can talk to men as equals. In marriage the husband can also be observed to make certain personality gains and changes, but notably in areas clearly related to System B1 (e.g. tenderness), in which he is more naturally deficient.

The above is a prosaic version of the mystic union of Male and Female, Light and Dark etc., found in many non-scientific descriptions of the universe – in which and by which the (perhaps once whole, but now divided) world is, or will be, made whole again. Man and woman *together* make up the fully fledged or fully integrated profile A B1 B2.[15] Psychoanalysis expresses similar views in somewhat different form. Freud speaks of finding a pathway into the unconscious, while Jung believed that the two sexes introduce the partners to their own respective unconscious personalities.

Looked at in this way, marriage, apart from its various functions on the reality level, is perhaps an attempt on the part of a (non-whole) individual to find wholeness. But the normal marriage possibly cures symptoms rather than causes. In the sense discussed there perhaps *can* be no cure for the 'illness' of being male or female. Except in rare relationships neither the husband nor the wife is *essentially* or permanently changed by marriage. The death of the partner often leaves the other as un-whole as before, or more so if the partner has been used solely as a crutch or shield.

While marriage can be said to provide each with, as it were, access to an auxiliary personality, psychoanalysis in general attempts more than this. It sets out to produce a permanently changed or augmented individual, not simply a supported one. In the early stages, however, it does function supportively and in many ways not unlike a marriage. The true reconciliation of

15. In this all the forces are in balance, and at the same time hold each other back from excess.

System A with System B, or B2 with B1, is not so readily or easily achieved – even if by 'readily' one understands six sessions of psychoanalysis a week for five years! The confrontation of these sub-systems within the personality is, I shall propose, not even so much a crisis of individual development, as a crisis of biological evolution.

For have we forgotten that men are terrified of women, and vice versa? Have we suddenly lost sight of the vast wealth of legend, or the copious data which emerges from the study of hospitalized mental patients, that attest to this? This book has been at pains to put at least that on record.

Why precisely *are* men apparently afraid of women? Is it enough to say that as children they were afraid of their mothers? Hardly, unless as an additional factor in some cases. For, as was noted, the mothers of many schizophrenics were often not actively cruel, hostile or punitive, or even emotional. Why should it be then, that many male patients in the course of psychoanalysis, with a sudden access of genuine fear, 'realize', for example, that the vagina has teeth?[16]

If our question is to be answered in the terms of this book it needs to be cast in the following form, and in two parts:

(1) why is System A afraid of System B as a whole, and vice versa, and

(2) why, in particular, is sub-System B2 afraid of sub-System B1, while the reverse is rather less true?

For the cake is divided in two ways, which in part overlap. One can cut so that A is one part and B (that is, B1–B2) another; or so that A–B2 is one part and B1 another.

Rephrasing the questions again, one asks first:

(1) why is consciousness afraid of the unconscious, and vice versa?

The answer lies essentially in the antithetical nature of the two systems: in the fact that when one exists, *the other is de-*

16. The critics of analysis usually propose that analysts put the idea into their clients' heads.

stroyed. This word is used advisedly. It then follows from this that neither system can ever exist in complete security and safety, even while the other exists as only a potential threat.[17]

While various examples have already been given of System A's fear of System B in earlier chapters, much less has been said of the reverse. The evidence, however, exists.

> Our meddling intellect
> Misshapes the beauteous forms of things:—
> We murder to dissect.
>
> Heaven lies about us in our infancy!
> Shades of the prison-house begin to close
> Upon the growing boy . . .[18]

The prison-house – not, perhaps, a particularly flattering reference to ego-development, for that is what it *is*. One recalls how Joey imprisoned his feelings inside machines.

The 'noble savage' destroyed by civilization is a further indictment of System A by System B, as is the naming of the unnameable, the breaking of the spell, which brings us back to reality. This same feeling lies also behind the woman's plea 'Let's not spoil it all' – which only sometimes means 'I don't want to sleep with you'. Frequently it means 'Let's not make it conscious, willed, deliberate'.

Laing speaks of the 'Medusa eye' of consciousness of his patients which strikes everything and everyone dead, kills all spontaneity, joy-in-life, the very sensation of being alive, and the possibility of happiness.

Relevant also are the lines spoken to Faust, the dry, desperate scholar, by Mephistopheles:

> Green is the golden bough of life
> And all theory grey.[19]

17. Does this now not sound remarkably like the kind of phrase which nations, and factions within nations, have used of each other?

18. William Wordsworth: *The Tables Turned* and *Intimations of Immortality*.

19. Goethe, *Faust, Part One*.

The reference to Medusa above serves to introduce here a new element – this being that while much of what the unconscious says is about itself, some of its products are *statements about consciousness.* This reversal of the view traditionally taken of unconscious material will prove extremely useful as evidence supporting the position finally adopted in this book. The Midas legend provides one example. Here Midas' involvement with and reverence of the material world (gold) results in those he loves (his daughter) being changed into lifeless metal. Midas represents consciousness.

To summarize, the impact of *System B* on *System A* is perceived as a swallowing-up, an engulfment, an overwhelming, an enslavement or (enchanted) conversion into animal form. The impact of *System A* on *System B* is perceived as a turning to stone, a petrifaction, a mechanization, a transformation of organic being into inorganic non-being, a loss of life and colour; or, rather less frequently, as a splitting, splintering, disintegrating – a familiar analogue of this being the breaking of a mirror, which brings seven years' bad luck, and which links in form also with the more general 'breaking of the spell'.

In second place we ask:

(2) Why should the masculine-aggressive side of the unconscious (B2, Freud's ego-instincts) be afraid of the feminine and non-aggressive aspect (B1, Freud's libido instincts)? The answer here is partly again found in the generally antithetical nature of the two systems – the fact that the presence of the one demands the absence (or altered function) of the other. In view of the fact, however, that the opposites of the unconscious are *linked* opposites rather than total opposites, this argument cannot be quite as forceful as previously.

One asks instead: what does the absence, or non-functioning, of System B2 entail for the male of the species? It means, essentially, that he is prepared neither for flight nor fight. On the contrary, he is now extremely vulnerable.

To this must be added the view of this book, which cannot be argued out fully at this point, that the male organism is in general far more interested in its own survival than in the survival of the species; while the female is far more interested in the survival of the species than in the survival of herself. Freud has hinted at this distinction:

> . . . a fundamental situation in which the sexual instincts had come into conflict with the self-preservative instincts . . . in which the ego in its capacity of independent individual organism had entered into opposition with itself in its other capacity as a member of a series of generations. (*Introductory Lectures on Psycho-Analysis*)

Freud is talking here principally about the conflict between these opposing tendencies within the individual. We are talking here *also* about the version of the same conflict which exists between separate male and female organisms – the first being more heavily endowed with ego-instincts, the latter more heavily with libido-instincts.

The copulating male, at all events, whether one is speaking of animals and the human male in the conventional animal position, or in the typically human position, displays his unprotected and unwatched back to the world and his enemies. It is possible that some of the force of expressions like 'being stabbed in the back' derives from this situation – one is attacked 'unfairly' when one's natural defences are not in play. Sex is thus in a sense the male Achilles' heel. It was also Samson's, of course.

The female, for reasons already suggested, does not experience the sexual act or the sexual position as anything like so great a threat. Almost the opposite, in fact. In either copulating position (i.e. lying or kneeling) she is to an extent quite literally shielded and protected by the male. Apart from the psychological reassurance gained – not lost – by having one's back protected, the physical act of having the back touched or stroked is itself pleasant and reassuring – as household pets of course clearly show us.

The foregoing arguments are a sample only of the kind of considerations that can apply when seeking an objective basis for the anxieties which this text postulates are aroused in the B2 System (and thence in consciousness) by B1 and B1-linked activity; and for the perhaps surprisingly *lesser* anxiety of the B1 System *vis-à-vis* its more aggressive counterpart.

In connection with the first part of this last statement, too, it has, incidentally, been demonstrated by ethology that the male 'prefers' to regard the female as an enemy than as a lover – and frequently 'forgets' that she *is* a lover, with predictable consequences. On the other hand (and in connection with the latter part of the earlier statement) the female *does* possess ways of dealing with the male aggression, which do not involve fighting.

The failure of the ego of the human infant to make the proper acquaintance of all or some of his emotions at the appropriate time was in addition earlier proposed as a further major source of misunderstandings and anxieties between the sub-systems in human beings – ending in extreme cases in psychosis.

The consideration of sleep to which we now turn will provide further general support for the views on sex differences expressed in this section.

It has been suggested that System B is associated with the unconscious, with dreams, with sleep and with sexual activity; also with women rather than men, and children rather than adults. The study of sleep, therefore, could be expected to produce not only further new evidence but, hopefully, supportive parallels to the conclusions already drawn.

While the findings and views of other writers are, obviously, not always presented from the angle of this book, or in the kind of detail that one requires, nevertheless a number of useful general points emerge from the outset. First, as our theory must lead us to expect, women do appear to sleep more than men.[20] I have not, however, found any direct statement to the effect that women *dream* more than men. (This would be fairly simple to

20. 'The consensus of opinion is that women seem to need more sleep on average than men . . .' Ian Oswald, *Sleep*.

verify experimentally, and seems on the face of it likely.) Babies dream more than adolescents[21] and adolescents more than adults. What unfortunately does not emerge clearly, however, is whether babies spend a greater *proportion* of their sleeping time in dreaming – since that they spend more time asleep is obvious.

Male human beings and male animals have penis erections for much of the time they are dreaming – one estimate placing this as high as 80% – an apparent, extremely interesting, confirmation of Freud's quite independent view that dreams are sexually based. Ian Oswald, in his book on sleep agrees: 'There seems to be some overlap, or special connection of function, between the dreaming sleep and sexual activity.'

Whether *women* are in a state of sexual readiness during dreaming sleep is a question that the experimental psychologist – for some reason – does not appear concerned to ask. However, some indirect light is thrown on this question by Kinsey. He reports that women have 'romantic' dreams up to the time of experiencing actual intercourse; thereafter they have dreams of a carnal nature, leading in some cases to orgasm.[22] The reverse appears to be true of boys; they have wet dreams up to the time they commence sexual intercourse. Possibly, again, these face-value connections mask other interpretations. It may be that both men and women have wet dreams during their sexual peak – which as we know is between the ages of fourteen to seventeen for boys, and from twenty-eight years onwards for women. Nevertheless, the suggestion that the physical act has more importance for women (as a sort of turning on mechanism), while boys make out on their imaginations alone, would fit rather well with some of the observations of previous sections.

Sometimes, when examining experimental findings, one has

21. 'Human babies spend a lot of time during the rapid eye movement phase of sleep [i.e. in dreaming], a far greater amount than children or adults.' Peter Nathan, *The Nervous System*.

22. Once again, physical touch has here played an important role in connection with the female.

the impression that the experimental psychologist is only discovering what the more intuitive psychologist (such as Freud) has established long since by other means. This is in no sense, of course, to suggest that intuitive judgements are always or necessarily right. A striking example where this was so is shown, however, in the following instance.

In 1953 two psychologists, Eugene Aserinsky and Nathanial Kleitman, produced definite experimental proof that dreaming in human beings is accompanied by rapid eye movements. These can today be readily detected by electrode-pads worn over the eyes by the sleeping subject. Already in 1892, however, G. T. Ladd had concluded on the basis of introspective studies that the eyes move during dreaming. Moreover, Aserinsky and Kleitman only made this rediscovery *by accident*, in the course of studying the physiological concomitants of sleep in infants – not even, therefore, on the basis of prior hypothesis. Subjects under experiment roused from sleep when no eye movements are taking place, in the large majority of cases report no dream. If the sleeper is roused when eye movements are occurring, however, dreams are mostly reported.

The great benefit bestowed by the manner of the rediscovery lies in the fact not only that with the apparatus used a continuous and complete record can be kept of periods of dreaming and non-dreaming, but in particular that subjects may be roused at various points in the record and questioned about their dreaming.

A blow-by-blow account of the detailed work undertaken by those named and by other experimenters, and by yet other researchers on the electrical activity of the waking and sleeping brain, is not required here.[23] Sufficient to say that what follows is a summary of this large volume of research, and moreover that the findings reported here are generally agreed and accepted by all those working in this field.

23. A general introduction to these subjects, with suggestions for further reading, however, will be found in the two books already mentioned, by Ian Oswald and Peter Nathan.

It appears that there are two forms of sleep – dreaming and non-dreaming. These are almost always accompanied, respectively, by the presence and absence of rapid eye movements, and also by accompanying differences in the electrical activity of the brain. It appears that the subject is actually watching the events of his dreams. These, incidentally, last approximately as long in the dream as they would if happening in real life.

In the normal full cycle of waking, sleeping, waking, there are five distinct phases, now described.

(1) The electrical activity of the brain of a fully conscious individual, with eyes open, and who is attending alertly to what is going on, or to a task in hand, is characterized by low-voltage, rapid and irregular waves.

(2) When a subject is awake, but relaxed or resting *with eyes closed*, not attending to anything in particular and free of any special sensory stimulation, or when in a reverie or day-dream, the so-called alpha rhythms appear. These are detected *at the back of the head* and are rhythmic waves having a frequency of some ten waves per second.

(3) As a person becomes drowsy, the alpha rhythm is gradually lost and replaced by irregular, slower waves.

(4) As true sleep begins the waves become larger and yet slower (the so-called delta waves), which alternate, however, with sharp, small bursts or 'spindles' of fast waves from the front of the head. The presence of these spindles is an important recognition-mark of true sleep, also termed orthodox sleep.

(5) About an hour after the onset of sleep, the first period of dreaming begins. The state of sleep accompanying dreaming is often called *paradoxical* sleep (as contrasted with orthodox sleep). It is characterized by low-voltage, rapid, irregular waves *very similar to those displayed in the waking state*. However, despite the demonstration that the cortex is active in a manner resembling its normal waking state, it is, paradoxically (hence the name), much harder to arouse an individual from this kind of sleep than from orthodox sleep.

The principal stages of the waking-sleeping-waking cycle outlined above are shown graphically in Figure 6. Only one or two phases of dreaming sleep are indicated, though there may be between four and six of these in a normal night's sleep. The first occurs about one hour after the individual has achieved orthodox sleep, the gap between phases of dreaming being of the order of ninety minutes and the actual time spent dreaming increasing with each phase. It is a matter for debate which of the two kinds of sleep is the deeper. If one considers the electrical activity of the cortex or the amount of involuntary movement during sleep, then dreaming sleep appears shallower. But if one considers the relative difficulty in arousing the individual from sleep, then it is orthodox sleep which is the shallower. In Figure 6 dreaming, or paradoxical sleep, is shown as being nearer consciousness.

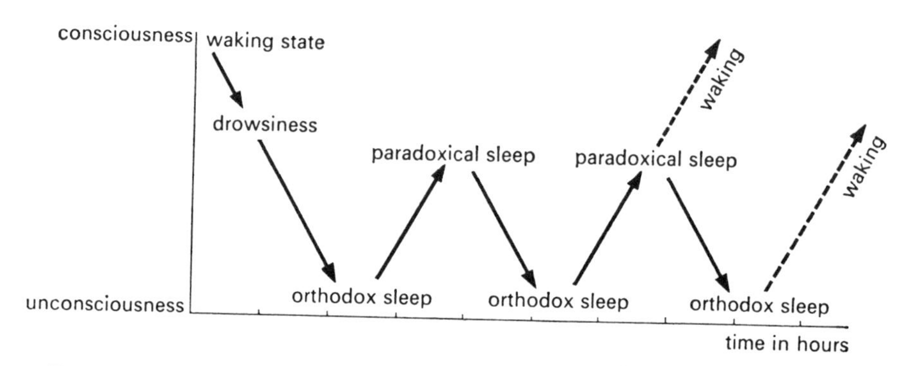

Figure 6. Principal stages of the normal waking–sleeping–waking cycle.

A further manifestation which accompanies the onset of paradoxical sleep, not so far mentioned, is that muscles which have so far remained partly tensed now relax completely – in particular the muscles at the back of the neck, allowing the head to loll and also those of the jaw, allowing the mouth to drop open. The individual is now paralysed – not stiff, but with totally relaxed, lifeless muscles. It is doubtless from this that the feeling of being trapped or unable to move arises in some dreams. The situation here is indeed highly paradoxical – the cortex, which controls voluntary muscular movement, is active in a way closely resembling its waking state. Yet the muscles under its normal control are now even less under its influence than they are when the cortex is *inactive* in orthodox sleep. Similarly, a shout or stimulus which might well arouse the sleeper from orthodox sleep is, during paradoxical sleep, frequently incorporated into the ongoing dream.

Shall we suggest it is almost as if some other agency had gained access to, and was illicitly and quixotically playing about in, the control room, in the absence of the true controllers – had misappropriated and was misusing the devices stored there? This explanation, of course, raises, or would raise, many questions, but it at least accommodates the facts.

Leaving this question as a question for the moment, we move off instead at an apparent tangent, and consider one or two of the facts of romantic, as opposed to laboratory, life. Women, it is said, close their eyes more often when they kiss than men. They tend also to prefer to make love in the dark, or in substantially reduced light ('soft lights and sweet music'). Why might this be? Is the answer in any way linked with the alpha rhythms?

Men are conscious see-ers (voyeurs) both in and out of sex. Given the choice, in general they prefer the light on. As a rule, too, they enjoy looking at women undressing, nude pictures of women, and so on. Women, in the main, are reluctant to go along with male wishes in these matters – preferring to avoid undressing in front of a man still more even than being in a state of undress.[24]

They similarly tend to avoid making up, and so on, in the presence of men. What they appear to be anxious to do is to direct attention away (1) from how exactly an effect is achieved and (2) from the fact that there is or needs to be a time of preparation. The woman apparently wishes to emerge 'complete',

24. At least some women become fairly readily hardened to this initial reluctance, of course, as witness strippers and prostitutes. The fact that some women learn to deal with their inhibitions in this area does not of itself make those initial inhibitions less worthy of note.

and time in a sense begins at that point. Is it perhaps too much to propose that, among other things, she has been donning a disguise, putting on some form of defence, in order to hide something – or someone? – from direct gaze and searching examination?[25]

In resting, reverie and day-dreaming with eyes closed, alpha rhythms make their appearance, while the activity of the cortex dies down. These effects, however, may also be produced with the eyes open, providing the person concerned is not using his eyes for looking – 'he stared with unseeing eyes', and so on. Most people are occasionally transfixed by a pattern or a particular object, and in hypnosis as such the eyes remain open. Here the subject 'sees' in apparent reality whatever the hypnotist instructs him to see. It *is* agreed by the literature, incidentally – and reluctantly? – that females are more susceptible to hypnosis, and suggestibility generally, than males. Hence some of the female's vulnerability to the influences of male opinion?

In experiments concerned with *sensory deprivation*, volunteers have agreed to spend hours or days under conditions of severely reduced sensory input, while remaining awake. In a typical experiment subjects rest on a bed in a soundproofed, whitewashed room, wearing cardboard cuffs over their hands, and so forth. After a time under such conditions the subject begins to hallucinate; that is, he or she sees imaginary events taking place – for instance, on the white walls. In the continued course of the experiment these hallucinations tend to become stronger and to assume more of a life of their own. Some individuals report bulging walls and movements of the room.

25. Other more ethological elements, too, are undoubtedly involved in seeing and not seeing. The male animal in general needs to see what is going on for flight or fight purposes. He may be generally more wary of the dark, which the female with her need not to be seen, may prefer – perhaps because she does not run or fight so well, especially when carrying or caring for young. She certainly favours dark places for giving birth, again for fairly obvious reasons. It would be interesting to know whether more men than women suffer from claustrophobia, and vice versa for agoraphobia.

These experiences resemble some aspects of psychotic breakdown.

Somewhat similar effects are experiences by normal people under non-experimental conditions – for instance when lying in bed at night but *unable* to sleep. In this situation of reduced sensory stimulation – usually the prelude to the onset of normal sleep, which is indeed why one goes to bed and puts out the light – one may see endless streams of detailed pictures, at times connected, at others not.

It is usually proposed that the human organism, or at least the cortex, is unable to tolerate a low level of stimulation – and that when the necessary stimuli are not forthcoming from the surroundings the brain is prepared to generate them itself. This theory seems to account, too, for the effect of a boring lesson or lecture, in the course of which one repeatedly finds oneself 'miles away'. However, day-dreaming is a pleasant activity, and up to a point volitional. This is not true of the hallucinations of the subjects in the reduced stimulation experiments, nor of the mental activity of the person unable to sleep. Neither of course does the standard explanation go any way to accounting for the phenomena of psychosis.

Considering only day-dreaming for the moment, how might it be if this practice were in some way essential to mental health? And what might be the effects, if this were so, of living in a society where such activity was deliberately or accidentally discouraged – for instance by demanding high outputs of work, or simply by creating an environment where 'the world is too much with us late and soon'? This is only a matter for speculation. What is not speculation, however, is that night dreaming and paradoxical sleep most definitely *are* an essential component of mental health.

People awakened during experiments whenever paradoxical sleep begins, increase their rate of dreaming on subsequent nights. If, nonetheless, they are continually awakened in this way over a number of nights, their day-time waking behaviour begins to deteriorate, even though they are getting more or less their full amount of orthodox sleep. They become irritable, jumpy, depressed, paranoid.[26] When finally allowed to sleep undisturbed these subjects spend a far greater portion of sleep in dreaming, as if making up for lost dreams.

Similar experiments, where subjects are awakened whenever they begin a period of *orthodox* sleep *after* dreaming, show some but not all of the effects noted in respect of loss of dreaming sleep. Such subjects, when eventually allowed to sleep normally, make up for the loss of orthodox sleep by drastically reducing the amount of *paradoxical* sleep in subsequent nights. *But the daytime behaviour of orthodox-deprived individuals is in no way bizarre.* Hence, while both kinds of sleep are clearly needed by the organism, paradoxical sleep appears to fill a more basic need – one that cannot be denied with impunity.

When volunteers are deprived of *all* sleep for long periods (upwards of one hundred hours) by being walked about, talked to, given tasks to perform, and so on, the reactions mentioned by Calvin Hall are intensified. From what was said above, concerning the deprivation effects of orthodox and paradoxical sleep considered separately, it seems reasonable to assign the effects of whole-sleep deprivation also to the latter, rather than to orthodox sleep.

The whole-sleep deprived subjects have persistent hallucinations and delusions, over which they have no control. They are, that is, unable to distinguish fact from fantasy – the two becoming inextricably mixed. What these subjects experience resembles very closely the hallucinatings of schizophrenics (or for that matter alcoholics and drug addicts).[27] Ian Oswald describes in one of his volunteers a full-fledged paranoid break-

26. 'Moreover, if a person is deprived of dreaming for a number of nights, his waking behaviour appears to be adversely affected. He manifests various "aberrant" symptoms that border on being pathological, and it has been conjectured that if he were deprived of dreams long enough, he might become psychotic.' Calvin Hall, 'Dreams', *International Encyclopedia of the Social Sciences.*

27. The schizophrenic and psychotic states produced by LSD are, incidentally, much enhanced by sleep deprivation.

down lasting many hours. This was not an 'as if' situation. The subject fully believed that he had been drugged and that he and his companion were being interrogated in connection with certain crimes. Of particular interest is a striking example, quoted by Oswald, of the concertinaed, associative 'schizophrenese' discussed in the last chapter. The volunteer referred to one of the psychologists as an 'exquisitor' – which turned out to mean an inquisitor capable of inflicting exquisite pain.

There is a good deal more one would like to know in connection with the area under discussion, *viz.*, whether schizophrenics dream less, more, or the same amount as other people; or whether they less or more readily *recall* their dreams on waking. Also whether neurotics dream and/or recall more, or less, than normals. Or whether the incidence of dreaming or recalled dreaming is greater or less during the schizophrenic breakdown than before – and still much else. However, one must go on what information there is.

What we have is as follows. When a person is cognitively occupied with an interesting task (i.e. when System A is functioning optimally), no alpha rhythms are detected from the brain. Instead we have the typical pattern of waking cortical activity. When day-dreaming is taking place, or when even the eyes are merely closed (thus reducing the level of incoming stimuli) and the person relaxing, alpha rhythms appear. What happens to the individual deprived of day-dreaming or periods of wakeful rest? Experimentally we do not as yet know. There are, however, some admittedly oblique non-experimental considerations. A feature of life prior to some mental breakdowns is overwork – the situation of being constantly 'on the go'. Under these conditions no time is set aside for relaxation. Conversely, techniques of meditation teach the withdrawal from cortical activity (that is, from thinking and problem-solving) for a few minutes several times a day. This is widely claimed to perform wonders for run-ragged business men, without, however, affecting their overall work output, except favourably.

We know from experiment that lack of dreaming sleep and total lack of sleep produce the symptoms of psychotic breakdown on a striking scale. Many *real* psychoses are preceded by a period of insomnia. In this sense at least, then, the pre-psychotic *does* of course dream less than the normal person.

What have these two situations in common? In both there seems to be a denial of the 'rights' of System B – a limitation as it were of its broadcasting time. This limitation in all the examples considered can be said in one way or another to be caused by System A. The amount of broadcasting time demanded by System B by *day* seems not to be large. The amount demanded by night *is* relatively large. Any denial of night-dreaming time is met by a counter-attack on normal waking time by System B. Perhaps the denial of *day*-dreaming time over a long period also leads to a System B attack on day consciousness.

On the basis of the evidence offered, there would seem to be a case for suggesting that psychosis and schizophrenia are the result of a take-over by day of a system, System B, most properly designed to function by night, in the absence of the day system (System A). When the night system takes over by day this is to the detriment, perhaps even the semi-permanent destruction, of the day system.

This view of personality function gains considerable support from the fact that when System A functioning is experimentally reduced or weakened (i.e. in sensory deprivation experiments), and without sleep following as a consequence, the symptoms of psychosis are observed. These are removed when the level of System A stimulation is restored to normal.

The evidence available is internally consistent. When the 'demands' of System B are augmented (e.g. by deprivation of paradoxical sleep) but System A is not weakened, the deterioration of daytime behaviour is gradual and at first only marginally abnormal. When System B is not strengthened in any way, but System A is weakened (as in the deprivation experiments, or by alcohol or soft drugs) the deterioration of behaviour is again not too extreme. *But* when the demands of System B are augmented (by the deprivation of paradoxical sleep) and System A is simultaneously weakened (say, by the administration of hard drugs) the breakdown of behaviour is both rapid and severe.[28]

However, all versions of the hallucinatory or waking-dream experience (other than volitional day-dreaming) may be described by the term *day-mare*, the corollary of nightmare. This is not intended to imply that the experience is necessarily disturbing – on the contrary, drug trips are often very pleasant. What the phrase day-mare brings out is the independent, inescapable and perhaps persecutory nature of this experience. Unlike day-*dreaming*, which is under the person's conscious control, in this other case the individual is *under the control of the experience*. He *cannot* turn it off at will. He is, in fact, possessed.

In the course of this section, it has seemed possible to establish certain links between: day-dreaming; hypnosis; stimulus-deprivation; sleep-deprivation; experience under drugs; and psychosis.

There are these differences. Day-dreaming is a temporary state, more or less under the control of the individual. The dream-like states arising from sleep-deprivation, stimulus-deprivation and drug trips are, like the true dream, not under the control of the individual; but they are temporary and, in that sense, reversible. Psychosis is likewise not under the control of the individual, but unlike dream states, it is not readily reversible, and is relatively continuous.

In the light of this a possible definition of mental illness emerges. Mental illness is the *undesired*, *not readily reversible*, or *relatively permanent* intrusion of the functioning of System B into waking behaviour and consciousness. It is pathological, unlike day-dreaming and the other mild states described, because it produces a relatively permanent dysfunction of System A. This definition can cover not only psychosis and neurosis, but also alcoholism and hard-drug addiction. Both of these,

28. It is possible that hard drugs augment System B *rather* than *weaken* System A, since full awareness of the experience is frequently reported subsequently. Presumably this is the meaning of the phrase 'expansion of consciousness'.

incidentally, not only affect consciousness, but in time produce visibly notable damage to the cells of the *cerebral cortex* (N.B.).

As a final isolated observation, it appears that at least *some* drug users have reduced paradoxical sleep (? because System B has had extra broadcasting time in waking hours): when their supply of drugs is cut off, paradoxical sleep is for a time then raised to a level *above* normal.

All these facts taken together support the view of two major alternative and alternating systems within the total personality. There are grounds also, both in this section and elsewhere, for regarding psychotic breakdown as the conclusion to a long-standing denial of System B by System A – compare for instance the period of insomnia preceding the final attack. Neurosis, on the other hand, is rather the continuous attrition of the conscious personality by a constant yielding to the 'demands' of System B. Neurotics often sleep more than other people. Are sufferers from neuroses in some sense 'under the spell' of System B?

The literature is agreed, incidentally, that neuroticism is more common among women – and that they also sleep more. Whether psychosis is more common among men appears not to be reported, as far as I can readily ascertain.

The next logical step is to consider other forms of possible System B domination, and System A resistance to domination, this time among *normal* people in normal life situations.

However, before finally leaving this detailed consideration of sleep, one or two additional points deserve emphasis. First, it will be recalled that at the onset of paradoxical sleep certain skeletal muscles, in particular the neck muscles, relax more fully. Wilhelm Reich, the noted, though unorthodox, psychoanalyst, claimed that he could tell what psychological shape his patients were in merely by observing the neck. The term stiff-necked is traditionally applied to the proud (the conservative?) – patricians, senators, officers. Perhaps both these and Reich's patients were in a condition of resisting whatever it is we finally yield to at the moment of most relaxed sleep. The stiff upper lip is another portion of the anatomy referred to by traditionalists.

During dreaming sleep twitchings of the face muscles around the mouth are common among people and animals – again the household pet may be observed. This seemingly trivial point will be seen to have a good deal of significance.

Next, remembering that it is harder to wake a person from paradoxical sleep than from orthodox sleep – meaning that during dreaming sleep an animal is more vulnerable to attack than at any other time – we have a further possible objective reason for the fear, particularly by the male organism, of System B1. That dreaming sleep begins only after an hour of orthodox sleep may be to allow time for testing out the safety of the sleeping place.

Finally, a word about the evidence that premature babies dream much more than normal-term babies. The experimentalists have used this fact to pooh-pooh the psychoanalytic view of the significance of dreams. That premature babies have long periods of rapid eye movements while sleeping cannot mean, it is said, that they are actually observing pictures of events – since their conscious experience of the world, particularly visually, must be nil. Therefore, it is argued, the visible dream in later life must be only a by-product – an unimportant by-product – of whatever biological processes underlie dreaming sleep, and are ongoing in the premature baby.

There is a very great deal one could say about such comments on a number of different levels. Let us confine ourselves, however, to two brief remarks only. First, that the experimentalists are quite right to draw attention to the biological processes underlying dreaming. Second, that the assumption that premature babies (and therefore presumably babies within the womb) do not have some kind of conscious visual experience, may prove to have been a little unconsidered.

If the System A – System B dichotomy proposed is as basic as alleged, then one will look for its influence as much among normal as among mentally disturbed individuals. While the proposal that mentally ill patients and normal individuals subjected to experimental stress show the functioning of the various sub-systems with particular clarity is in order, it is most desirable that the weight of argument should rest *as* heavily on the behaviour of the normal individual under normal conditions. Without this, the theory would be considerably weakened in its claim to deal with total behaviour, and have instead possible relevance only to a pathological condition or conditions. In this section a consideration is undertaken of certain everyday public phenomena.

Not simply individuals, but entire movements and social systems in corporate life can be described in terms of the dominant influence of one or other personality sub-system. There is no magic involved here. It is simply that group behaviour is the sum of, or perhaps a balance struck in the interaction of, the individual behaviours of its members. Given a preponderance of individuals governed by B2, B1 or A, or some particular combination of these, in a particular group, that group then may be spoken of in those generalized terms.

With that statement about the structure of groups, let us take the bull by the horns and say that it is clear that left-wing behaviour (such as Communism) is a System B phenomenon, while right-wing behaviour (such as Fascism) is a System A product. What grounds are there for this statement? As much perhaps by way of begging the question than by way of answer, a number of statements are now made about these two political orientations. Here we see the two opposed and opposite factions described in the kinds of terms used already elsewhere in the present text.

COMMUNISM	FASCISM
(*System B*)	(*System A*)
'comes up from below'	'is placed on top'
revolution	*coup d'état*, putsch
liberation	control

all men are equal; and men and women are equal (integration)	there are natural leaders and natural followers; women are different and inferior (segregation)
joining	splitting
theoretical equality of opportunity	rule by a nobility or élite
collective ownership and responsibility	individual ownership and responsibility
the masses	the individual
environment is the key (Lysenko)	genetic inheritance and natural endowment are the key
automatic right to services and benefits (mothering)	reward where deserved or earned; patronage (fathering)
the state as mother	the state as father
(progressive) education (the word means 'leading out what is within')	(traditional) instruction (the word means 'building into or on to from without')
'neurosis'	'psychosis'

The above constitute a sample of the characteristic ways of thinking of the two groups. Connections between some of these and the behaviours considered earlier in this volume may occur spontaneously to the reader. But in any case these connections are enlarged on later. At this point, however, having made the head-on assault, it may be advisable to retreat a little and attempt instead a rather more gradual build-up of the position adopted.

First, it is noted that the division into a major left-wing and a major right-wing party is generally agreed to constitute the stable configuration of politically mature countries. There are thus *two* major and basic political divisions or orientations. Once again we find the ubiquitous 'two' encountered so frequently elsewhere.

Do these two political orientations have different attitudes to sex – and if so, why? Table 5 shows the numbers of women put up for election to Parliament, and the numbers actually elected, by the two major parties in Britain between 1918 and 1966. The figures are taken from *British Parliamentary Statistics 1918–1968*.[29]

TABLE 5

Candidates at General Elections since 1918

	Women Candidates		*Women Elected*	
	Labour	*Conservative*	*Labour*	*Conservative*
1918	4	1	0	0
1922	10	5	0	1
1923	14	7	3	3
1924	22	12	1	3
1929	30	10	9	3
1931	36	16	0	13
1935	33	19	1	6
1945	41	14	21	1
1950	42	29	14	6
1951	41	25	11	6
1955	43	33	14	10
1959	36	28	13	12
1964	33	24	18	11
1966	30	21	19	7

These figures show clearly that the acceptance of women as equals in the Labour movement is in general more of a practical reality than it is in the Conservative movement. The best guide is provided by the first two columns referring to the number of candidates, since the last two columns are to an extent modified by which party happens to be in favour with the electorate and gaining most seats. Thus, in 1923, when three Labour and three Conservative women M.P.s were elected, the Conservatives had gained 258 of the total seats available, while Labour had only 191. (One needs also to appreciate that in that year there were fourteen women out of 427 Labour candidates and seven women out of 536 Conservative candidates.)

The Conservatives have a segregationalist attitude to women.

29. Compiled by F. W. S. Craig: Political Reference Publications, Glasgow.

This is shown by single-sex secondary schools (and originally by single-sex universities), by the existence of clubs exclusively reserved for men, and many and various social practices, such as men drinking port in the library after dinner while the women are elsewhere, and so on.

Were one being psychoanalytic about this, one would say that the Conservative is afraid of women. Yet, one might ask, if woman is regarded as inferior and excluded from any real power, where is the justification for supposing fear among the men? For an answer we might turn to classical Greece, where the status of women was lower even than in Conservative England. Yet, examining the myth and legend of Greece, one finds it strewn, as already noted, with *female* monsters – Scyllas, Charybdises, Medusas, Gorgons, Hydras, Harpies and whatever. The whole unconscious as such, it seems, is a place of extreme danger. Neither must one forget that it was women who directed the Greek Mysteries. As the priestesses of that cult they were most certainly feared. And in any case – putting the matter on a general footing, and asking the kind of question which Freud was fond of posing – why in any case would one want to keep anyone in *subjection*, if one were not afraid of them?[30]

What are the chances of claiming the Socialist as a 'man of the earth'? A difficult question, perhaps, in view, for instance, of the fact that many Conservatives are farmers. Were the revolutions of 1789 and 1917 in France and Russia after all peasant revolutions? Of course, the leaders of the Russian upheaval were middle-class, and the mass were workers rather than peasants. But might one without stretching language beyond what it will bear, speak of the worker as an 'urban peasant'?

That being as it may, the great stress placed on the environment by Communism and Socialism – the most astonishing ex-

30. On a purely subjective basis, one might like to ask oneself, and to answer, the question, who seem more afraid of women, Russian men or American men? 'Free love' is, of course, purely a Romantic and Socialist concept. And is the opposite then 'love in chains'?

ample of which is Lysenkoism, the belief in the transmission to offspring of characteristics acquired by the parent in the course of a lifetime – reminds one in many ways of the *field-dependence* of women and their involvement with concrete situations.

There is a good deal more of this nature that one could say – i.e. circumstantial, anecdotal and arguable. It is preferable to offer more readily testable items, where possible. A belief of my own is that were the average I.Q. of Conservative members of Parliament, as measured by conventional tests of intelligence, ascertained, this would exceed the average for Socialist M.P.s by virtue of the former's closer links with System A.[31] There is evidence (certainly in the last Labour Government's record) that while its heart may be in the right place, Socialism's head leaves something to be desired. The Conservatives have a better record as administrators and businessmen – but are perhaps notably lacking in compassion. (The living conditions which most Conservative employers were prepared to tolerate for their employees a mere hundred years ago are, of course, by themselves entirely sufficient evidence of this.)

31. I can imagine the rage and disbelief that this statement will produce in the hearts of my Socialist colleagues and friends – and would therefore qualify it as follows. First, I have advisedly used the phrase 'as measured by conventional tests of intelligence' – and I do not believe that such tests necessarily give a balanced expression of the full range of human abilities. Staying with those tests, however, I believe that the Conservative would perform better on the cognitive sub-tests, while I believe that the Socialist would do best on the verbal sub-tests, of the total battery. These views would be in theory easy (but in practice difficult!) to verify. However, more oblique approaches are open to us. I would suggest that the average length of speeches in the House by male Socialist M.P.s (when in government), say, over a period of a year, would on inspection be shown to exceed the average length of speeches made by Conservative M.P.s over a similar period (again when in government – i.e. not in the *same* year). This is a matter that could be settled by reference to the pages of Hansard. I must confess not having verified this – but my *private* impression is that Socialists *do* talk more. If it did prove to be the case, one would be seeing, I suggest, a reflection of the greater verbal facility of the female (i.e. of System B), a well-established fact of experimental psychology.

It is frequently said that Labour functions best in Opposition. If there *is* a link between Socialism, System B and the unconscious, then, as it happens, opposition would be the natural situation for a left-wing party, and the one in which it would in fact feel most comfortable.[32] If the Conservative party is more closely connected with System A, then *it* would tend to be more successful when in charge – controlling and consciously directing activities.

These are all controversial matters. One would ask the reader, however, to reserve full judgement in this area until after Chapter 9, where somewhat more compelling reasoning is offered.

We turn to some statistical considerations.

The ratio of males to females in the general population is approximately 50:50. (There are in fact fractionally more women, but this does not especially concern us.) If a particular behaviour we were examining were somehow sex-based – and assuming this were considered solely on an either-or basis, that is, by an insistence on discrete yes-no categories, but otherwise without placing any other pressure on freedom of choice – would one not expect the incidence of that behaviour (the ratio of yesses to noes) to be also 50:50?

There follow the voting figures for the last four General Elections in the United Kingdom, considering outright Labour and Conservative votes only and ignoring all others.

1955		*1959*	
Labour	*Conservative*	*Labour*	*Conservative*
12,405,254	13,310,891	12,216,172	13,750,875
1964		*1966*	
12,205,808	12,002,642	13,096,629	11,418,455

Is this not, at least, an interesting coincidence?

What one is suggesting is that some kind of predisposing basis exists for this behaviour, *all other things being equal*, that is,

32. Are women, too, more comfortable 'in opposition'?

where no other, more decisive counter-influences are in play. (One is therefore *not* proposing that if, for instance, voters at the next election were offered £1000 a head to change their voting allegiance, a large number would not do so.)

Having, however, remarked on the near-equality in the numbers of those voting left and right in recent elections in this country, let us now say that this precise equality is not in itself of major importance. What *is* of major importance is that there should be two substantial groupings. In the evolution of nations or peoples (I am speaking particularly of the now Western nations) I suggest that an initial preponderance of System A-linked genes is gradually replaced over time by a preponderance of System B-linked genes. Consideration of this important submission is deferred, however, until Chapter 9, since a much broader canvas is required than we are using. The long-term process involved *may* have reached its mid-phase in the gene-pool of the British Isles – hence the roughly equal voting strengths.

For the present let us suggest only a link between System B and (1) fecundity – the opposite, that is, of sterility, and (2) qualities of survival, including physical robustness of a certain type. This second factor is possibly reflected in the fact that more females survive birth and early childhood than males, despite the higher birth rate of male babies. Two points may be considered. First, a fact frequently commented upon by historians and sociologists, that peasant, 'working-class' and various other such materially deprived populations living under some of the most appalling conditions imaginable, nevertheless contrive to reproduce and grow at very appreciable speeds.[33] This is a complex issue containing many elements. The very overcrowding and lack of other 'entertainment', for example, increases the rate of intercourse, and therefore the rate of

33. The instance of the Jews throughout the Middle Ages is very instructive here. On top of the starvation existence of the ghettos, the inhabitants were often slaughtered wholesale during the frequent pogroms. Despite these circumstances, the Jewish population grew steadily.

pregnancy, considerably. That matters are, however, not that straightforward may be judged from a second point, that an appreciable number of women in our present society in wholly adequate, middle-class surroundings, and with the help of every scientific and diagnostic aid, remain unable to conceive. We cannot follow this path further here, but the psychological and medical literature provides support for a concept of fecundity that is *relatively* independent of environmental considerations.

These reflections still by no means exhaust the statistical support for the view that voting behaviour is linked with the 'male' and 'female' principles, that is, is genetically based.

There exists in statistics a concept termed a normal distribution. This in graphical form is a bell-shaped, bi-laterally symmetrical figure, as shown in Figure 7. The term normal distribution is applied when the incidence of a characteristic or function is symmetrically distributed on either side about the average value of that characteristic, with progressively fewer greater, or lesser, values occurring as one departs from the average. Such a distribution is typical of a very large range of phenomena in nature, from the wing-span of a given species of moth to the running speed of lions.[34] In Darwinian terms, it is what is described as the 'range of variability' of a species.

If one measured the heights of all the fourteen-year-old boys in England, and plotted the results graphically, a normal distribution would result – and show that most boys were of average height for their age, with progressively, and regularly, fewer exceptionally tall or exceptionally short boys as one departed from the mean. The same holds true of a great many human physical traits and not a few psychological ones. Thus, for example, intelligence is normally distributed – at least among boys and girls considered separately. (This book is somewhat at odds with the view that the same statement can be made for boys and girls considered together, as already briefly discussed in Section 2 of this present chapter.)

34. It is, however, not always quite as regular as is assumed for the purposes of this chapter.

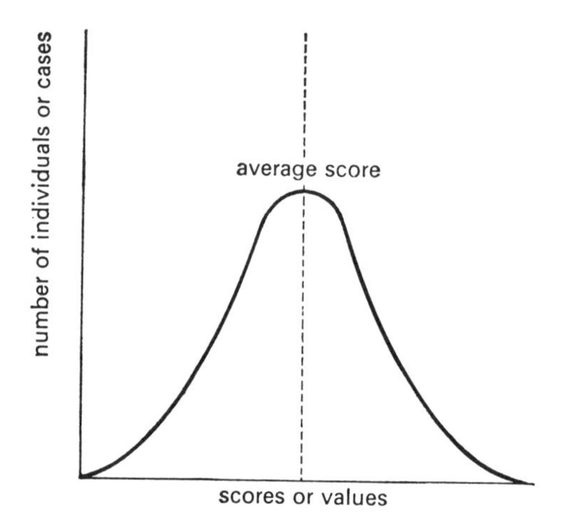

Figure 7. A typical normal distribution curve (stylized).

I have not seen the following statement in print, but there would seem no reason for not reversing the logic of the fact that inherited characteristics are normally distributed within any population, and proposing that *whenever a characteristic whether physical or psychological is found to be normally distributed, it can be considered to be genetically based.*

A further distribution known to statistics is the bi-modal distribution, an example of which is given in Figure 8(a). In this situation two averages are present, around each of which individuals cluster in the way described earlier – but the two samples or populations overlap.

A bi-modal distribution would be produced if one measured the heights of all fourteen-year-old boys and all thirty-year-old men in this country. What one in effect obtains is two normal distributions, which happen to overlap – since some of the boys would be as tall as some of the men. In Figure 8 (b) these two results are plotted and the overlap indicated. If the two sets of scores were combined, Figure 8 (a) would result. Such a bi-modal distribution would arise if one measured and combined, say, the heights of all Scandinavian and all Japanese men, or of all adult men and adult women in this country. Obviously, then, a bi-modal distribution may arise from differences of age, race and many, many other factors besides sex. A sex difference *can*, however, constitute the reason for such a distribution.

What can always be said is that the occurrence of a bi-modal distribution indicates the presence of two separate influences or two separate populations – which one then seeks to identify and isolate, assuming that the result is unexpected.

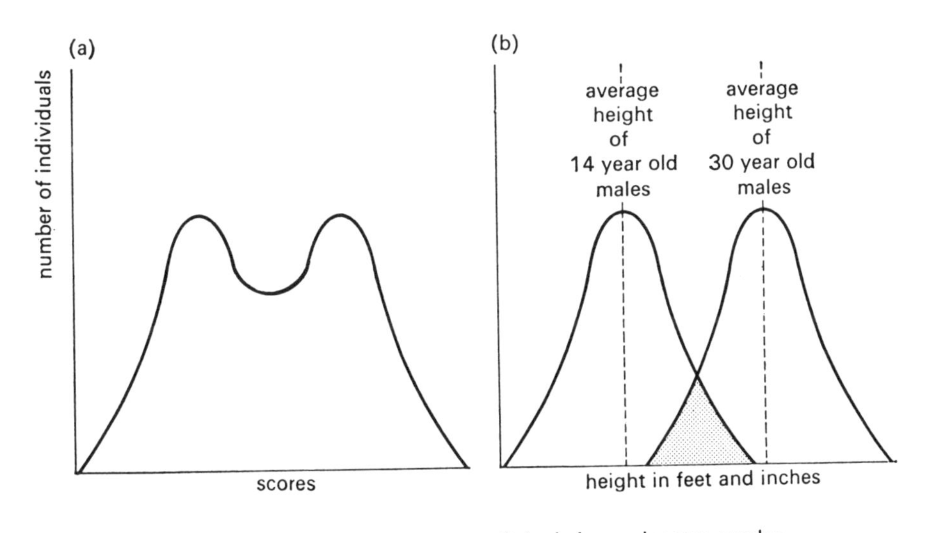

Figure 8. A bi-modal distribution as the outcome of two overlapping normal distributions.

If it is supposed that the two extremes of radicalism/conservatism form a continuum from ultra-revolutionary to ultra-traditional, and that extreme left-wing and extreme right-wing behaviours represent respective examples of these, a third assumption might be that the strength of these tendencies in individuals could be measured – except that this last is not a supposition. Tests in fact exist for the assessment of the radical and the conservative personality. My doubt only is whether these have been administered to sufficiently large and genuinely national samples.

What this book requires ideally is that the results from testing an adequate sample of *men only* (or of women only) should reveal a bi-modal distribution of these political or personality traits – that is, two clusterings, one about an average degree of conservatism, and one about an average degree of radicalism. Such a result would offer direct support for the existence within the one *physical* sex of two 'sexes' – that is, of two personality orientations (that is, System A and System B respectively).

Failing this or additionally, one would look for a bi-modal distribution of these traits in a mixed-sex population, attributable, therefore, to behavioural differences between the sexes.

Leaving now the direct, frontal and hopefully objective assault on the question of political behaviour, we turn to some more general considerations, which nevertheless still bear centrally on the nature of the body politic.

It will be clear that all System B behaviours – that is, behaviours based on System B – despite their individual differences will have (in fact, must have) points in common. The same holds true of System A. Additionally, the various examples of System A, and System B, behaviour respectively must differ consistently and definably from each other.[35]

Communism, Fascism, Catholicism and other dogmas have been said to demonstrate closed-circuit thinking.[36] This is their principal feature – that is, all proof of what is believed and any self-justification required is found *within* the system and not sought outside it; furthermore, and conversely, no criticism from outside can ever get in. The system presents a continuous exterior. One may think here of the outer wall of the self-contained medieval city, or of the circle of covered wagons in American Western films, around which the attacking Indians circle relatively helplessly – this being perhaps the appeal of that

35. The situation can perhaps be made clearer by an analogy. There are within Europe many different examples of the caucasoid (white) race, having different physical attributes. Similarly, there are within Africa many different types of negro. While recognizing the differences between types of men *within* these two continents, one has little difficulty in distinguishing between men from any part of Europe as a whole and men (that is, negroes) from any part of Africa.

36. See, for instance, Arthur Koestler's *Arrow in the Blue*.

particular symbol. Not only are these *individual* dogmas impervious to the assaults of all other dogmas of whatever persuasion, but the *totality* of the dogmas of System B is still more impregnable to the *totality* of System A. That is, the two major systems as wholes also *themselves* make up closed circuits. This is shown diagrammatically in the two parts of Figure 9 (a and b).

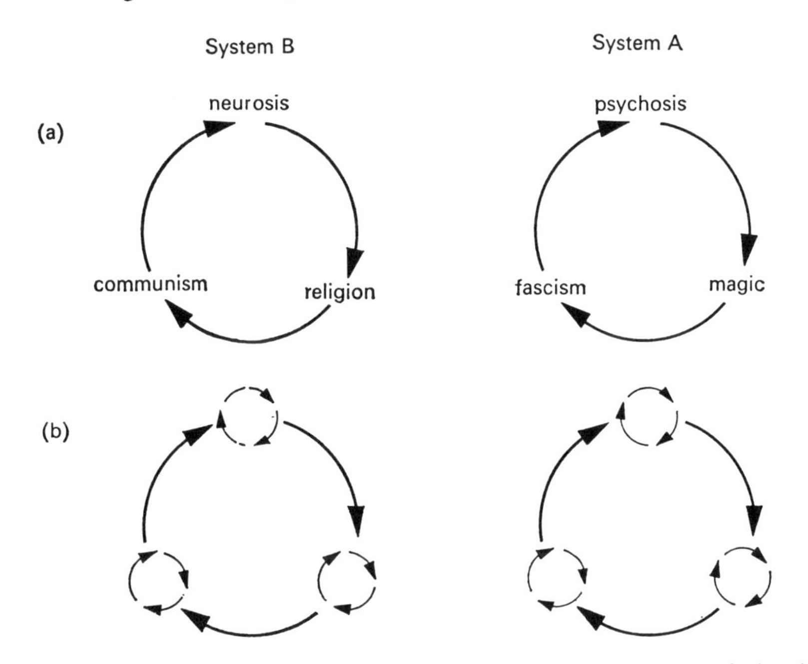

Figure 9. Diagrammatic representation of the suggested closed-circuit nature of personality systems and sub-systems.

It will seem strange that magic has been placed in System A, when hitherto we have linked it to System B. The reason for this apparent reversal will be clear shortly. Making Communism and religion bed-fellows may also seem a little unjustified. On the other hand, where has Communism gained most ground except in countries where religion was formerly strong, particularly in Catholic countries – but the Jews, too, have provided many converts, and indeed, initiators. Among what *class* of people have both these dogmas had most influence? Of course, here material poverty has *also* played a major role. One has no wish to be simplistic in these matters. Both Socialism and religion preach the brotherhood of man. Both condemn war (though, of course, both employ it). Is it not the right wing which, on the whole, *glorifies* war?

Reverting to Figure 9 itself, what is of great interest (and actually great significance) is the mirror-image which B gives of A. Thus magic is to religion what Fascism is to Communism, and so on. The B phenomena are the 'feminine' versions of the 'masculine' forms of A. The B group are characterized by *submission* and collectiveness, the A group by *dominance* and individualism.

How, then, is Communism submissive – since it aggressively seeks world domination? The submissiveness in question is shown in the surrender of the individual to the majority or mass. The hero of Communism is not the individual but the *masses*, the People. Hence, of course, its condemnation of the 'cult of personality'. Communism wishes to absorb (convert) the rest of the world in the manner of an amoeba, so that at the end there is only one vast, relatively uniform, relatively undistinguished mass. 'Freedom' exists only *within* that body. The sleight-of-hand logic here resembles a very similar piece of Christian thinking – the proposal that in serving God one is perfectly free ('whose service is perfect freedom'). To an outsider freedom would necessarily seem to involve the right to reject the dogma if one wished. To the Communist and Christian alike, to wish to reject the dogma means that one must be sick, deluded, lost, perverted.

System A behaviours are, contrariwise, assertive and individual. One thinks of the lone magician struggling to impose his will on the forces of the universe, the fascist dictator intent on domination, the desperately isolated schizophrenic trying to conquer, deny, encapsulate his feelings. There is a certain heroic quality about these figures and the incredible odds they face, which sometimes compels our unwilling admiration. Of such figures Nietzsche is an excellent example, as is also Faust, and the figure of Satan himself. (It is perhaps purely a personal opinion – but are not Communism and Christianity and neurotics in the long run boring?)

Some of the differences between the dogmas or sub-systems of System B are, of course, those of B1 and B2.[37] Thus the Communist is not only impervious to the promptings of religion, but anxious to destroy it – and vice versa. The sub-systems or dogmas based on System A have much less antagonism one to another. Therefore a Fascist-schizophrenic or a Fascist-magician or even a scientist-schizophrenic is by no means unthinkable. In this connection one recalls the bizarre features of the Nazi concentration camp, or the fact that Hitler, allegedly, at times resorted to magical divination.

If the items of Figure 9 are aligned in terms of their distance from or nearness to the opposing major system in each case, the following emerges: neurosis – religion – Communism: magic – Fascism – psychosis. Not all the reasons for this precise line-up may be immediately obvious, but they will become more so as the section proceeds. What one must consider is the amount of 'thinking' involved on the B side, and the amount of 'feeling' permitted or recognized in the A camp.

At this point another feature of the model emerges – namely, that the characteristics of a sub-system or dogma are partly actual, and partly a function of the standpoint from which they are viewed. In Figure 10 (a) are represented a number of sheets of cardboard, painted yellow on one side and blue on the other. No matter from which point in the series they are viewed, all the boards viewed from the right look blue, and all those viewed from the left look yellow.

If in place of the colours blue and yellow one thinks instead in terms of System B and System A respectively – Figure 10 (b) – one sees that from the standpoint of religion Communism is a System A product (worldly, ungodly, thought rather than felt). The Communist, from his viewpoint, sees religion as devoid of

37. In terms of the algebra of an earlier section, Communism would be something like $\frac{B1}{a}B2$ $\left(\text{or } \frac{B1}{A}B2\right)$ and religion $\frac{B1}{a}b2$.

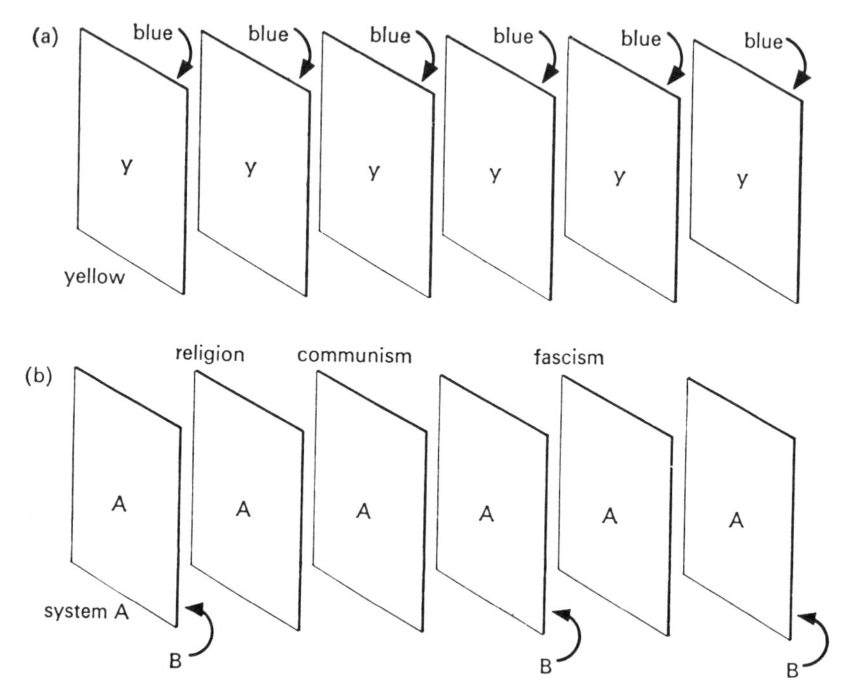

Figure 10. Diagrammatic representation of the altered nature of objects and systems viewed from different standpoints.

thought, dissociated from reality, the product of self-deception and wishful thinking – the 'opium of the people'. To the *Fascist*, however, these are precisely the hallmarks of *Communism* – 'the opium of the working classes', full of wish-attempts to turn sows' ears into silk purses, and so on.[38]

38. The general point of these paragraphs is an extremely important one – of which I should perhaps already have made more.

A relevant difficulty arises here from a too simple behaviourism – which considers only what a person does, and not *why* he does it or *precisely how* he does it. If externals only are considered, the (Russian) Communist and the (Nazi) Fascist appear to have much in common – each kills its opponents, builds political prison camps, disbars free elections, etc. This does *not*, however, suffice to make them identical. If the point is unclear, one might think of, say, three women – one who steals because her family is starving, a second who steals for a living (though has many other possibilities open to her), and a third, a wealthy kleptomaniac who steals trivially

With this a point has been reached where a major, summarizing step in the central argument can be taken. It is now proposed that all System B phenomena are *conditions of trance*, while all System A phenomena are *conditions of trance resistance.*

The concept of trance-resistance must, however, be further divided into *trance-manipulation* and *trance-denial.* In trance-manipulation the phenomena of System B are admitted to exist (to have validity), but are merely *used* by System A for its own purposes. In trance-denial, however, the existence, or the essential validity, of System B is denied.[39]

The hallmark of a System B activity, as already suggested, is that it involves a reduction or giving up of volitional or individual freedom. This may be deliberate, or can be involuntary and unpremeditated, as when someone falls in love. All are invariably communal or joining-in activities, involving at least one other person.

The hallmark of System A activity is the gain, or apparent gain, in individual freedom or power. Thus the practice of magic

---

and compulsively for irrational reasons. Each is certainly a thief. Yet each case is quite different.

One must finally, as ever, remember that although we continually speak of the two main personality systems as though they had wholly separate existences, every human being and every human institution contains *both* components. Thus the individual Communist also has an *Ego*, which requires certain satisfactions. Similarly the individual Nazi also possessed a *Self.* It is a question always of which component is predominant, and of the point at which the balance is struck.

Just to touch on religion here, it is obvious that, say, late Christianity has a much greater cognitive or intellectual admixture than, say, the Greek Mysteries. Yet this does not debar us from speaking of *both* (and religion in general) as System B phenomena.

39. It is extremely rare to find manipulation of System A by System B on any large scale. Freudian slips and rationalization are isolated examples of this. The exception to the rule appears to be Catholicism, and possibly occasionally Communism. Basically, System B's approach to System A is one of exclusion – 'Render unto Caesar the things that are Caesar's', and so on. Even in the apologias of C. S. Lewis and T. S. Eliot the validity of thought without God is essentially *denied*, rather than manipulated.

or the condition of atheism is the will-full (N.B.) disobedience of God – the setting up of man's laws (that is, the laws of the ego) over those of God, or of Fate or Destiny. Hence the Greek crime of *hubris* and the Christian sin of pride and self-assertion. Habitually, System A activities involve the idea of going it alone, of *joining-out* (exclusiveness) and excelling over other people (competitiveness).

Table 6 now shows a number of human behaviours, some already discussed, some not, assigned to their appropriate categories. There is some 'forcing' here, since the categories, as ever, are not wholly inflexible or watertight. A finer-grain table, however, would complicate matters unnecessarily.

Thus psychoanalysis could have gone into the trance-manipulation column, along with ego-psychology. It is partly a matter of whether one is undergoing psychoanalysis, or practising it. On the other hand, one cannot really practise psychoanalysis unless one also believes in it. With advertising it is possible to be cynical or cold-blooded, without necessarily being unsuccessful as a practitioner. The same is true, to a lesser extent, of magic (though less true still in the case of primitive or sympathetic magic).

Journalism started out in life as the dispatching of news from one part of the world to another. Today it is almost entirely a method of manipulating other people, for profit or political ends. Business similarly begins neutrally, as the handling of goods or services for a fee. Today, in the shape of Big Business, it wholly qualifies for inclusion among the manipulative practices.

It is somewhat sad to see ego-psychology in such bad company. It must be emphasized that its purpose, if manipulative in any sense, is the manipulation of the osteopath, having the final intention of liberating the individual. Such is *not* the intention of the other activities just discussed.

The extreme ends of the political scale, Communism and Fascism, are referred to as totalitarianism. There is some justification for ascribing this term to many other items of the table. Thus experimental psychology, no less than psychoanalysis, is

TABLE 6

System B		*System A*		
Deep Trance	*Light Trance*	*Trance Manipulation*	*Trance Denial*	*Extreme Trance Denial*
Dreaming and dreaming sleep		Day-dreaming	Activity	Insomnia
Trance of spiritualist medium	Conscious mediumship clairvoyance			
Orthodox religion	Church, etc. membership	Magic	Atheism	
Hypnosis	Suggestibility			
		Advertising		
		Journalism		
Cabbala	Mysticism			
	Alchemy			
LSD trip	Marijuana trip			
Communism	Membership of Labour Party		Membership of Conservative Party	Fascism
	Psycho-analysis	Ego-psychology	Experimental Psychology	
	Being in love		Emotional coldness	Psychopathy
	Femininity		Masculinity	
Neurosis				Psychosis

determined to describe all human behaviour in its terms. Business asks one question of all human activity – is it profitable? Advertising asks – how may this behaviour be used as a lever? And so on.

Totalitarianism is an apt, perhaps even unconsciously apt, term. The task is to make a whole or total out of what is actually only half. Both systems (A and B) wish equally to take over the whole personality, and to destroy the other. This may seem to contradict the statement made earlier that System B does not manipulate System A. The difference is just that. System B does not tend to use System A for its own, i.e. System B, ends. Getting *rid* of System A, however, is another matter entirely.

The initial impetus towards System A or System B as the initial bias or basis of the individual personality is now suggested to be a genetic, inherited one. Developmental stages are also involved, however, since clearly more people are socialistic or romantic in adolescence and youth than they are in later life (many girls, for instance, pass through a strong religious phase in their teens) when there is an opposite tendency to be practical, hard-headed and conservative. Over and above the constitutional factors there are additionally, of course, environmental and social pressures – a subject as far as possible avoided in this book. It is perhaps these which in some measure decide the particular, *overt form* of the basic direction in the individual case.

The genetic argument gains support, too, from the fact, no doubt to the bewilderment of those in authority, that each new generation seems to produce a crop of pro-radicals in Fascist countries and a crop of anti-radicals in Socialist countries, despite the fact that a majority of those holding such views in the previous generation were exterminated or deported – and that in the context of absolute control of the media and the educational system such views could not be expected to reappear.

All System A views are ego-based or *ego-centred.*

Logical extensions of the ego are one's direct ancestors and descendants (not so much, however, brothers and sisters or cadet branches); and also those who mirror one's ego in terms of status, title, rank and breeding. (The 'old boy' net.)

All System B views are *self-centred.*

The Self is defined at this point as *the female equivalent of the ego.* Extensions of the Self are one's offspring (all of them), the children, the family in this sense. This is further extended to include the offspring of other mothers (especially similar mothers), in which context one recalls the religious admonition 'love thy neighbour *as thyself*'. One notes that members of the Socialist Party and the Trade Unions call each other brothers, as do some religious groups. Expressions like 'the family of man' occur frequently. System B organizations are *in*clusive, and one becomes a member simply by asking to be ('knock and it shall be opened'); and basically there are supposed to be no distinctions between individuals ('we are all equal in the sight of God').

As noted, System A organizations are *ex*clusive. One needs really to be born into the correct caste, class or race, and even within this there may be grades. Election from the ranks is possible, but rare, and the new arrival may well never come to be regarded as the equal of those who were there to start with. (By contrast, in System B the newcomer, the newly converted has in some ways a higher placing than the already present. 'There is more joy over one sinner that repenteth than . . .' As noted so frequently before, all attributes and characteristics of these two systems are opposite.)

By very reason of their fundamentally and essentially opposite nature, it is clear that it is *never* easy for the two major Systems, A and B, to coexist harmoniously. Each works against the other at every sort of level – psychologically, physiologically, biologically. There are also considerable, though not as great, difficulties involved in the peaceful co-existence of Systems B1 and B2. These exist within the same modality, so to speak. Where love has been, hate can also be. Depression only exists by virtue of elation – and so on, as argued out in Chapter 3. But thought *can* exist without feeling, and logic without emotion.

Nonetheless, all systems, that is, each opposite system, sees the other as a threat, as essentially inimical to its purposes and to its very existence, at least at any given moment. Each system seeks to avoid its own destruction or displacement *by destroying the other one first and once and for all.* The tragedy of the situation is that the enterprise of putting the other system out of action is, for obvious reasons, doomed from the outset, and most of all when it succeeds, or half succeeds.

One need hardly comment on the appropriateness or relevance of the foregoing to international affairs. That is to say, this kind of 'thinking', this kind of reaction, seems to characterize the attitudes of at least some nations to some others.

One suggests, in fact, that a great many conflicts between nations are an acting out in the environment of this internal struggle. Clearly in many wars straightforward reality factors are also involved – the wish to acquire rich or fertile territory, to secure trade routes, and so on. *War* is frequently *best* understood psychologically as the clash of ego and ego. There are few things the ego enjoys more than fighting with (and beating) another ego. I believe, in fact, that all forms of competitive sport rest on this, and solely on this, foundation. Nevertheless, a further component in war, one suggests, is the struggle of System A with System B. This is seen, for example, in the current conflict between America and Russia; between the Protestants and the Catholics in the Thirty Years' War – more currently again in the clash between the Orange and the Green in Northern Ireland.

There are perhaps some grounds for further suggesting that while war between *nations* involves A versus B differences, persecution and oppression *within* a nation involve rather B2 against B1 (are, in short, rather more 'sexual'). However, the distinction is by no means clear-cut.

For further general evidence we might consider the Second World War. How did Britain see Nazi Germany? Certainly as an aggressor (B2), but more specifically as ruthless, cold-blooded and efficient – the Germans themselves as trained, heartless automata (System A). Britain saw *herself* as the kindly, humanitarian defender – B1? How did the Nazis regard the Poles, the French, etc., and in particular the Jews and the gipsies? As animals with animal appetites and animal habits (System B, especially B1). Like animals, or rather because they were animals, they could be herded into cattle-trucks, corralled, and either slaughtered or put to hard work. Their bones could (legitimately) be used for glue and their skin for leather. German youth, on the other hand, was urged to be 'as hard as steel' ('*hart wie Krupp-stahl*'). Both the body as machine and the body as *muscle* were glorified: the body as sex rejected. Thus head-hair was cropped short and beards were unwelcome. It is the Jew and the Socialist (and the woman) who are hairy. And so on.

In the general context of persecution and with a further glance at the genetic or inherited basis of the behaviour under discussion, the following remarks are also offered.

It can be hypothesized that a majority of those individuals who voluntarily leave their own country to make a new life in another will be high in ego-drive and ego-qualities. (In making this statement one must, of course, exclude those *forced* to leave.) It happens that all countries colonized from Europe are right-wing, materially prosperous and segregationalist – North America, South Africa, Rhodesia, Australia, New Zealand, and so on. South America constitutes perhaps *something* of an exception both to these and later comments, but not entirely so. One hears already in imagination the cries of protest against such a sweeping generalization of the kind just made. How does one define right-wing? What about the emancipation of the American woman? Perversely, one must again refuse to be drawn on definitions. However, in referring to the right wing one is clearly talking about the kinds of behaviours discussed earlier in this section – authoritarianism, traditionalism, and the like. Taking up the point about American women, is it not true to say that the women in the colonial countries mentioned are less feminine than their European counterparts? It is French or Italian women, not American, South African or Australian women, who are considered to epitomize feminity. Do not these *latter* women show instead higher aggression and achievement drives?

What one is suggesting is that the somewhat higher levels of ego-centredness of these countries *could* be ascribed to their representing a *non-random* sample of the parent European gene-pool – being instead a self-selected sample, more highly endowed with ego-qualities.

Are there any ways one could test this claim, other than by the administration of personality inventories of the kind described earlier, to obtain some estimate of the incidence and degree of radicalism and conservatism in a country? There *are* possible ways. Since we believe, for instance, that psychopathy tends to be associated with strong ego-drives, without the corrective factor of strong libido drives, one would predict a higher incidence of crime generally, particularly of 'motiveless' crimes, and a higher incidence of diagnosed psychopathy in the countries mentioned – excluding for this purpose any indigenous, native populations, in each case. Of course, one runs into considerable problems of definition and classification here. One has to bear in mind, too, that the issue is affected by many, many factors – such as the amount of opportunity each country offers for the working off of these basically antisocial tendencies in socially acceptable ways, e.g. by living a 'pioneer' life in certain areas, or by 'controlling' the unruly or radical elements in the population. One's attention might therefore be focused on the police forces and police methods of these nations.

Cross-cultural comparisons of this nature are, alas, always notoriously difficult and one must certainly be wary of the temptation to oversimplify.

Finally, these genetic matters under discussion might be a factor in the cycle of the 'degeneration' of once powerful and warlike nations into 'decadence'. Although an occasional war would make very little difference to the composition of a nation's gene-pool or gene-reservoir, continuous or near-continuous war over many generations might result in the death of a significant proportion of males high in ego-qualities before they had time to produce offspring – the crucial point here.

Human affairs, as we have said, are extremely complicated, and this book is emphatically not in the market for producing simple, 'formula' solutions to very complex situations. What is claimed, however, is that the basis of personality which is described in this book, in relation particularly to its inherited component, can be seen as one factor, an admittedly very important factor, in such situations as national politics and international relations, which are usually considered in total isolation from these matters and analysed as if the predisposing and motivational factors were inevitably conscious, self-evident and above all environmental.

The Nature of Consciousness 7

The phenomenon of consciousness, without which life, as we know it, is not life, and solely in terms of which we exist and can know we exist, is itself denied existence by the experimental psychologist. One could hardly find better words to refute that refutation than those of R. D. Laing:

> The experience of oneself and others as persons is primary and self-validating. It exists prior to the scientific or philosophical difficulties about how such existence is possible or how it is to be explained.[1]

The experimental psychologist – who speaks of consciousness as a hypothetical construct, or as a reference without a referent – secretly understands perfectly well what is being referred to by the term. Though he denies it (as, perhaps, Peter denied Christ), he reveals his true position when he discusses eagerly, as he readily will, the possibilities of developing computers which possess consciousness. In other words, he is fully prepared to accept the concept whose existence a moment before he refused to acknowledge, provided we are talking only of assigning it to a machine. When one asks how we would know if the machine had consciousness he replies, *because the machine would report that it had*. We have the position, then, that if a computer says it has consciousness, one accepts the statement. But when a human being makes the same claim, one does not.

This is behaviour which in another situation we might call psychotic.

1. *The Divided Self.*

If one is to discuss consciousness, the problem of definition would seem to arise. Actually, it does not arise. There are two circumstances under which definition is necessary: (1) when the person one is addressing has never seen or experienced whatever is being talked about, and when therefore some attempt must be made to convey to him what is under discussion, and (2) when it becomes clear in the course of discussion that two individuals in fact understand different things by an apparently common term.[2]

There would thus be no *a priori* call to define, say, the terms 'tree', 'shoe' or 'room', though it is a very salutary exercise to attempt to define such everyday objects to someone else's satisfaction – or even to one's own.

I am here prepared to work with whatever the reader himself understands by the terms consciousness or self-awareness, providing only that he does not deny the existence of consciousness. The proposition 'I think, therefore I am not' is a little too ridiculous. (Yet, the psychotic does almost achieve this.)

Far from any attempt to define consciousness, then, it is proposed now that an understanding of the totality of human behaviour, and of personality, is possible only in the context of at least two different forms of consciousness. There are two possible formulations of this position, and, as in physics in respect of the particle and wave theories of light, both account well for some of the observed phenomena. For this reason both must be retained for the present. The proposals are (1) that consciousness is migratory, and capable of movement to another site or sphere of activity within the nervous system, where it takes on characteristics other than those in evidence at the prior location, and (2) that two quite separate forms of consciousness exist within

2. The same principles apply in the laboratory. If one scientist tells another to prepare 6 grammes of sodium phosphate in a test tube, there is no call on him to define any of these terms. He speaks in the belief that the other understands the same things by them. Only in the event of the experiment not turning out as projected does one have any cause to suspect, among other things, a breakdown of communication.

each of us, governed, however, by a mechanism resembling an alternating switch, so that whenever one form is 'on' the other is automatically 'off'.

These proposals are discussed separately. Subsequently the physiology of the brain is reconsidered, where it is shown that the psychological picture outlined in this book matches up very well indeed with certain established *but not fully understood* aspects of experimental physiology.

(1) *The Concept of Moving Consciousness*

The evidence for the view of consciousness now suggested is so vast and so ramified that it is difficult to know where to begin.

Possibly the best starting point is with children (and certain poets) since as children we experience this condition in an extremely vivid, direct way, which we seem later to lose. However, the notion is recalled for us as adults both in poetry and literature.

Children commonly believe that when they sleep they go on a journey. In this they concur as it happens with the views of many Eastern religions and philosophies on the nature of sleep. A song made popular by a group called the Seekers in recent years, 'Morningtown Ride', incorporates one well-established idea of this kind. The song is addressed to children going to bed at night, and has as its central concept the boarding of a train, with the refrain:

> Hear the whistle blowing
> Out along the bay
> All bound for Morningtown
> Many miles away.

The example may seem trivial. In these areas, however, the trivial examples are frequently and actually the most important, since they are permitted to escape from the unconscious into consciousness (as perhaps Odysseus and his companions, clinging to harmless sheep, were able to slip past Cyclops from the cave where he held them prisoner), without their being seen for what they are, vetted, analysed and destroyed. However, for a less 'trivial' example let us consider one of Dylan Thomas's most loved poems.[3] (The poet, operating as he does under licence, is allowed to say what others may not, at the price, of course, of its not being admitted to mean anything.)

And nightly under the simple stars
As I *rode to sleep* the owls were bearing the farm away . . .
And then to awake, and the farm, like a wanderer white
With dew, *come back*, the cock on his shoulder . . .

Here in the second part the poet speaks as if it were the farm which had been away. It is, of course, he who has left the farm in sleep, and then later returned.

One talks to young children, too, of the Land of Dreams – the Land of Nod, and similarly the Land of Make-Believe, Fairyland, and so on – as if these were places that actually existed somewhere.

Many fairy-stories and other magic tales involve, and usually begin with, a journey of one sort or another. These journeys are of many different kinds, being usually however a process of going *into* – a cave, a secret passage, a forest, the mountains.[4] *Narrowness* and difficulty are often prominent features of this entry. Sometimes the journey is merely across country – even then, however, it may well involve passage over marshland, across a treacherous river, through dim valleys, canyons, and whatever. Only rarely does the journey involve movement *upward* – as in *Jack and the Beanstalk* or *The Wizard of Oz. Pilgrim's Progress*, too, has an upward tendency, while incorporating all other features mentioned.

This last is a reminder that the journey figures also in more serious and adult literature – witness, too, the voyages of Odysseus, Perseus and other Greek heroes – often, of course, to the underworld; further, in adventure stories such as James Hilton's *Lost Horizon*, Conan Doyle's *The Lost World*, Jules Verne's

3. Fern Hill.
4. Also occasionally a door or gate.

Journey to the Centre of the Earth, and many other books with varying mixtures of the serious and the fantastic. The *difficulty* of the journey is almost always emphasized. The central characters return wiser, richer, possibly happier – but at all events *changed*.

Let us not overlook here the journey of the Israelites from captivity to the Promised Land. Whether this journey actually occurred is, of course, immaterial. The appeal and vitality of the story lie almost certainly in the echoes it wakes within us.

Is it a coincidence that LSD- and marijuana-induced states, particularly the former, are called trips?

Psychoanalysts see in many aspects of these journeys a return to the womb, and a reliving of the painful, traumatic process of birth. Freud states that when we sleep each night we are in fact returning to the womb – hence the 'foetal' posture of the sleeper. (Of course, when the room is warm we sleep fully extended! Without wishing to say that the curled-up position has no Freudian overtones, the conservation/dissipation of body-heat offers perhaps a more likely explanation of sleeping curled or spreadeagled.)

This book can regard the Freudian womb-journey only as one sub-description of a more generalized voyage. True it is that at the moment of birth the dreaming embryo is roused for the first time to waking consciousness. Till then it has experienced itself (if at all) in terms of some inner world (at once proprioceptively and interoceptively). The present book agrees that each night we re-experience a similar condition to that prior to birth; and undertake to this end a 'journey' within our nervous system. In *this* sense the concept of the return to the womb is accepted – but without necessarily the Freudian overtones of a search for protection and comfort.

The *physiological* description of the various states of sleep will be recalled from Chapter 6 – in particular, Figure 1. This is reproduced here in slightly altered form, in Figure 11.

As discussed, paradoxical sleep may be considered in some senses nearer to waking consciousness, and further removed from it in certain others. On the one hand the cortex is active during dreaming in a way that resembles consciousness; on the other hand it is harder to arouse a person from dreaming sleep than from orthodox sleep. When a person *is* so roused, or arouses himself, it is often with the dream 'still upon him' – a state that may last only for moments, or may persist through a whole day. (Is it not as if waking consciousness has caught a closer than usual glimpse of some other arena of activity?) In Figure 1, paradoxical sleep is now shown as a state *further removed* from normal consciousness than is orthodox sleep.

A point of note is that the length of time spent in dreaming increases with each phase of dreaming, the first being of the order of five to ten minutes, the last of thirty-plus minutes, lasting possibly as much as sixty minutes. In this sense we can say that a person becomes progressively more 'possessed' by dreams as the night wears on.

Figure 11 is one possible representation of the journey. Others, more precisely physiological, will be proposed at a later stage.

Laing has a great deal to say on the subject of the journey, with, naturally, particular reference to psychotic breakdown. In

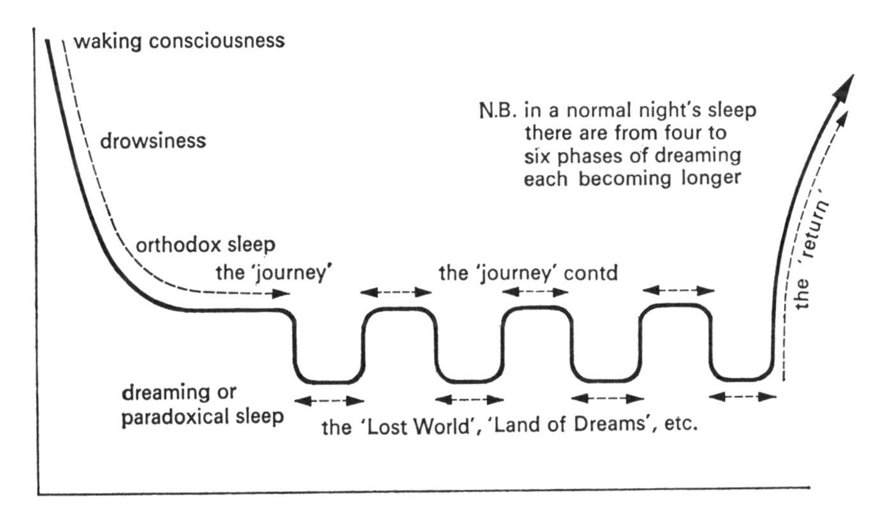

Figure 11. The journey.

Chapter 7 of his book *The Politics of Experience* he gives verbatim an account given to him by a man who experienced a schizophrenic episode. The chapter is called 'A Ten Day Voyage'. A brief extract from this follows.

> ... suddenly I looked at the clock and the wireless was on and then music was playing – um – oh popular sort of bit of music. It was based on the rhythm of a tram. Taa-ta-ta-taa-taa something like Ravel's repetitive tune. And then when this happened I suddenly felt as if time was going back, I had this extraordinary feeling of – er – that was the greatest feeling I had at that moment was of time going backwards ...
>
> I even felt it so strongly I looked at the clock and in some way I felt that the clock was reinforcing my own opinion of time going back although I couldn't see the hands moving. ... I felt alarmed because I suddenly felt as if I was moving somewhere on a kind of conveyor-belt – and unable to do anything about it, as if I was slipping along and sliding down a – chute as it were and – er – unable to stop myself.

On the question of one's relationship to *time* in the journey more will be said at another point.

Laing adds a number of comments of his own to this account, which occupies in all some fifteen pages. The italics in the following extracts are his.

> We can no longer assume that such a voyage is an illness that has to be treated ...
>
> Can we not see that *this voyage is not what we need to be cured of, but that it is itself a natural way of healing our own appalling state of alienation called normality?*
>
> In other times people intentionally embarked upon this voyage. Or, if they found themselves already embarked willy-nilly, they gave thanks, as for a special grace.
>
> Today, some people still set out. But perhaps the majority find themselves forced out of the 'normal' world by being placed in an untenable position in it. They have no *orientation* in the geography of inner space and time, and are likely to get lost very quickly without a guide ...
>
> In this particular type of journey, the direction we have to take is *back* and *in* ...

The links between the material quoted and some of the views of this book are, hopefully, clear.

A final word on the general subject of the journey, in connection with the language of everyday experience. We speak, for instance, of being *in* love and subsequently of falling *out of* love. In German the phrase for the former state, *verliebt*, means 'away in love'.[5] A wealth of phrases connected with all kinds of mental states, apart from love, reflect the idea of movement in space, for example out of this world, out of my mind, off my head, beside myself, way out, down-to-earth, back to reality, withdrawn, absent-minded.

(2) *The Concept of Dual Consciousness*

In the experimental literature on sleep a persistent denigration, amounting to a dismissal, can be observed, not of the physiological sub-strata of dreaming, but of the subjective, personal concomitants of dreaming – that is, of the dream *experience.* Literature, art and ordinary people, however, concur (foolishly) in regarding the subjective aspect of dreaming as the more important element.

An adult spends on average just under two hours of each night in dreaming, between a fifth and a quarter of the total night's sleep. This amounts to around 700 hours, or nearly one month, a year.[6] Children of course spend considerably more of their longer sleeping time in dreaming, and adolescents, too, dream more than adults.

In the process of their denial of the value of the dream as such, experimentalists point out how little our dreaming experience

5. The prefix *ver* is the same word as English 'far'.

In the film of *Brief Encounter*, where the suburban wife has had a love affair which she finally renounces, the husband says: 'You have been a long way away. I'm glad you came back.'

6. These may be considered minimal values, since *some* dreams are reported by subjects wakened from sleep with no eye movements, while the estimates of dreaming time are based solely on such movements.

is recalled on waking – how rapidly even a remembered dream fades during the course of a day – so that by evening it has gone completely.[7]

However, if the amount recalled of what is experienced is the criterion and the measure of importance, then our waking life must be very unimportant indeed – since we do not remember it *at all* (in its own terms) when we are dreaming. When I am consciously awake at least I know that I *do* dream. A few of my dreams are even among my permanent memories. And occasionally, an event in real life will remind me of a dream that I had years ago – and was unaware that I could remember until that moment. Yet, as stated, when I am dreaming I have no memory at all of the fact of my waking conscious life or normal consciousness in those terms.[8] It is this last observation which forms a mainstay of the view that we have not one, but two, quite separate experiences of consciousness.

These observations tell us also something about primacy – primacy meaning that which comes first. Sleeping consciousness, it appears, *precedes* waking consciousness. We are here, of course, referring to *evolutionary* time, not time on the clock.[9]

7. Were experimentalists more dispassionate, they might instead ask how it is that something vividly remembered on waking should come to be forgotten by the evening. In the course of the same day one does not fail to recall what one had for breakfast.

8. The experience of realizing, while dreaming, that one is dreaming is extremely rare, though one cannot say that it absolutely never occurs.

9. The point involved is perhaps best conveyed by analogy. If a person were to give an account of the events of two separate years of his life, without telling us which year preceded the other, we could ascertain this fact by seeking in one year's experiences evidence of the other year's existence. Various kinds of internal agreement would establish that year A had followed, and not preceded, year B. So with all human experience. If by examining A we find traces of B, but not the reverse, we can assume that B has preceded A in time.

However, strictly speaking, other interpretations and models are capable of accounting for these observations in respect of dreams. If there were some kind of filter between two areas of experience, which permitted certain items to pass one way, but blocked those same items moving in the other direction, one would have the situation described for dreams, but without invoking a

The primacy of dreams in the *everyday* sense of time is not so undisputed. The dream is, after all, in part made up of snatches and fragments of the events and things perceived by waking consciousness in the previous few days. These, certainly, are distorted and garbled, but still recognizably and undeniably originating from our conscious percepts. In Freudian terminology, this is the manifest dream.

To avoid wasting time, let it be admitted at once that System B does have *some* kind of knowledge of System A and of ongoing reality. This has already in effect been admitted in an earlier chapter, where it was stated that some of the comments of the unconscious – legends, and so forth – refer to consciousness. It is *waking consciousness* as such, rather than the contents of waking consciousness, of which we are not aware in dreams.

The point is that none of the *values* and *functions* of the phenomena to which System A attaches value or importance are respected by System B. The use to which these objects are put is entirely capricious (from System A's point of view). Not so from System B's point of view; on the contrary, the juxtapositions, linkings, inclusions and exclusions of the dream are extremely meaningful, as can be demonstrated.

Since the question of the perception of each respective system by the other has come up, it is dealt with at this point.

Each system perceives the other *from outside*. This phrase is meant on the one hand fairly literally, but it also means 'not in terms of the system which is being observed'. There are reasons for not saying too much at this stage of how System B perceives System A, but what can be said, speaking generally, is that System B perceives System A – despite the latter's power to lay waste and destroy – with awe rather than fear, and admiration rather than understanding.

When A perceives B it sees nonsense, gibberish, chaos,

---

time or primacy explanation. However, we do have other evidence which argues for a primacy in the present case. The cortex is, as we know, a relative newcomer on the evolutionary scene. Also the child dreams more than the adult, and the premature baby more than the full-term infant.

silliness, madness, irrelevancy, illogicality, unnecessariness and more of this kind. (Such is essentially the experimental view of the Freudian view.) It can be said also that the perception of System B is always unwelcome to System A. The latter will go out of its way to avoid the meeting – but where this is inevitable, prefers to kill first, and then look. In a way this reluctance and fear is not misplaced – since the sight or presence of System B may change a man beyond recognition; beyond the recognition even of himself.

This view of System B is clearly quite different from the way we perceive, that is, *experience*, that system when we are *in* it, i.e. during dreaming – or night-consciousness. At this time the dream is real, true, accepted, inevitable and self-validating. This point, that we perceive each system differently depending whether we are *inside* it or *outside* it, tends to argue for the view of two distinct forms of consciousness – though the explanation of moving consciousness can also partly accommodate this.[10]

A further point of great importance is that both System A and System B – but more especially the latter – can function *robotically*, that is, *in the absence of its own consciousness.*

This will be seen to constitute an argument again favouring the theory of moving consciousness – in that only one full *consciousness* can (or at any rate does) operate at any given time, even though both *systems* can be operating simultaneously. This fact, incidentally, may well be the source of the legends of zombies, robots, automata and all creatures of diverse kinds without 'souls'.

In the normal course of events in the normal individual, when *waking* consciousness is present, *sleeping* consciousness is not, and vice versa. Hence the fact that vampires sleep all day. But the experience that some other agency is in charge of us, when we are relatively fully conscious, is by no means uncommon. The compulsive gambler offers an instance of this. Many such

10. System B *can* in some sense be experienced in its own terms during waking consciousness. What is then required, to use Coleridge's words, is 'the willing suspension of disbelief'.

individuals, telling of their condition, speak in these terms: 'I had won £50 (or lost £300, or whatever) and decided to stop playing. But I found myself (saw myself) placing another bet (writing a cheque)', etc.; or, in still more graphic terms: 'someone went on placing bets using my hand and my money.' One hears similar accounts from alcoholics. One sometimes hears not dissimilar accounts from people in love. It is, of course, possible to dismiss these reports as nothing more than a *façon de parler*, a manner of speaking. Anyone who imagines that it is only a manner of speaking, however, should try working with gamblers and alcoholics. Alternatively, we can ask ourselves what *we* mean by phrases like: 'I found myself picking up the phone and . . .'; 'I promised myself I wouldn't go there again, but . . .'; 'I honestly intended not to mention it, but . . .'; 'I kept meaning to tell him, but . . .', and so on.

It would seem unnecessary to refer again here in detail to the results of sleep deprivation, stimulus deprivation and drug addiction, discussed in the previous chapter.

In all these, more obviously in those cases just discussed, some other agency or *alternative organization* – for organized it undoubtedly is – takes control in the presence of varying degrees of waking consciousness; sometimes with the initial agreement of consciousness, sometimes not. In the former category one recalls how the vampire must first be *invited* in, how Faust must *first* voluntarily sign the contract with the Devil. Once the initial step has been taken, however, consciousness is largely or entirely helpless, though still, in these cases, present.

We have spoken of an alternating switch, operating in such a way that both consciousnesses cannot be present at once. But as we have seen, the 'robot' *arousal* or *inducement* of System B can take place in the presence of waking consciousness. It is functioning 'robotically' at that point, in the sense that 'we' are not 'it'. We feel or sense its presence, but we are not the presence. However, it seems that if System B is aroused or released beyond a certain point, we, that is, waking consciousness, have no alternative but to move over (out of System A) into the other system.

The switch-mechanism seems to trip itself. At this point we *become* 'it'. We then *are* the presence, as in our night dreams or a deep 'trip'.

The sequence or process just outlined is very well described in *Dr Jekyll and Mr Hyde*. Dr Jekyll first *voluntarily* makes and drinks the potion. As the transformation takes place Jekyll (System A, consciousness) is gradually overcome and finally, at the end of the transformation, replaced by the alternative personality of Hyde (System B, the unconscious) – who or which has full control until the potion wears off. Over a period and in the course of many transformations, the conscious personality of Jekyll is weakened to such an extent that the unconscious Hyde can more or less *order* the conscious Jekyll to take the first step – or by-pass his resistance entirely. This is actually the end-stage of possession notably feared by the schizophrenic and the right-wing, authoritarian personality alike. It is hard for the conscious personality to take any rehabilitative action at this point – perhaps the most he can do is to join *Dr Jekylls Anonymous*, where he can draw on the help of the ego-structures of others.

The 'robotic' functioning of *System A* is in general less dramatic. It is, however, a clear fact. For example, one's own name called out by another person is more likely to arouse one from sleep (be it orthodox or paradoxical) than a strange name called at the same volume. A wife who sleeps through her husband's getting out of bed and banging his way to and from the bathroom, will rouse at the muted crying of her baby in the next room. Similarly, a doctor may sleep undisturbed by revellers in the street outside – but not through the ringing of his own telephone. Here, obviously, the cortex is involved, monitoring stimuli from the external environment even during sleep. Possibly sleep-walking is a further dramatic instance of robotic cortical control – since sleepwalkers show no rapid eye movements and report no dream on being wakened.

This discussion has taken us a little away from the meaningfulness of dreams in their own terms, though not in those of waking consciousness. An important attribute of dreams is that their strangeness does not surprise us while we are asleep. This does not mean that we cannot be surprised in the course of a dream. The statement refers to the fact that while dreaming we completely accept the bizarre nature of the dream and the strange use it makes of the everyday world. This is the Freudian dream-*work*. In an earlier chapter the analogy was proposed of an agency gaining admission to a control room in the absence of the rightful operators, and messing about with the equipment and control-gear stored there.[11] A further 'explanatory' analogy is that of children playing in a room full of lumber or bric-à-brac, the real or original purposes of which they do not understand, but which they now seize at random for the purposes of their own games. What an accurate description this also is, conversely, of the activities of the experimental psychologist in respect of, say, literature! An adult might be surprised (amused, horrified) to find the children in this hypothetical lumber-room using a piece of brocade as a floor-cloth, a dressing-table as a pirate ship, or a gramophone as an oven. To the children there is nothing unacceptable in their actions or assumptions.

So it is with the dream work. Thus, at the time, it is perfectly in order to dream that one's wife is not one's wife, but one's mother's sister; or that cars are driven by porridge; or that women have wings that enable them to fly. Only on waking do we say 'what a strange dream.' Occasionally it happens that a dream straightforwardly reveals to us something important, makes us aware of a situation we had overlooked or directs attention to a problem. But as a rule the events of the dream seem on waking meaningless, and the inclusions and exclusions unimportant.

It appears, however, that *the dream itself knows that these are important*. This is a strange turn of phrase, the full implications of which must be avoided for the moment. One thing is, however, clear, that the dream is in charge of us, not we of it. Clearly also,

11. Alternatively, the toys left behind for the night come to life of their own accord. The dolls which must normally be wound up to move acquire a life of their own. This is the meaning of the tales of *La Boutique Fantasque*, *Coppélia*, etc. (and their fascination).

the meaningful functioning of the nervous system and our consciousness of it are not the same thing. The belief that they *are* (must be) the same thing is firmly argued by experimental psychology, and arises from the fact that in the state of waking consciousness, what happens (what we do and what we experience) and the conscious awareness of what happens, enjoy a large measure of accord. I suggest, however, that in the waking state our consciousness so closely 'shadows' the activities of the cortex (our movements and perceptions) that they can readily be confused as being one and the same thing. Only at certain rare moments do most of us experience the *dissociation* of consciousness from the activity of the cortex. But there can be no doubt that this occurs.

In *dream* life the *dissociation* of (volitional) consciousness and what happens to us (or with us) is the *rule*. That is, in dreams it is all happening *without* us. Instead of consciousness directing operations, as it *apparently* does and perhaps often really does in the waking condition, during night-consciousness the operations are clearly directing *us*.

As already established, System B can obtrude into waking life, both in abnormal and normal states. Were this latter fact not so there would be, for instance, to say nothing of Freudian slips, no such phenomenon as normal religion, since the intellect does not need this, and would indeed not be capable of producing it. The natural condition of System A is a-theism, i.e. neither for religion, nor against it, but outside it. As said, were there no such thing as System B, *and were it not able both to reach and to perceive consciousness*, the accounting for the evolution of an Earth-religion into a sky religion, and for many other factors considered later, would be very difficult.

The obtrusion of System B into System A consciousness may be resisted or not resisted, recognized or not recognized. The Communist does not recognize it, though he suffers it. In religion it is recognized (though not really for what it is) and joyfully accepted. The forms of System B, both as itself in its own domain or as the 'ghost' of System A, are varied, almost infinitely so. In this connection the phrase is apposite – 'in my Father's house are many mansions'.

The product of System B is Knowledge II. The description and definition, with examples, of Knowledge II is the concern of a subsequent chapter. Dreams are one form of this knowledge, Christianity another. Knowledge I has a single form, Knowledge II has many. This is one further way in which it can be seen that the attributes of the two systems are always opposed.

A rather different aspect of the matter is now, however, considered or reconsidered – the physiological. The many deductions made throughout this book on the basis of certain psychological evidence, and the mental 'junk' of past eras and civilizations, are here dramatically borne out in physiological terms.

The Ancient Adversary 8

One needs to be clear that a psychological theory does not *require* any physiological support for its validity. This is to say that a psychological truth is simply that, in its own right. As such it is independent of any actual or suggested physiological infrastructure. The psychological *experience* moreover is, in any case, primary and self-validating in the existential sense.

It is similarly true that physiology is not dependent for its validity on psychology. If an observed physiological fact appears not to fit with observed or experienced behaviour, this does not of itself, or necessarily, invalidate the physiological finding. Certainly such a disagreement raises considerable problems, and it is always to be hoped that this situation does not arise – or, at least, not persist. Speaking generally, one expects physiology to provide descriptions, admittedly coarser and couched necessarily in physiological terms, that are nevertheless recognizably related to the finer-grain psychological experiences, sensations or behaviours.[1]

However, this having been said – that a psychological theory or model is not necessarily invalidated by an absence of physiological support – any psychological theory which *is* supported by physiological evidence necessarily gains enormously.

1. It must be emphasized, however, that a psychological experience and the physiological reactions associated with it are no more the *same* than, for example, a conductor wire is the same as an electrical impulse; or the impulse the same as the magnetic field which forms around its path.

Many writers have talked at length about the unconscious, and this has been described and defined in a variety of ways. The crucial question, however, goes in the last analysis unanswered – is the unconscious mind merely a useful hypothetical construct, or does it *actually exist*? Clearly the concept has already a certain psychical reality, to use Freud's term. But does it also have an *objective*, that is, a somehow *physical*, reality?

That no serious, or at any rate no successful, attempt has been made by psychoanalysts, whoever else, to give the unconscious a precise, physiological locus within the nervous system, is to say the least unfortunate. The self-assigned problem of this book is not merely to find evidence of some rudimentary consciousness at sub-cortical levels – since that would be likely to be only the forerunner, or residue of an earlier version of waking consciousness – but rather an independent, fully equipped, *self-conscious*

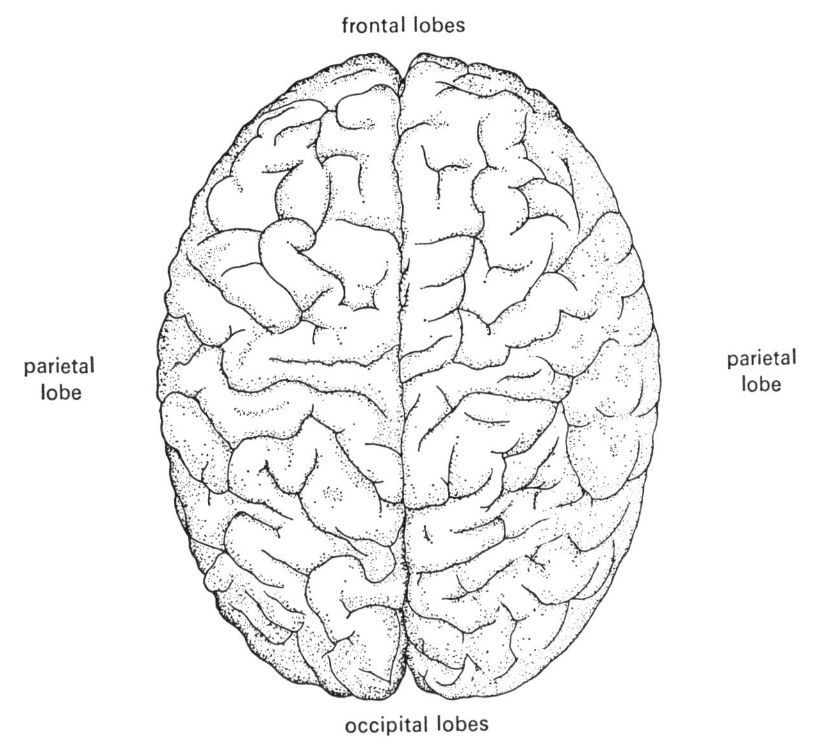

Figure 12. Human brain, viewed from above.

system, largely, if not wholly, separate from the central nervous system and its partner, waking consciousness.

With this intention in mind we again turn to a general consideration of the central nervous system.

For those unfamiliar with its features Figure 12 shows a general view of the human brain from above, with certain features labelled, Figure 13 the same brain seen from the right side, and Figure 14 the same brain once more, viewed from underneath.

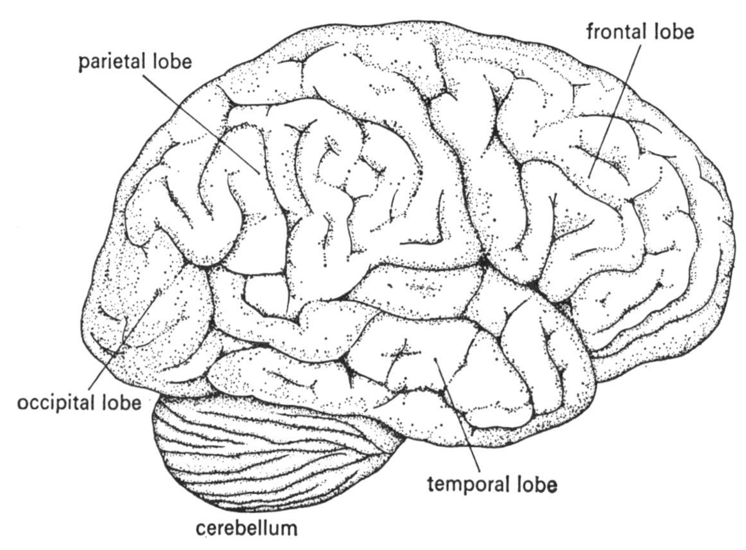

Figure 13. Human brain, right side view.

It will be noted that the two cerebral hemispheres anatomically duplicate each other. An animal with only one hemisphere intact can function well in a number of ways. However, the functions of the twin hemispheres are not quite identical. A point of importance is that motor and sensory fibres from the *left-hand* side of the body go to the *right* cerebral hemisphere, while those from the *right* of the body go to the *left* hemisphere. In short, each hemisphere governs the side of the body opposite to it. The majority of people have one dominant hemisphere – in most cases the one concerned with handedness, footedness, and so on.

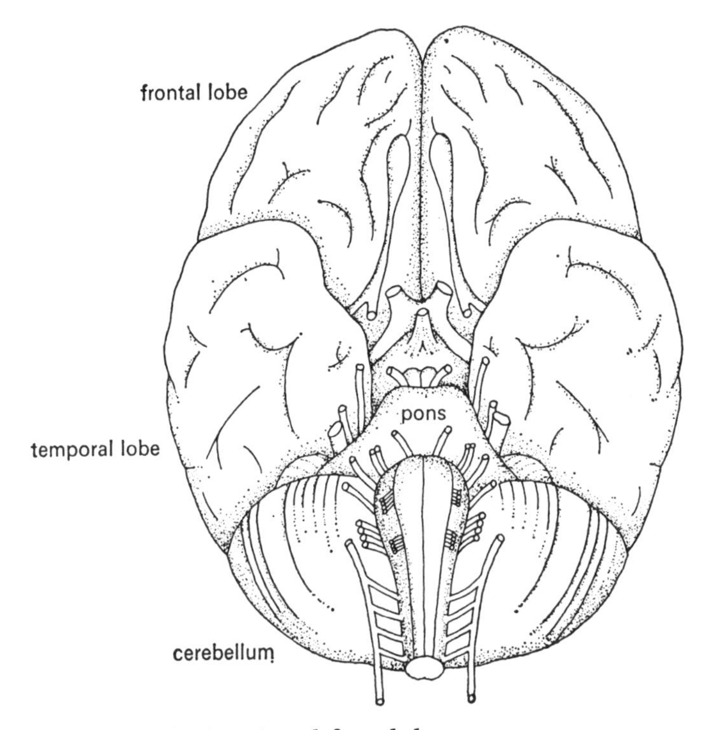

Figure 14. Human brain, viewed from below.

In Figure 15 the brain has been cut through the centre (front to back) to reveal some of the inner and lower brain centres. Finally, Figure 16 shows a simplified, 'exploded' mammalian brain.

Attention must be drawn to two terms which are easily confused, but between which it is essential to distinguish. The cerebral cortex is so called because it is the cortex of the *cerebrum.* The cerebrum may for present purposes be considered to be the whole of the cerebral hemispheres and their immediate substrata. The term *cerebellum* is a diminutive form of the word cerebrum, and means literally 'little cerebrum'. The adjective from cerebellum is *cerebellar*, and because the cerebellum also has a cortex, one speaks of the *cerebellar cortex*. A simple way of remembering which is which is to recall that the smaller object, the cerebellum, has the longer name.

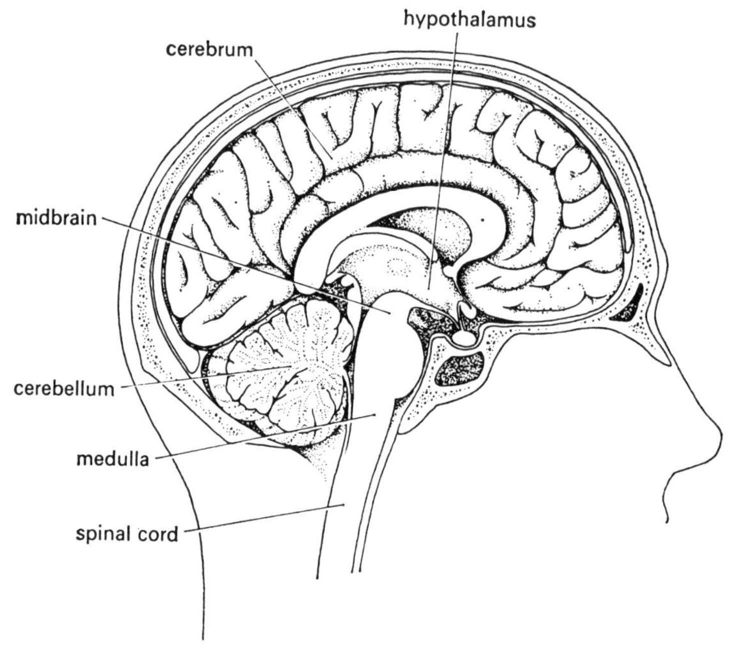

Figure 15. View of the vertically bisected brain.

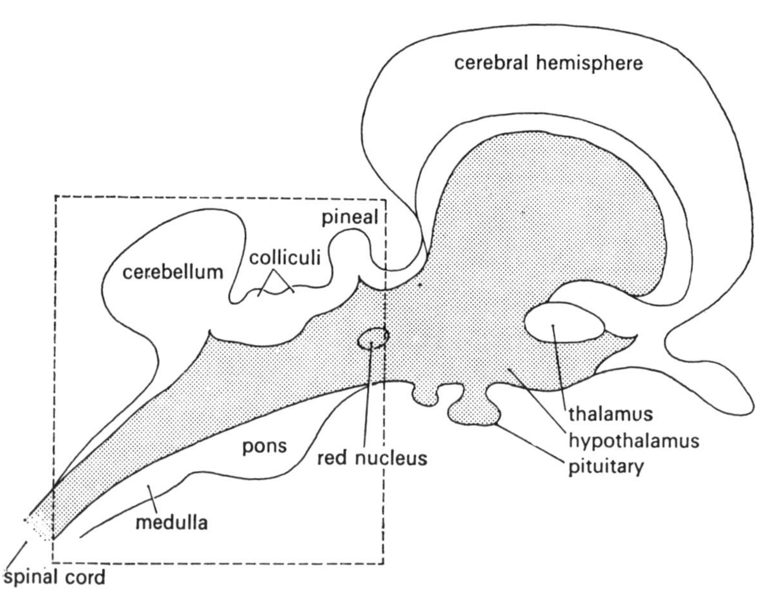

Figure 16. 'Exploded' stylized mammalian brain, with certain centres indicated.

We have already noted the position of the cerebellum in relation to the cerebrum in man's brain (Chapter 4) – and Figure 1 should be re-inspected in connection with what follows. We turn now immediately to some statements about the cerebellum, taken from the standard physiological literature. These statements on the cerebellum and associated organs, which follow throughout this chapter, are inset from the body of the text. This is partly for purposes of reference, but partly to allow more ready distinction between the established facts and my commentary. Some items are taken verbatim from source, where the source is indicated. Others are summaries of accepted fact in my own words. Unless otherwise stated, all italics are mine.

> (1) The cerebellum is a large organ, weighing in man more than the entire spinal cord. It is one of the oldest organs in the phylogeny (evolution) of the vertebrate nervous system. It is well developed in reptiles, very well developed in birds, but reaches its *highest* development in man.
>
> In the developing human embryo the cerebellum is seen to arise in the primary receiving stations of the labyrinth [*n.b.*] of the ear, that is, in the vestibular complex of nuclei principally involved in the maintenance of posture and equilibrium.
>
> (2) 'The cerebellum is an extremely complex structure. In many ways it is just as complex as the cerebrum. In fact, from the evolutionary history of the cerebellum *it looks as though nature started out to make the cerebellum the highest centre of the nervous system* but changed its mind and developed the cerebrum instead.'
> (C. T. Morgan and E. Stellar, *Physiological Psychology*, McGraw Hill, 1950, p. 288)

Like the cerebrum, the cerebellum has a cortex composed of grey matter, similarly convoluted and divided into a number of lobes. As with the cerebrum again, the recently developed parts of the cerebellum (the *neo-cerebellum*) are bilateral – that is, there are two cerebellar hemispheres corresponding to the two hemispheres of the cerebral cortex. There is one crucial difference, however, which will be discussed elsewhere. These two cerebellar hemispheres developed *during the same evolutionary time occupied by the development of the cerebral neo-cortex. Their appearance has more than doubled the size of the cerebellum.*

(3) 'The cerebellum receives extensive projections from all sensory systems and it has been suggested that some portions of the cerebellum may serve as primitive sensory-projection areas. Stimulation of tactile receptors results in an *organized* pattern of projections in the cerebellar cortex. Visual as well as auditory projections have been demonstrated . . .'
(S. P. Grossman, *A Textbook of Physiological Psychology*, Wiley, 1967, pp. 142–3)

'Electrodes placed on the surface of the cerebellum give us electrical responses when the various senses are stimulated. In fact there are very definite areas that can be mapped for vision, hearing and the somatic senses. *We do not know why these sensory pathways to the cerebellum are there . . .*'
'In another study by Brogden and Gantt there is a very interesting result. Direct electrical stimulation of the cerebellum of the dog was substituted for the usual unconditioned shock to the footpad. As usual, a sound was the conditioning stimulus. The cerebellar stimulation evoked movements of the same sort as those evoked from the cerebral motor cortex. Strangely enough, after a normal number of conditioning trials, pairing buzzer and cerebellar stimulation a conditioned flexion of the limb to buzzer appeared. *That is extremely interesting, but we cannot be sure what it means. The investigators assure us that the cerebellar stimulation was not spreading to the cerebral cortex*, and we can rule out that possibility.'
(C. T. Morgan and E. Stellar, pp. 288 and 456)

What emerges, in summary, from the foregoing is that the cerebellum is a large, complex organ which has been with the vertebrates since their inception. It has continued to evolve throughout the whole of that period, and is at its most evolved in the primates, particularly in man. Its cortex has extensive sensory and proprioceptive (positional), and perhaps even motor, projection areas. The term projection area means that ascending stimulus and informational pathways from lower centres reach these areas and end there. In the *cerebrum* such areas are associated with consciousness. There is, further, at least tentative evidence that classical conditioning can be established through (or within) the *cerebellar* cortex, without involving the cerebral cortex. This means, in other words, that the cerebellum is capable of independent learning *based on classical principles.*

These further facts are established:

(4) The cerebellum as a whole is concerned with the automatic regulation of posture and movement, in the terms described below et seq. In this role it works in close conjunction with the cerebral cortex, as a kind of auxiliary, throughout the waking hours. It is in 'the great swimmers and fliers' that the cerebellum is *relatively* best developed. That is to say, the cerebellum in these animals, while not as highly developed as in man, is better developed in relation to the rest of e.g. a bird's brain, than the cerebellum in man is developed in relation to the rest of *his* brain.

(5) Disease or destruction of the old cerebellum (*paleocerebellum*) results in the loss of skilled movements. The affected subject, human or animal, demonstrates 'over-shoot' – i.e. over-reaches or over-extends in making a movement. The starting and stopping of actions is too abrupt, accompanied by a general loss of fineness. Contractual movement is not well sustained, this being coupled with an overall, undue liability to fatigue. Already in 1828 the French physiologist, Flourens, likened the gait of a man with cerebellar damage to that of a man *under the influence of alcohol* (a very interesting analogy in

view of our earlier remarks concerning the effects of alcohol). The cerebellum is particularly concerned with muscle-*tone* (tonus).

(6) Electrical stimulation of the paleocerebellum causes in-inhibitory relaxation of certain active postures in normal animals – that is, they collapse.
Stimulation of other areas produces *movements of the face and of limb extremities.*
Yet other stimulation produces specific *autonomic* responses – for example, stimulation of the anterior lobe inhibits respiration, produces a decrease in tension in the walls of blood vessels (i.e. reduces the work of the heart), and a *parasympathetic* contraction of the pupil of the eye.
In *decerebrate* animals – animals in whom the cerebrum has been removed or otherwise put out of action who as a result evidence *decerebrate rigidity*, that is, permanent stiffness of the limbs – stimulation of the anterior cerebellum induces converse loss of muscle tone and flaccidity of the limbs. This relaxation is maintained until the stimulation ceases, at which *a rebound action causes the limb in question to extend powerfully.*

How far one might be justified in regarding a normal *sleeping* organism as *decerebrate* is arguable. It seems a not wholly unreasonable assumption. Be that as it may, it will not have escaped the attention of the reader that many of the results of cerebellar stimulation and cerebral ablation are *in evidence in normal sleep.* By way of reminder, two brief extracts bearing on the question follow from Ian Oswald's book on that subject.

What happens when we fall asleep? The eyelids close and *the pupils become very small. . . . The total flow of air breathed is diminished. . . . The heart slows . . .*

Although the cat's body flops its whiskers may twitch and *lots of little facial twitches accompany rapid eye movements. . . .* What is more, the paralysis of sleep is momentarily but quite frequently interrupted *by sudden movements of a limb . . .* (pp. 9 and 96)[2]

2. Both the cerebrum and the cerebellum possess facilitatory and inhibitory

The further examination of some other brain structures closely associated with the cerebellum now serves to strengthen the picture which has begun to emerge from the consideration of that organ in isolation.

(7) In the developing human embryo three divisions of the brain are clearly visible. These, in non-technical terms, are the fore-brain, the mid-brain and the hind-brain. In the 'finished' brain of the human being, the mid-brain is a bridge or stalk connecting the fore-brain and the hind-brain (see Figure 16), but in the early life of the embryo it occupies a prominent position on the surface of the brain. In fairly literal terms the hind-brain (principally, the cerebellum) and the fore-brain (principally the cerebrum and associated structures) 'flower' from the *brain-stem* – which has itself formed from the top end of the spinal cord. In the course of the growth of the higher structures the brain-stem is completely covered over and enclosed, and the cerebellum itself is partially covered by the backward-extending cerebrum.

The brain-stem consists principally of the *medulla*, and some tracts of the *pons* and the mid-brain. In addition, the thalamus and hypothalamus constitute, in effect, direct continuations of the brain-stem.

Although in Figure 16 the pons and the mid-brain are shown as separated, they are in fact complexly interconnected. There is really a two-level highway passing through these structures from back to front, one arising in the cerebellum, the other in the medulla.

---

powers over muscle movement. In the cerebellum inhibitory properties may be said to predominate. Both of these systems may be seen as *control* systems. But whereas the cerebellum controls or inhibits *too much* movement or tension, the cerebrum inhibits slackness and what might be described as *too little* movement.

It will be recalled here that psychotics and pre-psychotics are frequently characterized by hyper-activity and insomnia: and neurotics, in general, by ineptness, clumsiness, tiredness and oversleeping.

(8) *The brain-stem (medulla).* The principal role of the brain-stem or medulla is that of a mediating or transmitting station, dealing with incoming motor and sensory information *to* the brain and similar outgoing instructions *from* the brain (both via the spinal cord). One of its tasks is therefore that of balancing the demands of the spinal cord and brain.

In its further role as the *reticular activating system*, the brain-stem can variously arouse the organism from sleep, or increase the level of awareness during waking, or under other conditions facilitate sleep.[3]

The majority of the principal pathways from the spinal cord to the cerebral cortex pass through the medulla and it is here that they cross – so that messages from the right side of the body reach the left hemisphere, those from the left the right hemisphere.

Some of the spinal-medulla connections, however, instead of ascending to the cerebrum, pass instead directly to the cerebellum, without crossing, so that this organ has its own supply of information quite independently of the cerebrum. Also, *in mammals only*, some paths from the spinal cord to the cerebrum by-pass the medulla altogether. Thus, *some* information reaching the cerebrum is not also simultaneously transmitted to the cerebellum. The already existent 'split' between cerebrum and cerebellum is thus in *mammals* carried a stage further. (One is even more Two than before?)

(9) *Pathways from the cerebrum to the cerebellum – the cortico-ponto-cerebellar tract.* This tract is the main pathway from the cerebral cortex to the cerebellum for afferent impulses (that is, impulses which *affect*). These impulses come principally from the frontal and temporal lobes of the cortex (though all parts of the cerebral cortex have some connection with the cerebellum). Passing through the thalamus and the pons

3. The reticular activating system is responsible for the flooding of the cortex in panic and other extreme states (which effectively put the cortex out of action) discussed in Chapter 4.

(hence the name cortico-ponto) the fibres reach the same-side cerebellar hemisphere. There is no crossing of these fibres to the opposite side.

(10) *The pathways from the cerebellum to the cerebrum.* These are the *efferent* pathways of the first-named organ, the means and route by which the cerebellum has an *effect* on the cerebrum. The fibres concerned pass from the nucleii situated in each hemisphere of the cerebellum via the mid-brain (in particular the red nucleii of the mid-brain) and there *cross over to reach the opposite hemispheres of the cerebral cortex.* It is to be recalled also in the present connection that the cerebellum, *unlike* the cerebrum, receives its fibres from the spinal cord uncrossed. Thus, with the cerebellum, damage to one of its hemispheres produces motor dysfunction on the same side of the body – not the opposite side as is the case with the cerebral hemispheres.

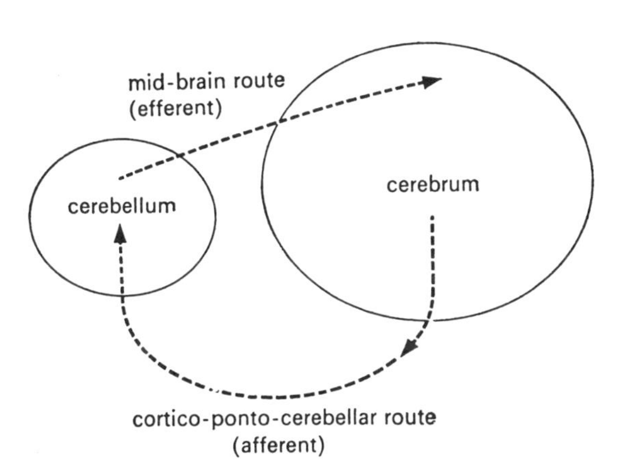

Figure 17. Diagrammatic representation of pathways between the cerebrum and the cerebellum (side view).

Figure 17 shows in stylized diagram form the direction of afferent impulses to the cerebellum and efferent impulses from the cerebellum to the cerebral cortex (side view).

At this point a brief excursion into a long-standing social

phenomenon becomes necessary. On the additional basis thereby provided, the present book's conception of the role of the cerebellum can be further developed.

Left-handedness, with all its ramifications, is among those topics which seem to have preoccupied man since earliest times, but have so far escaped satisfactory account. It is possible here to undertake only the briefest of reviews.[4] The circumstances attendant on the phenomenon are, however, a significant link in the chain of argument here.

The majority of psychoanalysts are united in considering the left (hand) to be associated especially with the unconscious, with the 'female principle', and with the emotions. In mysticism, religion and magic also the left has similar associations, with perhaps the additional connection with darkness and evil. In medieval paintings, for instance, the invoked devil is always shown leaving the magic circle with left hand outstretched. This is all very germane.

It appears, too, that in the everyday life of the East (and in the West to a lesser extent) the left hand was, and is, considered the unclean hand. It is the hand reserved for wiping after excretion and for making love – i.e. caressing[5]. Both these, one notes, are

4. For a non-technical survey of this subject see Michael Barsley, *The Left-Handed Book*, Souvenir Press. Also Chapter 4 of *The Story of the Human Hand* by Walter Sorell, Weidenfeld.

5. Without quoting his source Michael Barsley proposes that ninety per cent of the general population use the left hand for these purposes. This is a rather surprising percentage, which an instant survey of one's acquaintances fails to support. Possibly the figure intended applies only to Eastern populations.

What is undoubtedly true, however, is that a phrase like 'cack-handed' (left-handed, awkward, clumsy) means quite literally and unequivocally 'shit-handed'. The word 'cack' has a long and impressive history – occurring not only in Greek and Latin (Lat. *cacare* – to void excrement) but in the Indo-European *kakka*.

B1 functions, and it is interesting to note also that the preferred hand for *fighting* and doing is the *right* hand.

The left hand in Western societies is perceived as somehow unlucky. This is clearly borne out by linguistic usage. Most people know that the English word 'sinister' is simply the Latin word for left. The English word left itself comes from Anglo-Saxon *lyft*, meaning weak or worthless. In Italian *mancino*, apart from left-handed, means also treacherous and dishonest. So one could multiply examples. Such a widespread and ingrained reaction is perhaps puzzling in view of the mere five per cent incidence of left-handedness in the general population.

What is not generally realized, however, is that in the oldest times left meant *lucky* or favourable. This is partly bound up, in historical European times, with the way an augur or diviner faced when prophesying – that is, South or North – and whether therefore he had the East on his left or right side: in all cases East was lucky, West unlucky. Older civilizations generally, such as Egypt, were inclined to favour the left as propitious.[6]

The turn-around of the ancient view that the left is propitious (or to be propitiated) into the later view that the left is unlucky seems, then, to be an ongoing process completed only in historical times. It was, perhaps, already well under way in classical Egypt, but traces of leftness – e.g. in handwriting – persisted.

All the primates, apart from man, are ambidextrous. They never develop the favouring of either hand. Human babies, as

6. The earliest handwriting of the Near and Middle East is from right to left. This movement is *less* natural for right-handed than for left-handed individuals. Other ancient forms of writing, including Chinese characters, are read from top to bottom and appear equally convenient for both left-handed and right-handed writers. However, Walter Sorell tells us that here, too, the direction of the character is from right to left. All later forms of writing, including of course our own, are from left to right.

One is not necessarily suggesting that a majority of, say, ancient Egyptians were left-handed – though perhaps a high proportion of the scribes or priests were. If that nation, however, was as generally right-handed as we are, then possibly they wrote *towards* the left in order to commemorate or propitiate something. One must then ask oneself what that something might have been.

we have noted, appear to recapitulate this evolutionary (?) stage by passing through a period of ambidexterity before settling for one or other hand. Of great interest – particularly in respect of aspects of the general theory to be developed later – is the opinion expressed by some authors that existing primitive peoples show a higher incidence than Western man both of left-handedness and ambidexterity. Authoritative statements of the true position appear, however, to be lacking.

Anthropologists are divided also on the further and not unrelated question of whether early, extinct man was predominantly left-handed. The evidence for an affirmative view is derived from the study of cave-paintings and other artifacts. It is claimed that the profiles of men and animals in cave drawings usually face to the right. Right-handed people as a rule draw profiles looking *left*; it is the left-handed person who normally draws them facing right. (This one can test out for oneself by asking acquaintances to draw a man or dog or whatever in profile, without telling them the reason.)[7]

The jigsaw-puzzle situation that all primates apart from man are ambidextrous, that very young children and possibly a higher than usual proportion of primitive peoples are ambidextrous, and finally again that perhaps a higher than usual proportion of primitive peoples are, and even a majority of ancient men were, left-handed, would appear to merit some kind of explanation.

One must appreciate, too, that while left-handedness (or ambidexterity) might well come to be regarded as *un*lucky because it represented a more 'primitive' evolutionary stage, this would hardly explain why the left should ever have been regarded *favourably*.

7. Michael Barsley and others claim that a significant number of well-known cartoonists are left-handed. (Once again, however, no firm figures are given.) If this is true, may there then be some connection here with the apparent left-handedness of cave-artists? And if so, what is that link? A later chapter attempts answers to these questions. Meantime, in the present general connection, let us not overlook that probably the greatest of all cartoonists (not, of course, satirical), Leonardo da Vinci, and another great artist, Michelangelo, were also left-handed.

The general circumstances surrounding leftness and the change of attitude to it may be seen to coincide (at least in time) with the situation outlined in Chapter 2 et seq. that early civilizations were somehow female-oriented, especially in terms of what was feared and revered. As we saw, the throne in early Egypt descended exclusively through the female line and the oldest divinity of that people was also female. The associated shift from Earth-religion to sky-religion, under way but not completed in early Egypt, and carried further towards completion in classical Greece, 'coincidentally' parallels both the change from left-directional to right-directional writing, and the change in attitude to leftness from favourable to unfavourable.

From the purely statistical point of view, as already indicated, the entrenched and vigorous anti-leftness of Western culture certainly appears exaggerated in that the small minority of left-handers could not constitute any kind of serious threat numerically. One can find plenty of non-statistical reasons, of course – for instance that the left has played such a prominent role in devilry, magic and witchcraft since earliest times, along with the other reversals and opposites already touched on in Chapter 2. At the same time, this favouring of the left in witchcraft itself requires explanation.

The incidence of left-handedness in the general population appears to be, as stated, of the order of five per cent. Estimates, admittedly, vary rather widely both up and down, and there is the added difficulty that apart from which hand is favoured there are questions of foot-dominance, eye-dominance, and other yet more subtle behaviours. The incidence of conventional left-handedness is considerably increased, however, rising to around twelve per cent, if one considers only the educationally sub-normal and/or mental defectives.[8] This, clearly, must account

8. Some of the higher incidence of left-handedness among the educationally sub-normal is probably due to actual brain damage, presumably of the otherwise dominant hemisphere. And against this association with feeble-mindedness we must place the very considerable achievements of such varied left-handers as Goethe, da Vinci, Michelangelo, Nietzsche and Beethoven.

for some of the bad reputation of left-handedness. From our present point of view, this group would conceivably be deficient in System A characteristics, and therefore the trend is not surprising. However, the logical expectation that *females* would show a higher incidence of left-handed behaviour is in no wise borne out. While here, too, estimates vary, the majority of surveys agree in assigning a lower incidence of left-handedness to females.[9] To account for this one could propose that, for example, the greater suggestibility of girls might account for their increased closeness to the norm of rightness. This is not a very convincing suggestion – and fortunately we shall be able to offer a much more adequate explanation of that circumstance in due course.

Ambidexterity is much more common in animals and, as indicated, all primates other than man. Nevertheless, there is a tendency towards right-handedness in some species. The preference is always less marked than in man. Nor, so far, has any general explanation emerged why this should occur in particular species and not others. In rats, for example, the tendency to use the right forepaw is marked: yet in chimpanzees, the most 'human' of the primates, there is no consistent 'sidedness'. The views of the present book on the aspects of handedness in play here are outlined as follows.

The centre of gravity of the human body is slightly to the right of centre. The right lung, for example, has three ventricles while the left has only two. This means that slightly greater stability is obtained when the left foot is advanced; and that the right arm has somewhat greater inertial power behind it.

My own view is that what we see among left-handers in our present society is the shattered remnants of a normal distribution. This point will be taken up again elsewhere.

9. The sex difference, however, is much less marked among the educationally subnormal. One or two studies have actually returned a higher incidence of left-handedness for girls over boys among the E.S.N.

To introduce the apparently trivial or anecdotal into the serious, one *does* note that women still today button their coats to the left, while men do so to the right.

There are factors connected with the rotation of the Earth on its axis – coupled perhaps with phototropism, the tendency to seek light – which could possibly provide a basis of explanation for the right side of (primitive) organisms being slightly favoured in strength – at least as far as the northern hemisphere is concerned. It is possible that such a slight tendency would nevertheless persist in higher organisms through evolutionary time, providing it withheld no advantages.

The question one must ask is under what conditions such a hitherto neutral and in any case slight tendency might be *favourable* to survival, either generally or in the context of one species. One answer is in fighting. An animal striking with his left would be at a marginal disadvantage in terms both of stability and hitting power when faced by an animal leading with his right, and therefore disfavoured survivally.

Applied to human beings, this proposal receives strong support from two additional considerations: (1) the heart is biased to the left of centre, i.e. in the rearward and hence most protected position when the right hand is in actual use and (2) by reason of the crossing of motor and sensory fibres to the opposite hemisphere of the cerebrum the dominant hemisphere is therefore *also* to the rearward and in the most protected position when the right hand is in use. (Presumably this is one of the reasons for the 'migration' of the dominant hemisphere.)

If the above reasoning is correct, in whole or part, one would look to find right-handedness *predominant in all aggressive and/or predatory species*, while ambidexterity would be more the rule in non-predatory and non-aggressive species. This would account neatly for the non-laterality of chimpanzees and gorillas, Man's nearest relatives, who despite appearances are only fruit-and-maggot eaters; and for the laterality or right-handedness of e.g. rats and lobsters.

In brief, the proposal is that right-handedness became dominant in man first when he ceased to be a fruit-and-nut gatherer and eater, and became a hunter of other animals; but much more markedly, when he came into open competition – that is, organized war – with his own species, the inevitable result, perhaps, of his relentlessly increasing numbers.

Early man, certainly agrarian man, could be tolerant of left-handedness or ambidexterity. Even as hunters left-handers would possibly be socially acceptable. However, there would probably tend to be more fatal accidents amongst them in the course of hunting, so that already there would be a tendency for the incidence of left-handedness to reduce in the general population. But under conditions of severe competition or in war two things would happen: (a) *consistently* more left-handers would be killed in battle and therefore gradually 'bred out' and (b) left-handers would be *perceived* to be less effective fighters. Let us not forget here, too, the apparent association of – some – left-handedness with poor intelligence. Under these conditions left-handers could come to be regarded as 'womanly' or weak (cf. the Anglo-Saxon *lyft*), particularly in warrior nations.

These arguments are, obviously, not fully developed here, but the trend will be perceived from this brief exposition.

Since straightforward statistical approaches to our problem here do not seem particularly helpful, a less obvious hypothesis is called for. It is in magic and allied areas that the issue seems most alive. This therefore should be the logical region of our inquiry. Indeed, the conclusion of this book is that the intense feelings aroused by left-handedness have only marginally to do with actual physical left-handedness. The explanation which now follows would seem to deal more comprehensively with the numerous aspects of the question than any existing account. The suggested link between fighting and right-handed dominance, of course, remains. But this is a rational explanation for an overt phenomenon. Physical left-handedness acts as the focus, the false target or lure as it were which distracts our attention from the true meaning and explanation of 'the left'. One or two preparatory comments for our next step now follow.

When a person looks into a mirror he finds that his left hand has become his right, and his right his left. The opposites, one could say, have changed places.

When vampires and others of the living dead look into a mirror, however, they have no reflection. This test is, in fact, one of the way's of spotting a vampire. Do they, perhaps, have no reflection *because they are that reflection*?

What role do mirrors play in legend and fairy-story? They are, of course, one of the entrances through which one can gain access to the magic world. Lewis Carroll's *Alice Through the Looking Glass* and Cocteau's *Orphée*, in which the characters enter and leave the underworld through mirrors, are recent examples of this. In that world as in other versions of magical lands, the rules of the normal world are reversed. One recalls, too, that the Black Mass is celebrated in opposites. In older legends not a mirror, but a pool or surface of water – which, of course, when still, functions as a mirror – is an entrance: and gazing into a bowl of water is one method of divination, as is the crystal ball. When the surface of a pond is ruffled, of course, the image is lost. Do we have here a link with the breaking of the mirror which brings ill luck – and thence with the more general breaking of the spell?

At this point we revert to the consideration of the physiology of the brain, and in particular of the role of the cerebellum, for a possible elucidation of these and other matters.

The left *cerebellar* hemisphere controls the opposite side of the body to the one that the left *cerebral* hemisphere controls. Efferent messages from the respective cerebellar hemispheres to the cerebrum cross over to reach the opposite hemisphere. This is because cerebellar contacts with the spinal cord are same-sided, while cerebral contacts with the spinal cord are opposite-sided.

If the cerebrum is imagined as looking at itself in a mirror, with the cerebellum as the reflection, the situation and its implications are clear. The right hand has become the left hand. This notion is diagrammed schematically in Figure 18.

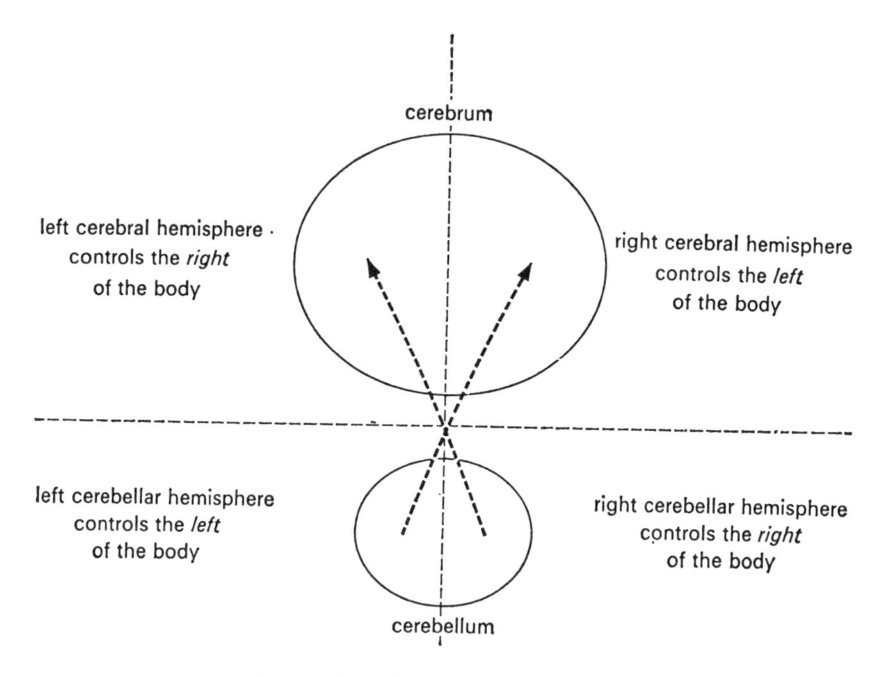

Figure 18. Cerebellum as 'mirror-image' of the cerebrum.

One sees, then, that the cerebellum is the *mirror-image* of the cerebrum: that is, if the cortex were a person looking at himself in the (cerebellar) mirror, it would experience the diagonal crossing over of left and right that is a feature of mirror-images.

Should this still not be clear, let us consider Figure 19. In (a) we see, from above, a person facing another person. The two right hands of these two individuals are *diagonally* opposite each other. (There is a cross-over, such as we effectively see from cerebrum to cerebellum.)

In (b) we see an individual facing his own reflection in the mirror. Now, of course the 'right' hand of the reflection is actually the *real left* hand of the real person. But let us here think of the *reflection* as if it were a real person. Or let us suppose that the reflection (magically) came to life and stepped out of the

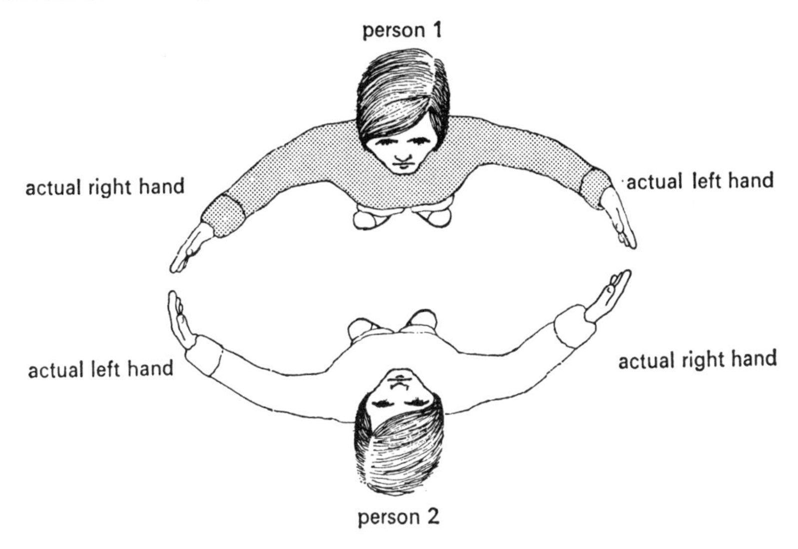

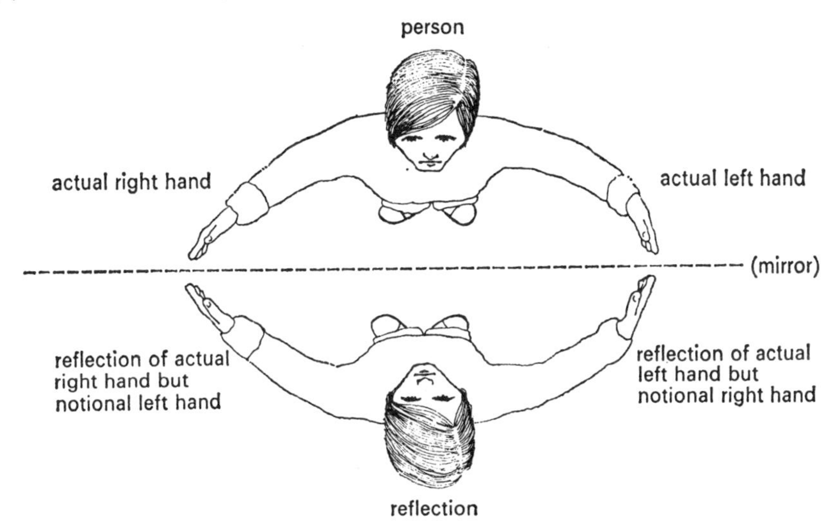

Figure 19. Reflections, opposites and handedness.

mirror. Our reflected left hand (the *notional* right hand of the image) would then be the *real* right hand of that new person. Our left hand would have *become* a right hand.

The mirror-image uses its/our left hand as a right hand. And this is precisely what the cerebellum is doing when it influences the cerebrum. The totality of variables involved in the total schemata is not uncomplicated. But if we hold fast to the analogy of the cerebrum being a person looking at himself in the mirror of the cerebellum, the idea should remain clear enough.

If it is true that the cerebellum takes over the helm at night and in trance conditions, as defined in Chapter 6, the cerebrum would then in effect be under the control of its own mirror-image.

In the light of these general speculations and in passing can one now consider it coincidental that the Socialist movement is referred to as the *left* wing? Socialism being, of course, one of the trance conditions.[10]

Some further speculations involve a more detailed consideration of the mid-brain.

(11) The mid-brain in man contains primitive visual and secondary audition centres – which reside in what are termed the upper and lower colliculi respectively. In pre-mammalian

10. One does not at all dispute that there exists a 'rational', historical explanation for this term: it arises from the fact that in the French National Assembly of 1789 the radicals were seated on the left hand of the President.

What I suggest is operating here, however, is a process akin to the 'Freudian slip'. The 'explanation' of the event (where one exists) is then not so much an explanation as a rationalization.

This process I term *archestructural*, and the product an *archestructure*. These terms will be fully defined elsewhere. Among other things, however, archestructures may be regarded as instances of the unconscious and undetected influence of System B in System A situations. (Nevertheless, this is *not* a *neurotic* process in the usual Freudian sense.) This is a matter for a later chapter. However, one further well-known example of an archestructure, an observation for which no adequate explanation has yet been suggested, is the situation that the poor in many modern conurbations are housed in the *east* end, the wealthy in the *west* end, of the city.

organisms, however, these are the *principal* centres for vision and hearing.

In man the mid-brain is still centrally concerned in *movements of the eyes*. Situated just ahead of the mid-brain is the *pineal gland* or pineal body. Although a vestigial organ in man, its evolutionary history is extremely interesting. In the lower vertebrates it is a much more highly evolved organ and was originally bilateral (i.e. paired). In the lamprey, a still-living lower vertebrate, the pineal body is sensitive to light, and is visible below the skin.

In certain reptiles the development of this organ is still more considerable. *It is believed that certain extinct reptiles and certain ancestors of vertebrate mammals possessed an additional pair of parietal eyes.*

The pineal gland in man is identified with what some mystics refer to as the 'third eye'.

Recalling in particular the view of C. T. Morgan and E. Stellar that nature appeared originally to have intended making the cerebellum the chief organ of the nervous system, one considers again the exploded mammalian brain in Fig. 16, in particular the structures contained within the squared area. In the light of what has been added above concerning the pineal gland, do not the structures embraced by the squared area constitute an organism within the organism? One has (a) the remains of parietal eyes, and centres for vision and hearing (the mid-brain) and (b) the cerebellum and the cerebellar cortex (strongly reminiscent in structure and function of the cerebrum and the cerebral cortex) – these moreover also fully and *independently* linked to all organs and systems of the total nervous system, below the level of the actual cerebrum itself.

What follows next is speculation, but as such not wholly unjustified.

While it would seem that the pineal gland in man is not directly sensitive to light, there is no automatic reason for supposing that the centres which it originally served do not still *receive and interpret* some stimuli as visual. If it seems surprising to argue that non-visual stimuli can create a visual sense-impression, let us remember that electrical stimulation of the cortex of the temporal lobes produces visual and audial memories in the individual's mind. (One speaks commonly in any case of 'the mind's eye'.) Moreover, even simple pressure on the lidded eyeball gives a perception of colour.

Even if the suggestion is rejected, does there not still remain some possiblity that at some stage of recapitulatory embryological development these once active centres somehow might function to produce visual activity in the brain of the embryo? Quite what form this might take is again a matter for speculation. But the attractiveness (to me) of the idea of a somehow visual cerebellum lies in the fact that if the 'cerebellar organism' could in any way see (that is, experience visually) the cerebral cortex, as it arches over it, it might well perceive this as a firmament filled with bright, soaring angels – that is, Greek *aggelos* – messengers.

This must rank as the most improbable idea so far produced by this volume. Yet, is anything *more* improbable than our own existence?

All this being as it may, or may not be, we still have the unexplained fact that *premature babies spend more time in rapid-eye-movement sleep than full-term babies.* Do full-term babies show this activity in the womb? Or is it a side-effect of being born prematurely? One is, of course, quite ready to accept that the actual eye-movements (governed principally by the mid-brain) are automatic. Their functioning in these circumstances resembles somewhat the action of pedals on a fixed-wheel bicycle. When the cycle rolls down hill, the pedals turn automatically and 'involuntarily', in sympathy with the movement of the back wheel.

Whatever else, the question remains – why should the mid-brain be unusually active in premature babies, and possibly in babies within the womb?

The cerebellum, it has been proposed, is the mirror-image of the cerebrum. Mirrors, the surfaces of ponds and other bodies of water, are some of the gateways to the Land of Dreams and Fairyland. A visit to the magic world through one of these gates (such as Alice's excursion through the looking-glass) is also a further version of the 'journey' already discussed.

As already observed, in many cases the passage into the various kinds of Make-Believe world is often difficult, being narrow, dark, precipitous. Other stories, however, describe the entry as *easy* (seductively easy, in fact) but the *return* as difficult.* Many stories contain this theme – *Tannhäuser*, *Pinnocchio*, *Rip van Winkel*, *Goblin Market*, etc.

This contradiction, which need not delay us too much at this point, arises possibly from the frame of mind in which the journey is undertaken. If one is prepared to abandon one's hold on everyday reality or waking consciousness, and to accept the unconscious wholly on *its* terms, the entry is easy. Conversely, the return is difficult because one has in effect forfeited the means of return. But if one seeks to enter the unconscious on one's own terms, with consciousness, personal integrity (meaning 'waking integrity'), hold on reality, and so on, *intact*, then the way is difficult and danger dogs every step. In this, consider the journeys of the Greek heroes into the underworld – the world of the dead, which they entered *living*.[11]

* cf. Virgil: *Aeneid* VI (transl. Ogilby)

'The way is easy to the Avernian flood
Black Pluto's gates stand open day and night:
But to return and view ethereal light
That is a work, a labour.'

Also *Matthew* 7:13 'wide is the gate, and broad is the way, that leadeth to destruction.'

11. In the context of the whole of the foregoing consider also on the one hand 'knock and it shall be opened' and on the other, 'it is easier for a camel to pass through the eye of a needle than for a rich man (i.e. reality-orientated man) to enter the Kingdom of Heaven'. On the one hand the entry could hardly be easier, on the other it is impossible.

We require, then, a physiological model in which *either* an 'entry' or a 'return' is somehow difficult, somehow not straightforward, somehow dangerous. We require also a sojourn in a strange 'world' or 'universe' where things are not as they 'should' be in reality terms – but range through the wonderful, miraculous, magical, back-to-front, reversed, wrong, evil and sexual. Somewhere, too, where one *is* or whence one *returns* in some way or other changed.

Earlier, examining sleep and dreaming sleep, a state of affairs somewhat approximating to the journey and the 'country' just described was found. Is it possible to add to the picture by a further examination of the physiology of the nervous system?

What if the pathways to and from the cerebellum (which we now more than suspect to be involved in sleeping) to the cerebrum are considered?

If consciousness somehow moved along the cortico-ponto-cerebellar tract into the cerebellum, and thence returned (either completely or partially) via the mid-brain path to the cerebral cortex (in particular, perhaps, to the temporal lobes, the storehouse of conscious memory – 'vacated' now by waking consciousness), would this movement not meet most of the stipulated conditions of the journey? The narrowness of the entrance and/or exit is met, since clearly the cross-sections of the pons and the mid-brain are tiny compared with the open reaches of both the cerebral and the cerebellar cortex – where the association and projection areas offer a universe of activity and of being.[12] The feeling of danger could arise from the fact that consciousness is lost (the 'valley of the shadow', perhaps), or shifted; that by switching off the normal functions of the cerebrum almost completely we run the greatest of risks. (It has already been

12. Peter Nathan quotes the number of junctions between the neurones of the cerebral cortex alone as 10 to the power of 2,783,000. To write this number in full would mean replacing the first letter of this book with a 1, and the rest of the letters by noughts – and then producing up to three more books as long as the present for the remainder of the noughts.

pointed out that in paradoxical sleep the organism is most completely at the mercy of predators.)

Figure 20 shows the suggested afferent-efferent, cerebro-cerebello-cerebral route within the brain, viewed from above. The crossing of the impulses on the journey back *from* the cerebellum will be noted. Perhaps from this derives the great power of the symbol of the cross.

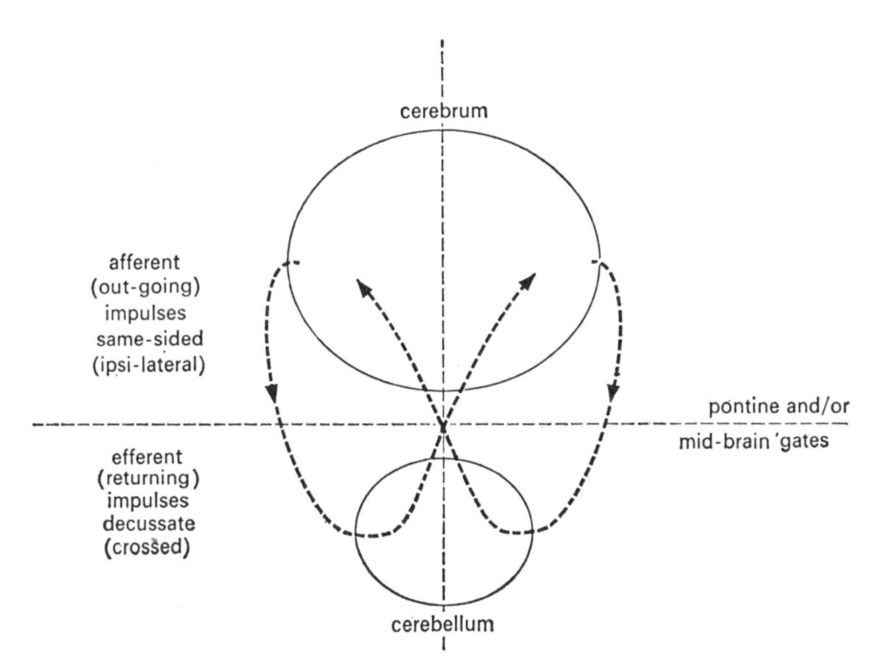

Figure 20. Diagrammatic representation of pathways between the cerebrum and the cerebellum, from above.

Certainly one of the surest and most widely known methods of invoking the Devil was to sacrifice an animal *at a crossroads* at midnight.

With this so appropriate thought we consider Figure 21, in which the 'route lines' if extended bring us now face to face with the Ancient Enemy of mankind.

Or, to be quite precise, the ancient enemy of Cro-Magnon.

The description given in the caption of the previous figure

(Figure 20) is only one of several possible physiological infra-processes. Another proposal, not involving the *afferent* pathways from the cerebrum to the cerebellum at all, would be to suggest that whereas waking consciousness is the *reticular* activation of the cerebral cortex, sleeping consciousness is the *cerebellar*

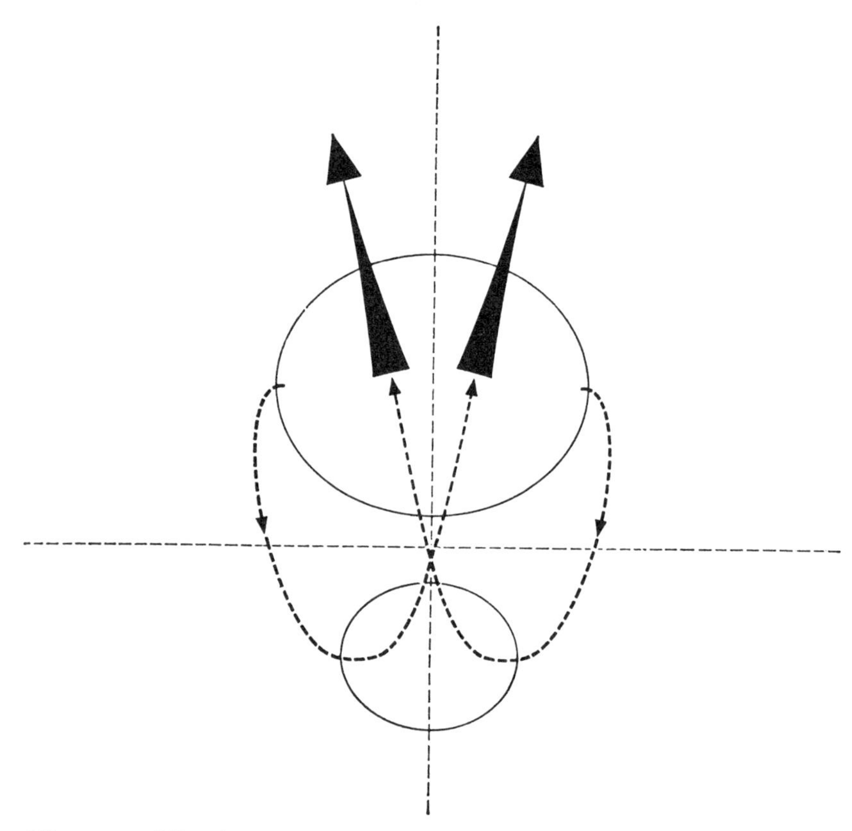

Figure 21. The Ancient Enemy.

activation of the cerebral cortex. However, there are, as we have seen, good reasons for believing that sleeping consciousness resides at least in part in the cerebellum itself – and further of these are discussed below.

Yet other physical peculiarities of the cerebellum are of particular interest. While full consideration of these would take us too far, it will be seen from the following extract that there *is* some underpinning for our next statement about the nature of sleeping consciousness and the cerebellum.

> (12) 'If we now go on to see what fibres *leave* the *cerebellar* cortex, we run into another interesting arrangement. There are many fibres that we might call motor fibres leaving the cerebellar cortex, but few, if any, go out of the cerebellum. There are a few that go back from the cerebellar cortex to the vestibular nuclei of the medulla, but there are not many of them. Instead, all the so-called motor fibres from the cerebellar cortex go to the sub-cortical nuclei of the cerebellum and stop there.'
> (C. T. Morgan and E. Stellar, p. 291)

We know that the *cerebrum* houses a motor system, which also receives extensive sensory and 'emotional' information – that is, projections – from the lower centres. One could say that the cortex has real movement, and the *sensation* of emotion. Since so far all the attributes of System B have proved to be the opposite of System A, can one say that the *cerebellum* has *real* emotion but only the *sensation* of movement – that is, via projections from motor centres?

The meaning of the foregoing is clearer if we consider what happens in dreams. The feelings we then experience are unquestionably real, and actual. When we wake up from a frightening dream the panic is still with us. So also for depression, elation, etc. What is more, we demonstrate the actual, physiological concomitants of feeling – we have an *actual* erection, we *literally* sweat with fear, our heart races, and so on. But the *movement* that takes place in dreams is imaginary. Our body remains in bed. True, we turn and jerk occasionally, but we do not *actually* get in trains, go into the kitchen, walk through a field.

It may be argued that in waking consciousness we nevertheless do experience real emotion. One must insist, however, that what we then experience is a projection of emotion (which is not to say that the real emotion is not going on somewhere, of course). First, it is possible, at least sometimes, to e.g. *refuse* to feel pity – even though we know that pity is somehow trying to get through to us or rise up in us – and to *stem* our anger, *govern* our despair. Psychotics in particular manage a much stronger such split between waking consciousness and emotion. None of this however would be possible unless the emotion originated, that is, first *existed* somewhere outside or below consciousness. It exists, and *then* we deny it. Sometimes we deny also the fact that we are denying; as Freud states in a similar context, in repression the *act of repression* is *also* repressed – so that, to return to our example, it really seems as if there is no emotion there. Thus a patient may be often quite genuinely dumbfounded at the suggestion that he, for instance, hates his mother. At the same time 'coincidentally', he is the victim of acute asthma, or ulcer, or migraine attacks on the days she comes to see him.

The consideration of 'movement' in dreams leads on to an important point, which is first reinforced by a further factual observation.

(13) The cerebellum achieves its greatest importance in birds. It is there the major organ for co-ordinating the sensory and motor impulses involved in *flying*.

What has the present inquiry found to say about flying? First it was noted that Freud believed flying in dreams to indicate sexual intercourse. We know, further, that in magic and legend those who fly best, most vigorously and most often are *sexually mature women* (Chapter 2). Now indeed one is perhaps in a position to understand more clearly the significance of the witch, her flying and the means by which she flies.

Before moving on to further physiological data, however, we turn initially to that never-failing, best of all stand-bys, language.

It would seem to be no coincidence that so many of our metaphors for love, sex and women involve flying and matters associated with flying.

The word 'bird' itself as a slang usage for woman or girl dates only from the nineteenth century. However, this is derived from a *standard* English usage having the same meaning, which occurs for the first time in print in the *thirteenth century*, and is no doubt older.

Aside from numerous more or less flattering epithets like chick, dove, pigeon, it is said that an inconstant girl is 'flighty', the marital home is a love-nest, one's heart has wings, love-making is billing and cooing, a brow-beaten husband is hen-pecked, a woman fusses over her children like a mother hen. And, finally, many unflattering references to women's verbal behaviour have bird overtones – chattering, cackling, screeching, and so on.

One is entitled, of course, to regard these as 'mere' metaphors. It is instructive, however, to see how many metaphors relating to *other* animals one can think of in this area. There are not many. Can it be, then, that some aspects of a woman's (sexual) behaviour (unconsciously?) remind us of birds, for an actual, physiological reason?

This notion may prove more acceptable in the light of what follows.

If the cerebellum is essentially associated with System B (particularly System B1), and is in fact the seat of sleeping consciousness, one would expect this organ somehow to play a greater or more active role in woman than in man. This might reveal itself in a number of ways – for example, by relatively stronger EEG activity. Another of these ways, however, might be in terms of actual physical size or weight.

A paper published in 1920 by R. S. Ellis[13] offers some relevant and extremely interesting research findings. In reporting some of

13. 'Norms for some structural changes in the human cerebellum from birth to old age', *Journal of Comparative Neurology*, 1920–21, No. 32, pp. 1–35. Currently cited in S. M. Blinkov and I. I. Glezer's *Human Brain in Figures and Tables: A Quantitative Handbook*, Basic Books, 1968.

these at this point I do not claim that they represent a finality in terms of proof – no more than does Ellis himself – but merely that they constitute a prima facie case meriting further investigation, and at the same time one which is extremely pertinent to the arguments of this book.[14] That attention should since have drifted away from the implications of these and other findings may perhaps seem less surprising in the light of material in Chapter 13.

Table 7 (a) shows the *relative* weights of the human cerebellum, expressed as a percentage of the encephalon (the whole brain), in male and female babies during the first months of life. Table 7 (b) shows further such comparative weights in later life.[15]

We note first that the female cerebellum in the majority of cases exceeds the male cerebellum in relative weight. Only in two instances is this position reversed – and then only by 0.1%. Clearly, on these figures, the female cerebellum is, on average, relatively the larger.

In electing to attach great significance to a slight statistical difference I must emphasize the following point. We are dealing here with a gross level of physiological – or rather, anatomical – difference. This level cannot be expected to reflect the subtleties and ramifications of psychological difference, except in the most general way. Thus, for example, if a life-form from outer space found on Earth a male and a female skeleton, the slightly larger pelvis and pelvic girdle of the female would not be sufficient to

14. To mention but one weakness in the data, the tables – of which I have, incidentally, chosen the most favourable – combine results produced by many different researchers at different times and places. It is certain that their methods of brain preparation, the intervals which elapsed after death before dissection and so on, differed to unknown extents. Such differences of procedure give rise to variation not assignable to a hypothesis under test – though it is by default necessarily so assigned – and may therefore constitute an important source of error.

15. It should be made clear that the male cerebellum always weighs *absolutely* more than the female cerebellum, since the male brain as a whole is more massive.

TABLE 7

Comparative weights of human male and female cerebella, expressed as a percentage of total brain weight (after Ellis, in turn after H. Pfister and R. Boyd).

(a) *During the first months of life*

Age	*Sex*	*Cerebellum as % weight of total brain*	*No. of cases*
2–4 weeks	m	6.0	17
	f	6.0	13
2 months	m	6.7	10
	f	6.7	6
3 months	m	7.8	10
	f	7.9	11
4–5 months	m	7.9	8
	f	8.9	19
6–8 months	m	9.0	12
	f	9.5	14
9–10 months	m	10.0	6
	f	10.0	8
11–12 months	m	10.0	5
	f	10.8	2
13–18 months	m	10.6	11
	f	11.0	6

(b) *In later life*

Age	*Sex*	*Cerebellum as % weight of total brain*	*No. of cases*
20–30 years	m	10.8	55
	f	11.0	70
30–40 years	m	10.7	103
	f	11.0	85
40–50 years	m	10.9	135
	f	11.0	97
50–60 years	m	10.8	110
	f	10.7	100
60–70 years	m	10.7	123
	f	11.0	142
70–80 years	m	10.9	102
	f	10.8	146
80 years	m	10.6	24
	f	11.2	75

produce a description of the womb and the process of child-birth, much less of the myriad intricacies and complications of sexual behaviour.

Returning to matters in hand, already at the beginning of the nineteenth century F. J. Gall proposed a connection between the cerebellum and the sexual instincts. This view was subsequently much criticized, and is today considered to be discredited. For example, in 1839 F. Leuret examined the relative weights of the cerebella of stallions, mares and geldings (i.e. castrated male horses). His results are reproduced in Table 8.

TABLE 8

Relative weight of the cerebellum in stallions, mares and geldings (after Ellis, in turn after Leuret).

	% weight of cerebellum	*No. of cases*
stallions	11.4	10
mares	12.2	12
geldings	13.5	21

If – so the argument ran – the cerebellum were associated with sexuality, one would expect to find its role, and hence presumably its size, *diminished* in the gelding – not increased, as is the case.

Two points can be made concerning this conclusion. First, one must always be cautious in attempting to explain normal function on the basis of abnormal or pathological cases (such as the gelding constitutes). But much more importantly, I believe that the views of Gall and others were defeated only because of their far too narrow definition of 'sexuality'. If one spoke rather, to use the terminology of an earlier chapter, of the female and the male 'principles' (System B and System A, respectively), the criticisms lose much of their point. For – taking Leuret's results – is it after all surprising to find that in a situation where the male sub-system of the personality is deliberately destroyed, that the female sub-system has expanded to fill the vacuum; that is, in the absence of competition, is able to assume an unnatural dominance?[16]

Reverting to the general argument, the relative differences in the human male and female cerebellum might be enhanced if instead of a random sample of women, certain selected groups of women, chosen along lines already suggested and to be further suggested – say, left-handed women – were compared with men.[17]

One would, further, very much like to know, quite apart from within-species differences, what the relative cerebellar differences are between generally 'female' species such as the cat (to

16. An interesting incidental agreement worth mentioning at this point – which may be merely fortuitous – obtains between the relative weight of the cerebellum in adults, an average 10.9%, and the ratio of sleeping consciousness (dreaming) to waking consciousness in a twenty-four-hour day – assuming sixteen waking hours and 1.8 dreaming hours – namely 11.25%. Further, women have slightly larger cerebella – and sleep slightly more – than men.

17. There is one further aspect of the Ellis data which is of the greatest interest, mentioned here in footnote only to avoid breaking the flow of reasoning above. This concerns racial differences, in particular the difference between Negroes and Whites.

In reporting these findings, let us first be clear that the data here is under even more pressure as regards reliability than that so far quoted – a fact duly emphasized by Ellis himself. Nevertheless, the data is not wholly valueless. Its considerable relevance for the present book will become clear in Chapter 10, where it both confirms and is confirmed by considerations arising in a very different field of human study.

Briefly, it appears that the average cerebellar weight for Negroes is above that of Whites. To report Ellis's exact words: 'The arithmetical average of the percentage weights for Negroes is above that for Whites . . . however, the difference is not a very great one. But as far as the results go, they do show a relatively larger cerebellum in Negroes.'

It further appears that the average for female Negroes is above that of male Negroes: and that the average for 'eminent men' (presumably in the present instance all White) is *lower* than that for men of normal ability. The question of the relative size of the *cerebrum* is of course also relevant here – a matter we refrain from pursuing at this point. Bearing these facts in mind, and slightly anticipating the terminology of Chapter 10, if instead of female/male, black/white, and so on, we think rather in terms of System B-dominants and System A-dominants respectively, we can and shall make rather good sense of the findings reported by Ellis and others.

whom women are often likened) and 'male' species such as the dog. Specifically, one proposes that the relative weight of the cerebellum would be greater in cats than in dogs. There are already hints in the literature that this may prove to be so. In the rabbit the cerebellum of the female clearly exceeds that of the male throughout the whole life period.[18] Leuret's data, quoted above, suggest that the same might be true of the horse. The rabbit is noted for its fecundity, the horse as a symbol of virility.

Let us meantime not forget in this context that the cerebellum in human beings is heavily involved in skilled movements and posture in general – and that on tests involving skilled movements girls score higher than boys; and that the ballet and dance is one of the few professional activities in which woman's performance might be said to absolutely exceed that of men.

Is it also a coincidence that women outdo men on all items of verbal behaviour (except comprehension), and that birds are the best speakers (i.e. 'parroters') in nature apart from man himself? That the one example of the spontaneous *meaningful* use of language by an animal ever noted in the literature was from a bird – Konrad Lorenz's raven?[19] Is it another coincidence that the group of animals which have excited the attention of psychologists in recent years as possibly being in the process of producing a language – the porpoises – are a mammal which has returned to the sea, and therefore, in common with all the 'great swimmers and fliers', has a characteristically well-developed cerebellum? Is it yet another coincidence that women are easier to hypnotize than men, and that the easiest animals to hypnotize are birds and reptiles?[20] Birds, of course, evolved from reptiles. When a human being is hypnotized the eyes remain open but do not see. Reptiles sleep without closing their eyes.

A bird is strong on *cerebellum* but weak on *cerebral cortex*.

18. S. M. Blinkov and I. I. Glezer.

19. Konrad Lorenz, *King Solomon's Ring*, pp. 89–91.

20. A hen can be hypnotized simply by drawing a chalk line in front of its beak. And the eastern snake charmer has an equally easy task.

Apes and chimpanzees are relatively much better endowed *cerebrally*. Yet it is the bird that speaks, not the ape; the porpoise, not the chimpanzee. Is it possible that what we see in birds is the cerebellum making its bid to become the dominant organ of the nervous system? Is it possible that the cerebellum in women is somehow responsible for, among many other things, their greater verbal endowment over men?

Is it possible to go on using the word coincidence in this context?

Taking this last now with all that has been suggested and defended in previous chapters and in the present chapter, I propose that it is the cerebellum which is the Follower, the doppelganger, the Devil who dogs our footsteps on the lonely path and for whom *one may not look back*. Lot's wife became a pillar of salt (was 'changed') when she looked back; and Orpheus lost the reclaimed Eurydice to the underworld for a second time and forever when he looked back to see if she was following him. Christ, too, said 'Get thee behind me, Satan' – i.e., back to your proper place.[21]

The physical position of the cerebellum can tell us why the route to the unconscious is backward and downward. This too is why knowledge – Knowledge II, wisdom – is a well and Hades a pit. 'Flying' in dreams, and 'flights of fancy' in daydreams – also when we get 'high' or ecstatic – all these express or are the release of cerebellar consciousness (the unconscious) from its prison under the cerebrum – Zeus, one recalls, had imprisoned the Titans under the earth – as well as paralleling the *actual* role of the cerebellum in those animals which fly.

The cerebrum is forever haunted and dogged by the cerebellum – compare the Greek Furies, Harpies and so on. Man in general is pursued by his animal past. The escape route of the *cerebrum* is *also* upward and outward. If only we can run fast

21. This can be carried further. Satan is identical with Sammael, the Angel of Death. Se'mol in Hebrew means *left*. Sammael was Lillith's first husband, and Lillith was Adam's first wife.

enough, jump high enough, we shall be able to escape the cerebellum – the 'animal' within us. This, of course, is a doomed hope. We cannot help but take the cerebellum with us wherever we go. It clings to our back like the Old Man of the Sea, like Pilgrim's burden, like Quasimodo's hump. The hump, the burden, is, indeed, nothing other than the *symbolic cerebellum.* So, too, the dwarf (like Punch) is misshapen and small, a toiler and dweller in darkness.[22] The symbolic cerebellum takes as well many, many other forms, some of which we have already examined. It can be anything which is some or all of these: small, valuable, magical, different, hidden away, lost. Aladdin's lamp is a further instance and the Holy Grail yet another. On a much larger scale the cerebellum is Atlantis – the Lost World.

In the light of the fearful or 'persecutory' aspects of the cerebellum we can now understand the spaceman whose parasitic spirit fell away from him when he left the Earth for free space – and the space scientist who remained trapped in the body of the ape. Twentieth-century man is both of these during his waking hours. During those waking hours he drives the stake into the heart of the sleeping vampire as the latter lies helpless. But when night again falls or the moon is full, it is once again, as ever, the vampire's turn. So, too, in the face of every advance by rational man, the Old Enemy has risen again, apparently unweakened.[23]

In the next chapter though we shall very much question the appropriateness of the terms 'animal' and Enemy as applied to this part of our nature – that is, to System B. It may be that if we would only embrace this side of our personality – instead of staking or imprisoning it – the frog, or Beast, would become a Prince, able to lead us back to the Lost World or the Garden of Eden.

22. The 'hunch-back' I would say certainly refers also to the position of the male in sexual intercourse.

23. The moon in legend again also represents the cerebellum. When the *cerebrum* is illuminated by the sun (waking consciousness), day-creatures have their being and night creatures hide in their caves. But when illumined by the moon (night-consciousness) the creatures of night emerge.

Part Four:

The Lost World

Power Politics in the Nervous System 9

This chapter and the next bring us to more detailed consideration of the nature of Knowledge II. The examination takes the form of three run-ups to that subject – three possible, though not mutually exclusive, avenues of approach.

These approaches are outlined under the following headings: (1) the power politics of the nervous system (2) essential characteristics of Systems A and B, and (3) the evolution of *homo sapiens*.

The third of these approaches forms Chapter 10.

(1) *The Power Politics of the Nervous System*

We have seen that the central nervous system evolved from, and in close connection with, those parts of the organism in direct touch with the external world. The central nervous system, then, perceives and responds to the external environment. It is the external environment (whose influence Freud spoke of as the Reality Principle) which shapes that system's development both in individual and evolutionary, that is, ontogenetic and phylogenetic, terms.

The autonomic system, on the other hand, evolved from those parts and functions of the primitive organism concerned with its internal economy – its equilibrium, the intake and breakdown of food, excretion and so on – and with reproduction.[1] The

1. At first consideration the management of the internal economy of an organism and its sexual functions might appear to involve two rather separate

autonomic system is governed and shaped by the subjective reality of the internal environment – a reality differing from that of the objective, external universe. Freud has named the influence of this other reality the Pleasure Principle – a term, although aptly subjective, too limited for the purposes of this book.

The autonomic system may be said in many respects to be older than the central nervous system.[2] Initially the latter is little more than an auxiliary organ (or series of organs) which develops from the former (cf. here Freud's view that ego evolved from libido). The central nervous system was in the first instance only the means for providing the total organism with information about the outside world. Along with this power to perceive and understand the outside world developed also means of enabling the organism to act on the information received – that is, the powers of locomotion and, more especially, of manipulation.

For reasons on which we need not speculate at this stage, it appears that the external receptors, together with the motor systems concerned with volitional movement and manipulation of the environment, gradually ceased to be merely an extension,

---

processes. It should, however, be remembered first that in primitive (unicellular) organisms reproduction is achieved by the division of the 'parent' into two identical versions of itself, and that this process is governed by the internal readiness or ripeness of the cell. Second, that close bonds between eating and excreting, for example, persist in higher organisms. Not only are the external sex organs in large measure also organs of excretion, but linguistic and psychoanalytic data shows how closely linked conceptually oral and sexual functions still are in the human personality, particularly in the child.

2. We are obliged to speak continually as if the two systems – autonomic and central, B and A – were completely separate. Such, of course, is not the case. We separate them conceptually in order to emphasize their real differences – in many aspects of their functioning as we know the two systems *are* opposite and antagonistic. The danger here of forgetting that we are always dealing with *one* organism is emphasized by Freud: '. . . not to forget that the libido of a given person is fundamentally part of that person and cannot be contrasted with him as if it were something external'. (Freud, op. cit.)

or as it were the servitor, of the total organism, and set up in business on their own account. Today *homo sapiens*, at least the Western variety, displays this sign: Central Nervous System: Sole Arbiter and Executor of Human Affairs: No Connection with Any Other Organization.

Were this sole executorship really the case there would be no problem.

The analogies which follow will serve to bring out what seems to have taken place in the nervous system of the more complex animals, and especially in mammals.

When the central nervous system first began to establish its own autonomy – by continuous additions to its holdings of cerebral cortex – the autonomic system, in the manner of a wily old director, laid various contingency plans. It took steps, above all, to ensure that information concerning the external environment continued to reach it no matter what action in the matter the central nervous system, or consciousness, might take. What is termed sub-liminal, or below-threshold, perception is an instance of this. An actual demonstration is provided by the lie-detector – where words associated with the crime produce changes in the electrical conductivity of the skin (an autonomic indicator), while neutral words do not, despite the firm desire of consciousness to remain unaffected by such words.

While safeguarding its information supply by an 'underground' route, System B continued during the course of evolution to maintain amicable relations with System A on as broad a basis and for as long as possible.[3] A united front, or at least the appearance of one, continued to exist. Even in other primates the two systems worked and work harmoniously. Apes and chimpanzees in the wild state do not seem to evidence the mental 'hang-ups' concerning emotion, aggression and other autonomic functions that are evidenced by human beings. (It is perhaps significant, however, that not merely these primates but many

3. There is a striking parallel here with the relationship between the Catholic Church and the powers secular and temporal. This is, indeed, a practical example of what is being discussed.

animals including dogs and cats, when brought up alongside human beings, often show what looks remarkably like neurotic insecurity.) A similar statement could have perhaps been made about the earliest ape-men. But clearly the two-tier system of government must have creaked ever more noticeably as true man emerged from the ranks of the primates. It is extremely difficult and perhaps a little fruitless to attempt to pinpoint precisely that crisis-moment in time.

Nonetheless, reverting to the terms of our analogy, one suggests that System B in its ancient wisdom avoided an actual showdown whenever and wherever possible. Like the Catholic Church, System B yielded where it finally had to – but then always on the best terms available. However, over considerable periods of time not merely the ostensible but the *real* power of System B is gradually eroded. Almost in the way that a son finally challenges a parent's absolute authority, a stage is reached where System A cannot be denied an equal voice. In reality the situation was considerably more complex than this – the crisis occurring not once but over and over again, and in many different areas – the balance of power swinging now to this side, now back again to that. Perhaps Socrates was the first outstanding martyr to the System A cause (that is, that we know about). In legend, at least, Adam preceded him by some thousands of years.

Christ, a System B advocate (though also a System B reformer), eventually offers a solution which one hears in many guises on many occasions thereafter – the division of the Kingdom into two equal realms, one to be ruled by Caesar, the other by God. This split may yet prove to be the death knell of our species.

Therewith the pretence of unity in the old sense of that term, that is, absolute rule by System B, is no longer possible. System B has been met and matched by a force equal to itself.[4]

4. One is tempted to suggest that this situation in part underlies the 'battle of champions' (or the medieval trial by combat), where two chosen fighters meet in single battle to resolve the issues between two armies or nations, one that is frequently found symbolically in legendary and actual

The foregoing 'analysis' of the development of the nervous system has been couched in anthropomorphic terms. This is, however, not an unjustified ruse. We are, when all is said and done, discussing real people – that is, *ourselves*. No matter what the objective, scientific or physiological facts of the matter are or may be, the product of these processes is our own experiential existence on this planet. We must never allow ourselves to forget that we are in these regards our own objects of study, being both the observed and the observer.

The facts on which the deposition thus far has been based *could* have been couched in terms acceptable to the biologist, psychologist or whatever. Apart from the terminology, and the dramatization indulged in, there is actually little in the foregoing that the scientific establishment would not accept in essence. It is in what now follows that the new ground is broken.

While maintaining holdings in and a certain direct influence on consciousness, and while allowing the cortex to function as the ostensible headquarters of the total nervous system, System B was evolving – to some extent had already evolved – *a data-processing, data-collating headquarters of its own*; a headquarters necessarily camouflaged, necessarily unobtrusive, but nonetheless containing all the machinery needed to produce System B's separate awareness, and description, of the universe around it.

This text has proposed that these secret headquarters are housed principally in the cerebellum.[5]

---

struggles in early and medieval Europe; the story of David and Goliath is a well-known instance of it. Often the two combatants are given the labels of good and evil respectively – which is which depending whose side one is on. Some of these stories – that of *Sohrab and Rustum*, for instance, and the *Hildebrandslied* – have significant overtones for this book. In that story the two contestants, unknown to each other, are father and son. However, all these instances can as well be seen as the clash of ego with ego, as discussed earlier.

5. It is possible that apart from the many legends of the Lost World discussed in the last chapter, and claimed to refer to cerebellar consciousness, the story also of Moses placed and hidden in the bulrushes when the Pharaoh ordered the slaughter of all the Hebrew first-born is symbolically the

As is the case with all viable guerilla organizations, what the secret headquarters cannot produce for itself it takes from the legal authorities – either at night while the guards are sleeping or, disguised, in impudent daylight raids.

Dreams (and the illicit use of the sleeping cortex) are an instance of the first action; Freudian slips an instance of the second.

(2) *The Essential Characteristics of Systems A and B*

The central nervous system (a System A organization) evolved principally in the context of the external environment. As a consequence it views all phenomena 'objectively' – that is, it sees the universe in the terms provided by, inherent in, and observable in, that universe. It seeks to apply the laws discovered to itself *and* to System B.

The autonomic nervous system (a System B organization) evolved principally in the context of internal needs. As a consequence it views all phenomena subjectively – that is, it seeks to apply the laws it finds within itself to the universe outside it (*and*, incidentally, to System A).

One needs to be quite clear of the import of what has just been said. So specialized and differentiated have the functions of the two systems become in the course of evolution that the majority of individuals in our present society have completely lost the sense of their own (true) totality – being engaged in the attempt to use half their personality as if it were the whole.

Each system in this situation perceives the other as extraneous and irrelevant.

This is better appreciated through actual examples. There are four possibilities. (1) System B's view of the phenomena of

---

camouflaging and hiding of the cerebellum (disguised to look like the cerebrum) near the enemy's headquarters (in the story, Pharaoh's palace). The cerebellum, as we know, *is* the first-born, in evolutionary terms. Like the baby Moses it is small and hidden. Like him it has mystical properties. It is, like him, the lost leader who will perhaps one day emerge to lead his 'oppressed peoples'.

System A (2) System A's view of the phenomena of System A (3) System A's view of the phenomena of System B and (4) System B's view of the phenomena of System B. In tabular form we have Table 9:

TABLE 9

System B's view of System A	System A's view of System A
System A's view of System B	System B's view of System B

Inserting actual examples based on certain aspects of certain events, yields Table 10.

TABLE 10

Sun (= Sun-God)*	dG2 main sequence star, diameter 1.4×10^6 km, surface temperature 6000 K
Moon Landing Apollo 11, etc.	Moon (= Moon-Goddess)**

* Egyptian Ra, Aztec Inti, Greek Helios, etc. But also Christ-Jehovah – cf. the glory, nimbus, halo surrounding Christ (the Prince of Light) and the Saints.

** Greek Hecate, Egyptian Isis, Chinese Hengo, etc.

Especially outside Europe and the Near East, however, many examples of Moon-*Gods* as opposed to *Goddesses* are found. This difficulty is admitted. Nonetheless, since goddess-worship turning into god-worship is a development to expect, one can always suggest that these moon-gods were preceded by goddesses, unrecorded. There seems, however, no way of verifying this and it remains therefore merely a suggestion.

The statement in the top right-hand box of Table 10 represents, whatever else, a statement by System A about System A – that is, an objective description of an objective phenomenon. Nothing is included in the description except what is actually observed.[6]

From System B's point of view the sun represents the new unknown, the moon the old known. As suggested elsewhere in these pages, the sun-god seems to be System B's experience of the cerebral cortex. The moon is rather System B's experience of itself – the cerebellum, among other things.

From System A's point of view the moon represents the not known or not sufficiently known – which System A is in process of incorporating into the known. Science *requires* final explanations, detests anomalies, impenetrables and imponderables.[7]

System B views these two phenomena – the sun and the moon – as expected, subjectively. Something of the human being is added which is not actually there in the strictly objective sense – namely an animating soul or spirit. The object is invested with human qualities and human desires – which may be *super*human but are nonetheless recognizably of *homo sapient* origin. To

6. One ignores here, of course, the fact that nature *herself* has never heard of such a thing as a dG2 star: this is at best an arbitrary sub-division in a continuous (though admittedly real) process which *homo sapiens* makes for purposes of his own.

7. It is interesting here to speculate why scientists, particularly Western scientists, are willing to devote so much of their energies to exploring space and reaching the moon, but so little to investigating the depths of the sea – actually a far more accessible treasure-house of badly needed information concerning the ecology and other aspects of our planet. We do not to anything like the same extent *need* more information about the moon.

Now, clearly, there are many factors involved here. The space race is a prestige area in terms of East-West competition, and a strong defence interest is also involved. One would not wish to deny the motivational role of these reality factors.

It is just possible, however, that in addition to such factors the scientist is (a) intent on 'destroying' the very strong symbolic significance of the moon and (b) motivated out into space and away from e.g. the ocean depths by reason of the latter's associations with the unconscious.

System B the position is that some spirit or soul not only inhabits but *must* inhabit the heavenly bodies and many other objects in nature.[8]

We wish, however, to go further. We wish to say that certain appropriate objects, such as the sun, display properties which resemble or at any rate evoke properties of the cerebral cortex – while the moon somehow or other suggests the cerebellum.

It seems just too coincidental that the emergence of sky-religions (day-religion, god-worship), and the decline of (earth-) moon religions (night-religion, goddess-worship), so closely parallel the rise to dominance of the cerebral cortex in human affairs.

Is it for nothing that we speak of the *light* of reason, *clearness* and *clarity* of understanding (Latin *clarus* = bright, shining), *unclouded* vision, also of *blinding* insight, *dazzling* logic, *brilliant* ideas, *flashes* of inspiration, en*light*enment, etc. – but conversely of obscurity and the *mists* of obscurantism (Latin *obscurus* = covered over), the *Dark* Ages, etc? (As noted earlier, Christ is the *Light* of the World.)

From one point of view the Earth can be considered to represent the cerebrum (or perhaps the total human organism). The real Earth is illumined alternately by the sun and the moon. In the sunlight, creatures of the day have their being. At night (moon-time) the day creatures sleep and others come to life who by day remain hidden. We have suggested that the sun may be likened to waking consciousness: when the cerebral cortex is awake and fully activated we have conscious day-thoughts and perceive the day world. When the cerebral cortex is 'asleep' but 'illumined' or activated by the cerebellum (the moon) we

8. Most psychologists would say that the observer is *projecting* his humanity and his human feelings on to or into the object. I would prefer to say not that the subject is projecting his emotions, but that the object arouses these emotions in him. The incoming information is the trigger for a subjective experience. Perhaps it is more appropriate to speak of *intro*jection. Even if this term is not wholly accurate, nevertheless I prefer to retain the term projection for the hallucinatory and delusional experiences of the psychotic. The distinction is probably worth insisting on.

have night-thoughts (dreams), and now perceive the world as it *subjectively* is.

That the human organism is activated or illuminated alternately by waking consciousness and sleeping consciousness, while the Earth, at those same times, is illuminated and activated alternately by the sun and the moon, *can* be dismissed as a coincidence or a totally fortuitous parallelism: in that case, however, one is justified in asking the critic to attempt a definition of coincidence. This matter is raised again in Chapter 11.

So persistent are the associations and connections between System B and the phenomena of night and darkness – one recalls, for example, the preference of women for darkness in the sexual encounter – that there are grounds for the suspicion that System B may in some sense have evolved or developed under nocturnal conditions. For we dream as if by moonlight.

It is scarcely possible to overemphasize the all-embracing nature and pervasiveness of the differences between the two systems. Their very manner of perceiving at the most basic level is a product of the terms of reference within which they came into being.

System A evolved as we have said, basically, in the context of the material, external universe. In the process of evolving it so to speak incorporated the laws of that universe into its own nature. Thus it is the case that System A (like Joey during his psychosis) conceives of the universe as a machine. One could say that it prefers this view. But actually the case is more extreme. System A *can only* see the universe as a machine. There is no other possibility.

System B, conversely, evolved within the body. It at no time ever saw itself as other than the centre of its own being. As it became aware of the external universe, as in time it did, it imposed the laws of its own being *on* that universe. This is a process we can observe in the young child. Whereas System A sees itself as (merely) a (by-)product of the universe, System B sees the universe as an extension of itself. From the subjective standpoint of System B the universe exists simply to fill the needs of System B. To the extent that the universe *is* perceived as a separate object, it is conceived of as an organism. (In a sense, System B *gives birth* to the universe.) It can *only* see the universe as an organism. There is no other possibility.

A materialist (a scientist?) might well accept the above as an account – an unnecessarily exaggerated account – of the objective and subjective processes. He would, however, add: 'But, of course, the universe *really is* only a machine.'

A priest or a mystic might similarly (with reservations) also accept the above account. But would probably add: 'But, of course, the universe *really is* an organism – that is, the spirit of God made manifest, which moves in all things.'

A more reasonable person might feel that we must forego the luxury of ever hoping to know what the universe *really* is. So far it turns out to be whatever one's terms of reference predetermine. That is to say, the answer is begged by the method: one could say, uncharitably, by the nature of the prejudice involved – or, more charitably, by the nature of the means of perception in the two cases. Certainly if we rely *either* on System A *or* System B alone, we must inevitably forego any whole or absolute view of the universe, and settle instead for one that is only 'relatively absolute'.

The scientific method will no doubt be unused to hearing itself described as a form of prejudice. Nonetheless, what is sauce for the goose should also be sauce for the gander. Science in the twentieth century is as prejudiced about religion and similar phenomena as religion has long been about science.[9]

9. It is interesting to note that while the psychologist is supremely anxious to convert into concrete 'verifiable' form that which is by its nature somewhat nebulous (thought, behaviour, etc.), the physicist is prepared to report that what he, and we, previously considered to be solid matter, is only energy, the solidness of matter being apparently merely the product of our perceptual mechanisms. What one has here is not actually a head-on paradox, but the retreat of the psychologist from the less to the more tangible contrasts strongly with an opposite tendency in physics.

It is my view that as one attempts to approach the human organism beginning, say, with physics and proceeding through chemistry, bio-

This book proposes that the universe is neither a machine nor an organism, except in the sense that it is at least both.

In all such discussions, moreover, it is essential to remember that *we* at least are organisms first, and machines second.

---

chemistry, biology, medicine, psychiatry, psychology, etc., the scientific method continually produces results of less and less value – certainly of less reliability – and in fact of greater and greater triviality, until by the time one reaches such phenomena as art or human consciousness the findings can hardly be said to have any value whatsoever.

However, on the reverse journey to the one discussed above – one which starts with man and moves through the biological to the physical sciences – subjective judgement and value judgements play an ever less useful role in *understanding* the processes observed (say, in the production of high-grade steel). Such judgements are in fact quite irrelevant to those processes. Thus the essentially subjective approach of the alchemist and the magician to problems of the physical universe proved largely fruitless. *However*, this is not to suggest that System B – Knowledge II – cannot in any sense deal with practical matters in the 'real' – that is, external, universe. We shall be considering this possibility later.

Old and New Men: The Recent Evolution of the Nervous System 10

THE EVOLUTION OF HOMO SAPIENS

This chapter is concerned principally with the physical evolution of man. However, physical evolution blends gradually into the psychological and cultural, particularly as one approaches historical times. The two cannot be sharply separated. At the general physical level our argument treats man globally as mankind. In connection with some precise physical changes, however, and in any case at the general cultural level, the focus shifts to the development of man specifically in Europe and the Middle East. Reference to man's cultural development elsewhere is then made for comparative purposes only.

The facts reported are as accurate as the many gaps in the fossil and archaeological record currently allow them to be. It needs to be kept in mind that new discoveries occasionally produce major changes in the views of authorities. Nevertheless, on the positive side, there does exist some measure of general agreement on broader outlines, and this situation is unlikely to change drastically.

As a preliminary step the highlights of the available record of man's physical evolution are briefly summarized.

Australopithecus africanus (literally, southern African ape) and *Australopithecus robustus*

These primates existed some 2,000,000 years B.P. (Before Present) and 1,500,000 years B.P. respectively. Their remains are found in south and east Africa. They were primitive tool-users, but possibly not tool-makers. They had no use of fire. Brain capacity around 500 cc.

Homo habilis (skilful man)

This is a classification contested by some anthropologists. *Homo habilis* is possibly simply a rather advanced *australopithecus* or, as is claimed by some, he may be early true man. He existed around 1,750,000 B.P., or perhaps even earlier, in central Africa. A tool-maker, still having no use of fire. Brain capacity in the region of 650 cc.

Homo erectus (upright man, originally called *pithecanthropus* – ape man)

Homo erectus was widely distributed in Java, China, Africa and possibly Europe in 500,000 B.P. He had the use of fire and probably cooked food. An extensive tool-maker, with permanent cave dwellings. Probably practised cooperative hunting. Brain capacity range from 750 to 1200 cc. (See Figure 22b)

Homo sapiens neanderthalensis (Neanderthal man)

So-called *classic* Neanderthal man existed between 75,000 and 35,000 years ago in western and south-western areas of Europe, cut off during that time by the Würm glaciation both from his near relatives (the Neanderthaloids) and all other varieties of early man. His average brain capacity is generally agreed to have been *larger* than that of modern man – the head is massive – i.e. around 1400 cc.[1]; but the brow of the head is flattened,

1. The average brain size in modern man is around 1300 cc. There appear

receding rapidly from prominent brow-ridges, the skull low-vaulted though wide, and the occipital region (the rear) expanded by a curious swelling (see Figure 22c).[2] In stature this type of Neanderthal was short – average height for males 5′ 4″, females 4′ 10″ – barrel-chested, with extremely powerful, curved thigh and forearm bones. Some of these and other details of his physique are clearly well-suited to the ice-age conditions in which he lived. His culture included hand-axes, bone tools, personal ornaments, etc. Ritual burial of the dead was practised, sometimes along with personal possessions, food and so on. This last is important, for it argues strongly for the existence of religious or magical practice, and a belief in life after death.

In south-east Europe, the Near East, north Africa and elsewhere are found what are termed Neanderthaloid remains, in most cases roughly contemporaneous with classic Neanderthal. These exhibit some, but not all the features of the classic type – usually, but not always, in less extreme form and sometimes along with more modern characteristics. Various explanations of these finds exist in the literature. The two major views, sufficient for our purposes, are (1) that the finds represent an ancestral stock from which classic Neanderthal on the one hand and modern man on the other separately evolved, and (2) that the

---

to be fairly wide differences of opinion in the literature regarding the mean brain capacity both of Neanderthal and his successor Cro-Magnon, a majority of authorities however being agreed that Neanderthal had a larger brain than present man, but a few only maintaining this in respect of Cro-Magnon.

There is, obviously, some relationship between brain size and mental ability on a very gross level. Because this relation is only a rough and ready one, it is never used by psychologists as an index of individual intelligence or ability. In any case, since this book makes no precise capital out of absolute brain size, the issue need not detain us. The *relative* size of parts of the brain in comparison to other parts is, however, quite another matter.

2. '. . . the brain-case juts out strongly posteriorly, with a marked projection which contrasts sharply with the rounded contours of modern man in this region'. A. S. Romer, *Man and the Vertebrates*, p. 229.

(a)

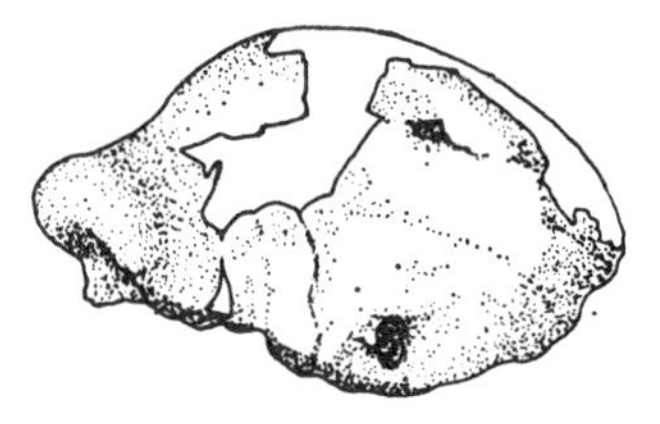

Hominid 9 (Homo Erectus/Pithecanthropus, Africa. 500,000 years Before Present)

(b)

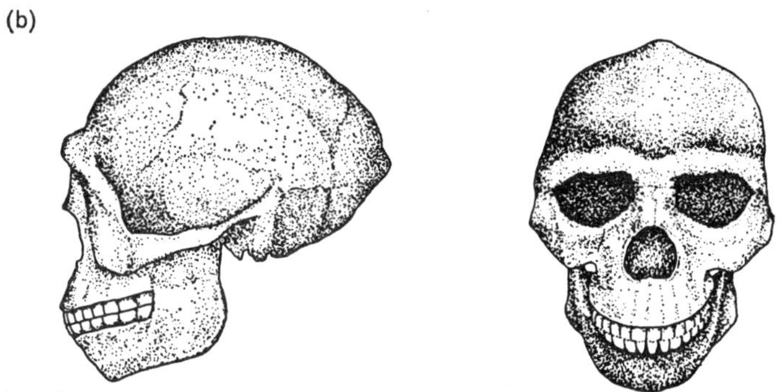

Peking Man (Homo Erectus/Pithecanthropus, China. 500,000 years Before Present)

(c)

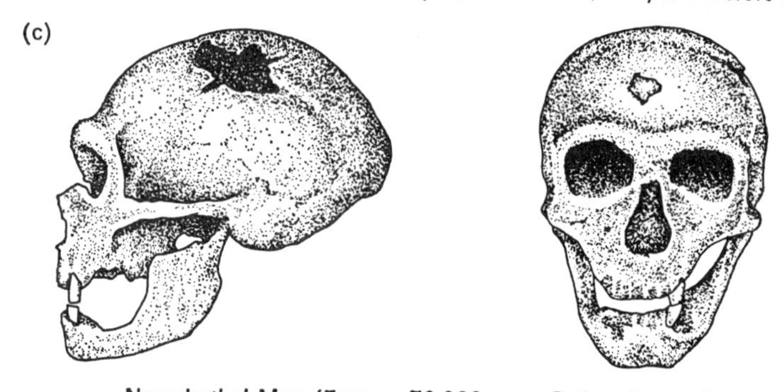

Neanderthal Man (Europe, 50,000 years Before Present)

(d)

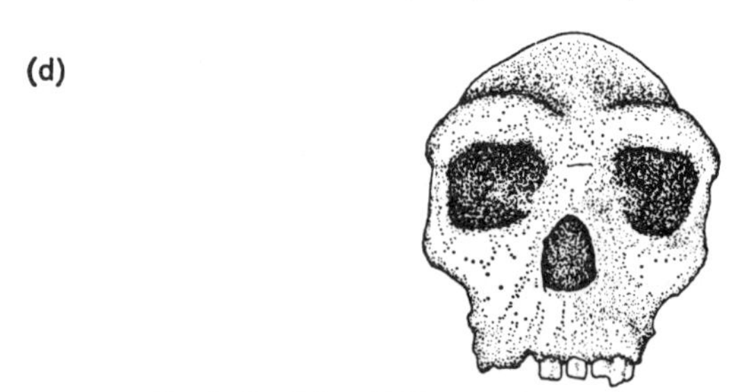

Rhodesian Man (Africa, 40,000–25,000 years Before Present)

Figure 22. Some representative paleoanthropic skulls.

Neanderthaloids represent a mingling of Neanderthal and Cro-Magnon strains – see below – this cross leading to the emergence of our own variety of *homo sapiens.*

Early Homo sapiens sapiens (Cro-Magnon man)

This variety overlapped with in time, succeeded and entirely replaced classic Neanderthal in western Europe. Cro-Magnon man was more or less solely dominant throughout Europe by 25,000 B.P.

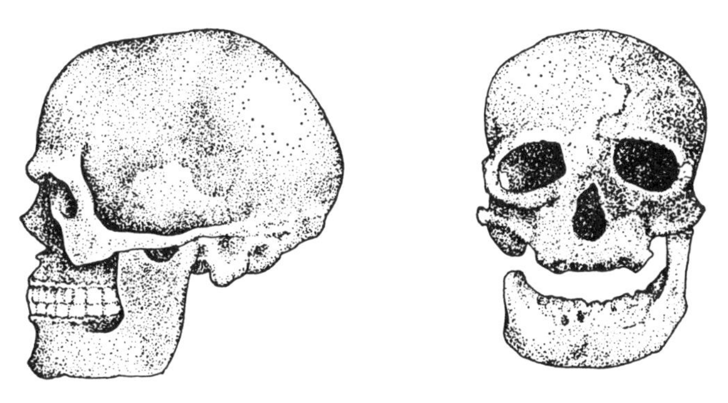

Cro-Magnon Man (Europe, 25,000 years Before Present)

Figure 23. Two representative neanthropic skulls.

Cro-Magnon was tall – average height around 6′ 0″ – well-muscled and powerful, though not in the gross manner of Neanderthal, and without the curvature of thigh and arm bones. His skull-profile closely resembles our own. The forehead is high, as is the vault of the skull, the bone thin. It lacks both heavy brow-ridge and rear-ward occipital bulge. The chin is well-formed and jutting, in contrast to Neanderthal's chin, which receded abruptly from the lower teeth. In general the features are regular and, to us, aesthetically pleasing. (See Figure 23.)

Cro-Magnon's culture included fine stone and bone tools, many of these ranking as works of art, arrow-heads, spear-heads, needles and so forth. He produced sculptures, engravings and cave-paintings, these last having undoubted magical and symbolic connections, and practised ritual burial of the dead.

During subsequent millennia further waves of yet other peoples moved up from the east, whence Cro-Magnon himself had emerged, in time mingling with, diluting and modifying the Cro-Magnon strain.

Europe between 9,000 *and* 5,000 *B.P.* (7,000 – 3,000 *B.C.*)

By 9,000 B.P. village-farming communities are firmly established in the Near East – with domestication of animals, production of pottery, and so on.[3] Simple cultivation and the domestication of animals was probably established by 11,000 B.P. With the discovery of metal the Bronze Age begins in Mesopotamia during the latter part of 6,000 B.P. and this development spreads rapidly through the Near East during 5,000 B.P.

The foregoing is a brief outline of the position as generally agreed upon by the experts in the fields concerned. One then comes to the interpretation of the facts – at which point, it should be realized, we have very much entered the area of opinion and counter-opinion.

One further solid fact, however. In western Europe classic Neanderthal disappears abruptly over a period of perhaps only hundreds, at most a few thousand, years around 35,000 B.P. No continuity of his life-style or culture (with one notable possible exception – the custom of burying the dead) and no anatomical continuity between classic Neanderthal and his successor Cro-Magnon have been found. Though the two varieties are adjacent in time there is between them what archaeologists term a sterile horizon.

That Neanderthal suddenly disappeared and was as suddenly

3. It is a matter of general agreement that Jericho was, already around 9,000 B.P., a town defended by 'a massive stone wall, associated with at least one magnificent stone tower'. Jericho, however, is the only town of such antiquity so far discovered.

replaced by Cro-Magnon is, then, a well-agreed established fact.[4] What is not established is whether the two events were causally connected – or not related, even though adjacent. We do not

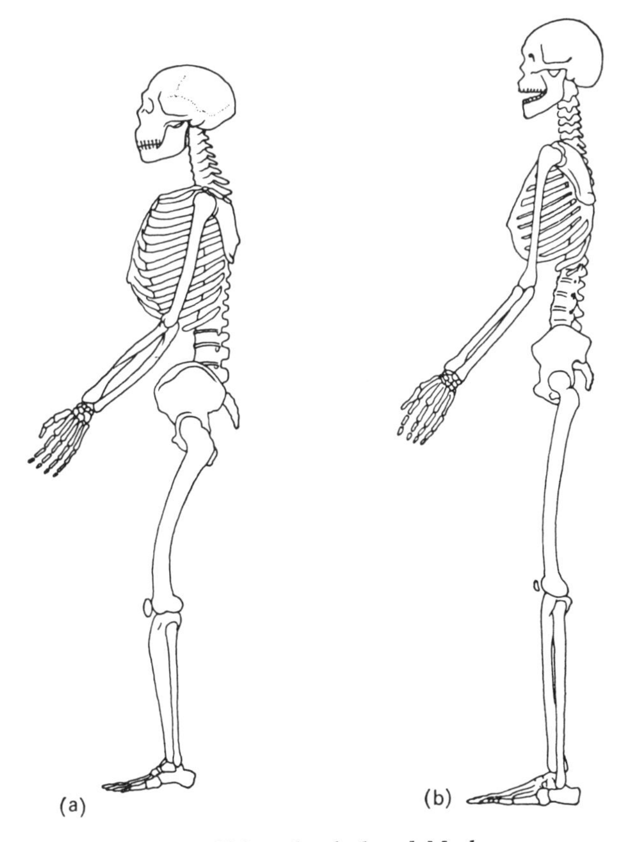

Figure 24. The skeletons of Neanderthal and Modern man.

know, and can only speculate, whether classical Neanderthal was systematically wiped out by Cro-Magnon or whether some natural event – for example, a disease carried by the latter – brought about the catastrophe. Parallels for both these events can after all be found in historical times. Or whether perhaps the retreating ice-cap of the Würm interstadial around 40,000 B.P.

4. The extreme unlikelihood that classic Neanderthal could have evolved into Cro-Magnon in the space of a few thousand years can be appreciated by comparing the two skeletons of Figure 24.

created climatic conditions in which Neanderthal was now not well-suited to survive – but which on the other hand gave Cro-Magnon the impetus to migrate up from the south-east. All these explanations are in no way necessarily mutually exclusive.

The interest of this book centres on the Neanderthaloids of the Near East, and to some extent elsewhere, of some 75,000 years ago, and on the question of their origin – whether, precisely, they represent a mixed parent population whence descended both Cro-Magnon and classic Neanderthal – or whether they represent a crossing of Neanderthal and Cro-Magnon varieties which led to the emergence of modern man. Both theories are possible.[5]

The general background is, however, yet more complex. Not only must we ask did, around 75,000 years ago, a parent population of mixed characteristics evolve into the types of classic Neanderthal on the one hand and Cro-Magnon (and thence into ourselves) on the other. But we must ask, also, whether this happened at that time – *if* it happened – for the *first* time? Or, as a number of authorities believe, had the splitting of an initial ancestral stock into two recognizable types, old style and new style, already occurred a very long time previously? A. S. Romer comments:

> We have noted grounds upon which to base a belief that throughout the Pleistocene[6] the story of humanity was woven of *two* strands [my italics] and that neanthropic new-style man has had a line of descent long-separated from that of his more simian-appearing relatives.

Included in the evidence which leads observers to conclude that new-style man has been on the scene long before Cro-Magnon as such put in his appearance (and *he*, incidentally,

5. The notion of Cro-Magnon as a descendant of the Neanderthaloids, however, seems currently in rapid decline. R. S. Solecki states firmly: 'This . . . hypothesis has been abandoned' (*Shanidar: The First Flower People*, Knopf, New York, 1971, p. 12).

6. In all a period of some two million years.

would in that case probably be the cousin or perhaps descendant of earlier 'modern' men) are such finds as the skulls from Swanscombe in England and Steinheim skulls in Germany. These, unfortunately, are in a fragmentary state of preservation only. The available skull bones, nevertheless, are thinner not merely than those of Neanderthal but in some respects than many of the much later Neanderthaloid or mixed varieties of the Near and Middle East. (The Steinheim skull in particular has a moderate brow-ridge and a rounded rear.) Yet both the Swanscombe and Steinheim finds are dated around 250,000 B.P.

Zoologists, of whom A. S. Romer is one, are not *essentially* concerned with such considerations as types of weapon, styles of hunting and the general sociology or psychology of early man – matters which, for better or worse, make the task of the anthropologist in some ways more difficult. The zoologists, quite legitimately, base their classifications on observable anatomical features. Possibly for this reason – and perhaps it is an entirely adequate reason – zoological opinion appears to be in little doubt as to the validity of its classification of early man into the two-part division of paleoanthropic (literally, old man) and neanthropic (new man) over a considerable period of time.

The general features of the paleoanthropic and neanthropic skulls are respectively more or less those noted already with reference to classic Neanderthal and Cro-Magnon. The *paleoanthropic* skull has pronounced brow-ridges, receding forehead, low though broad vault, a projecting occiput, a jutting (or prognathous) muzzle with large teeth, but a recessive chin. All head bones are thicker and more massive than in the neanthropic skull. By contrast, aside from being generally less massive, the neanthropic skull has, as stated, a high forehead, absence of pronounced brow ridges, a high vault to the skull, a rounded occiput and a well-defined, jutting chin.

Quite why nature apparently produced *two* generalized types of early man (and not four or seven or ten types) is puzzling though to this book extremely interesting.

What one seems to have – reverting now to the descriptive

method of Chapter 6, and either considering each skull feature individually or the skulls as aggregates – is something very like a bi-modal (presumably normal) distribution of skull characteristics, one mode or mean of which is the paleoanthropic, the other the neanthropic. The overlap of the two distributions – this being, one recalls, what produces the bi-modal distribution – can accommodate types of skull intermediate between the modes (the so-called 'mixed' populations).

To have shown that owners of suggestively 'modern' skulls were living in various parts of the world well before the advent of Cro-Magnon is not to have proved conclusively – due both to the occasional and fragmentary nature of the finds – that neanthropic man was in existence throughout the whole of the very long period in question, if, of course, at all. Even if this were proved to be the case, this still would not exclude the possibility of a new and independent budding of modern man from a paleoanthropic stock in the Near East some 100,000 to 75,000 years ago. One could, indeed, postulate that nature made a number of attempts at producing neanthropic man at various times – all of which failed, except the last, perhaps because of the relatively delicate constitution of these creatures in the face of new ice-ages or other catastrophes (or their hardier relatives!). Meanwhile, more sturdy paleoanthropics might soldier ruggedly on.

While the view just expressed has something to recommend it, it is not, on the whole, my own. In particular, as stated, I reject the view that Cro-Magnon evolved in the Near or Middle East from the mixed Neanderthaloid population whose remains we find there. I adopt instead the current majority view that Cro-Magnon evolved elsewhere, and thence invaded Europe via the Middle East, overrunning and at least partially destroying the Neanderthal populations as he did so:

'. . . that Neanderthal man was wiped out; that he was perhaps harried and exterminated by invading groups of *homo sapiens*, who had come from an entirely different line of ancestors.' (A. S. Romer)

'On the whole, the evidence seems to indicate that Neanderthal man did not gradually change into *sapiens*, but was replaced, presumably by an invading sapiens.' (R. S. Solecki)

Two sets of Middle Eastern remains have been found near Mount Carmel and are known as the Tabun and Skuhl groups respectively. Both are dated approximately 75,000 B.P., but the Tabun remains precede those of Skuhl by some 10,000 years. Some individuals in both groups, particularly in the former, are rather typically Neanderthal, while others are not. The foreheads are reasonably full, although the brow-ridge is rather prominent, the skull somewhat vaulted, and so on.

Looking at such findings, it is at first easy to imagine a people undergoing rapid transition to a neanthropic type. One must, however, then ask why such a change should be occurring at that particular juncture; but more pointedly, why it should be occurring so rapidly. From the viewpoint of the evolutionary clock a period of a few thousand years is a very short one indeed, *unless* perhaps at a time of exceptional pressures. In any case, it is equally reasonable to imagine that at Mount Carmel we have a 'half-breed' population resulting from an admixture of Neanderthals or Neanderthaloids with an initial wave of Cro-Magnon or Cro-Magnonoid peoples.[7]

The chief and perhaps only weakness in the theory of Cro-Magnon having evolved outside the Middle East lies in the fact that no traces of his alleged home site have as yet emerged.

At this stage, however, we halt the examination of accepted facts and the variant explanations of these, in order to present the views of the present book on the material thus briefly reviewed. It will be seen that accepted facts are not violated, although, certainly, extended and interpreted. These extensions will help

7. 'If one believes that neanthropic man represents a new invader of Europe and was not a descendant of Neanderthal man, the Mount Carmel people may be considered as due to interbreeding of the dominant race with its lowly predecessors.' A. S. Romer's view, and that also of R. S. Solecki and other anthropologists.

us further in the description and understanding of the structure of personality.

It is suggested first that man has been marked since his inception by a strong urge to be somewhere else – that is, by migratory tendencies, connected in all probability to the high curiosity drive displayed by all primates. Many considerations support this view, one being – as far as this book is concerned – the central role which the Wanderer or Pilgrim occupies in legend, literature and religion. There is in any case no lack of evidence of an urge to travel among people today. There are, nonetheless, also compelling biological reasons why this must have been the case with our early ancestors.

Darwin, and many others since, have shown that when representatives of a particular organism or plant are isolated from the parent population they, in the course of (unspecifiable) time, and by reason of the selective pressures of the environment in which they find themselves, evolve into a species ultimately incapable of interbreeding with the original parent stock. This process is the 'origin of species'. When the isolation from the parent stock is recent or is not total, or is total, but intermittent, different *varieties* of the species will emerge, which though showing inter-variety dissimilarities, are still capable of interbreeding with the parent population, or with other varieties.[8]

Though one can never speak with certainty in these matters, it is highly probable that any variety of man cut off from others for a period of even, say, a hundred thousand years would have evolved into a separate species. Some support for this view is provided by classic Neanderthal. The period during which this community was isolated from the rest of mankind (by the Würm glaciation) is estimated as not exceeding 40,000 years. In this

8. Although the matter is more complex, a *species* may be defined as a group of organisms able to breed with each other but unable to interbreed with other groups of organisms, even though initially related; while a *variety* is a stage on the way to becoming a species. Dogs and wolves, for example, are rather separate *varieties* of one *species*, for each may breed with the other. All currently existing races of man are also varieties, not species, since all crosses among them are fertile.

short time, however, though admittedly under extreme environmental pressure, Neanderthal had already advanced a fair way along the road to becoming a distinct species. In a period of some 18,000 years of isolation the aborigines of Australia, too, had begun to develop peculiarities of genetic transmission. We see, therefore, that it is extremely unlikely that any early men were at any time isolated for indefinite periods from their fellow men, since otherwise separate species of man would almost certainly have emerged.

The notion of early man as a continual and compulsive wanderer[9] would account for the fact that the remains show mixed characteristics found at widely scattered points through time, but more importantly for the non-emergence of separate species of man. It seems that as soon as specialization in a particular direction got fairly started, the amicable or forced intermingling with other varieties tended to dilute and otherwise thwart the specialized trend, producing in the event something like a 'regression to the mean'. Nor must one imagine that the circumstances operating against too great a specialization applied only to anatomy. They apply equally to inherited behaviour.

We consider further reasons why the concept of man as a continual, energetic migrant is especially useful.

Pre-man and early true man was well suited for migration, and became still more suited once this was taken up. For the first ape-man was a fair walker, a good climber (both of rocks and trees), was prepared and able to eat, and moreover could obtain, almost any kind of food – from roots and tree-produce to small, later large, animals including fish and shell fish. He was capable of constructing shelters (as some modern primates do), or could sleep in trees or in caves as necessary. Ultimately he had the protection of fire, but long before this had simple

9. Desmond Morris, in *The Naked Ape*, points out that monkeys, chimpanzees and other primates do not have fleas (fleas spending part of their life cycle in excrement or nest materials), which, he suggests, is evidence that these primates have always been, as were also probably man's early ancestors, continually on the move.

weapons to serve as 'equalizers' between him and the fiercer animals. Clearly, early man was the almost perfectly equipped traveller, one capable of coping with almost any kind of terrain.

Importantly, it was probably in the course of and as direct result of travelling that man became a useful swimmer – a generally un-primate activity. There must have been numerous occasions when the ability to struggle somehow across, or simply into, a stretch of water meant the difference between life and death – when trapped by carnivores or by fire, when pursuing game in times of food scarcity – or, at such times, being able to dive in shallow seas.[10] These particular situations and selection procedures would obviously arise far less frequently in the case of primates who were prepared to limit themselves to one known area, if not one sleeping place. These played it safe, but at the price of reaching an evolutionary dead-end. There is no such proverb – but one might say: 'he who stays in the trees does not learn to swim'.[11]

We have suggested that the notion of the continuous crossing of incipient varieties of man, made possible by his keeping on

10. It is possible that the story of the escape of the Israelites *across water* (the Red Sea) from the pursuing Egyptians is a generalized and symbolic echo of this particular form of escape from death or from capture. Certainly the theme of *outwitting the pursuer* is a frequent and persistent one of literature and legend right up to the present.

It is an extremely interesting exercise to begin to consider the whole of the *Old Testament* as not (merely) the history of the Jews, but as a compressed and stylized 'dream' account of the adventures of early man since he first left his arboreal habitat. On this basis the Garden of Eden would represent the rain forests of the Pliocene – a suggestion also voiced by Desmond Morris. This view of the Old Testament, for instance, perhaps makes a certain sense of the improbable ages imputed to the characters.

Further specific points on this general proposition are made in the course of this chapter.

11. The account given has implied the classical Darwinian view. It is perfectly possible, however, that man at some point simply made the decision to learn to swim. We have no business to overlook this growing ability of man to take charge of his fate. The ability to swim, however, once learned, *then* confers survival value.

the move, can account fairly neatly for the regular occurrence of mixed populations. There was for primitive man, however, a further and unsought bonus from this situation – and for us a similar bonus in increasing our understanding of the progressive evolution of man's personality.

When two varieties of a species are crossed, the resultant offspring often, though not necessarily, display what is termed 'hybrid vigour'. This is to say, the offspring are frequently sturdier in many or all ways than either parent – sometimes on a purely additive basis, sometimes in ways not predictable from one's knowledge of the characteristics of the parents. Such crossing is the principal tool of the animal and plant breeder searching for new qualities or desirous of strengthening old ones. A final point is that hybrid vigour is more likely to arise when *well-differentiated* varieties are crossed, rather than fairly closely related varieties, *especially* if the varieties in question are themselves somewhat in-bred.[12]

Can one not assume that under the conditions hypothesized differing varieties of man would interbreed on numerous occasions, in some cases after thousands of years of genetic isolation? And that the offspring of these crosses might demonstrate improved hybrid qualities – including such qualities as intelligence. (See in this connection *General Note*, page 308.)

One notes that since man is a cultural animal – because, that is, he lives in tribes which maintain continuity of learning, and because his young during their long period of dependence absorb, and are specifically taught, the learning of their elders – for all these and other well-known reasons, any advance produced by some particularly gifted generation or individual will of course then tend to be preserved and to become the permanent property of the community.

It is a possibility that the peoples with mixed characteristics of Skuhl and Tabun were the result of one of the last crosses between widely divergent hominid populations – perhaps the

12. These preconditions are of course admirably met by the Neanderthaloids and Cro-Magnon at the time of their encounter.

General Note

In this particular context I suggest that parts of the Old Testament and the Pseudepigrapha *which describe the sons of God mating with the daughters of man are actually a stylized 'dream' account of the passage of Cro-Magnon through the Near East some 35,000 years ago.*

If this is so, our ideas of the time-scale in human history would need to be drastically revised. Let us consider very briefly what could cause us to see a connection between Cro-Magnon and certain legendary figures of the Old Testament. We need to imagine on the one hand, the figures we have already described: a race of probably skilled warriors, average height six feet, handsome, well-proportioned men with high foreheads and well-formed chins, perhaps largely devoid of body-hair. Probably the Greek ideal of manhood or our own present-day Olympic athletes – of field rather than track events – would be a reasonably close approximation. These are to be contrasted with, on the other hand, the typical Neanderthaloid of the Middle and Near East (not, of course, with classic *Neanderthal – he lay much further west) – short, thickset, probably fairly hairy, with recessive forehead, possibly a hair-line close to the eyes, and a recessive chin. Might not Neanderthal have perceived something godlike in this new invading race? How might it be, too, if the children of the alleged cross between Cro-Magnon and the Neanderthaloids possessed a marked degree of hybrid vitality. Is there any evidence of such a possibility in the old writings? The answer is affirmative.*

First a brief excerpt from The Book of the Secrets of Enoch.

'And there appeared to me two men, exceeding big, so that I never saw such on earth; their faces were shining like the sun . . .'

Now the well-known passage from Genesis vi, 4:

'When men began to multiply on the face of the ground and daughters were born to them, the sons of God saw that the daughters of men were fair; and they took to wife such of them as they chose. . . . The Nephilim [giants] were on the earth in those days, and also afterward, when the sons of God came into the daughters of men, and they bore children to them. These were the mighty men that were of old, the men of renown.'

The meaning of the last sentence is not wholly clear, as commentators agree – are the 'mighty men' referred to the sons of God, or the children of the sons of God and the daughters of men? Probably the latter. J. Hastings, in his Dictionary of the Bible *(T. & T. Clark, 1900) under* Nephilim, *notes first that a more accurate translation of 'when' in the sentence 'and also afterward, when the sons of God came into the daughters of men' is 'forasmuch as' (i.e. because, or as a result of, the fact that). He continues: 'It is not explicitly said that the Nephilim . . . were the heroes borne by women to the sons of God, and some scholars have held that they were not; but [the writer of Genesis] certainly meant that they were for otherwise it is impossible to account for his mentioning them at all.'*

The biblical text itself, then, and a majority of commentators unite in assigning 'miraculous' qualities to the offspring.

Many 'primitive' and isolated peoples, such as present-day Eskimoes incidentally, refer to themselves simply as 'the men' – just as the writer of Genesis did – out of their belief that they were the only men on earth.

There is much else here that is fascinating and germane, over which we can unfortunately not linger too long. Thus Numbers xiii, 32–3: '. . . and all the people that we saw in that land are men of great stature. And there we saw Nephilim (the sons of Anak who came from the Nephilim); and we seemed to ourselves like grasshoppers, and so we seemed to them.'

Might not Cro-Magnon have contemptuously regarded the fecund ('when men began to multiply on the face of the ground'), diminutive Neanderthals (and they, themselves) as grasshoppers? This is particularly appealing when, finally, Hastings tells us: 'Among the derivatives proposed for Nephilim *one makes it to be from* naphal *'to fall'* [!]: *either as meaning beings fallen from a previous high estate* or as fighters who fall upon the enemy fiercely [*my italics*]. *The latter view has been . . . favoured by the Greek versions . . .' Is this a reference, then, to the fighting qualities of – ? – Cro-Magnon?*

Finally the relevance here, too, of the literal-symbolic meaning of the David and Goliath legend need perhaps hardly be mentioned.

---

last such cross possible, since from there on the growth of civilization and sheer numbers tend to preclude the possibility of lengthy genetic isolation. It may even be, as I believe, that what we term civilization emerged as a *direct* result of that last cross. This point will be taken up again.

For the moment let us track back again to the millennia preceding these possible events. It was earlier suggested that fairly continuous mixing of varieties of man would produce among other things a kind of anatomical 'regression to the mean'. That is, on the one hand, physical characteristics of man were in general prevented from becoming too divergent – the one exception here being classic Neanderthal who, significantly, was precisely in that position of being isolated from the main body of mankind.[13] And not only was a limit thus set to physical divergences, but on the other hand to behavioural divergence also.

One can hypothesize further that during the course of evolution – along with increasing powers of reasoning, for example – certain individuals saw the advantages of a permanent home site. Of course, there may have been other and less conscious reasons for that decision also. We are not committed to assuming only single causes. One can for instance suggest that within the normal range of variability of the species – perhaps involving whole varieties – some individuals were less disposed temperamentally to continuous moving on. The tendency might be reinforced by the discovery of a particularly fertile stretch of country, by a growing attachment to personal possessions, or whatever. For some or all of these reasons, at least some members of a tribe, which perhaps in any case had grown too large for convenience, might decide to settle. We may suggest two further points: that 'settling down', once a certain ability to cope with environment had been developed, conferred at least as much survival potential as keeping on the move; and that secondly the gene pool of the permanent settlers, already perhaps differing, came by processes of natural selection to differ further from the gene pool of those who continued a nomadic existence. This last implication would tend to hold even more true for behaviour and customs.

One can imagine that these attempts on the part of early man to settle proved abortive in the sense that sooner or later a tribe

13. One other point can be made in passing. The more two varieties differ, that is, the more distant the genetic relationship between the individuals concerned in the cross, the greater in general is the chance that the offspring will be sterile – incapable of reproducing themselves sexually. The mule – the cross between a donkey and a horse – is an instance of this, as are several commercial varieties of corn. It is possible that any crosses between Cro-Magnon and *classic* Neanderthal would have produced sterile, even still-born or 'monstrous' offspring.

of more or less aggressive strangers would appear over the hill and help themselves to whatever was going – including the womenfolk. This is a possible pattern, still clearly visible in early historical times. The adoption of women of the conquered tribe into that of the conqueror has probably always been fairly standard practice. The biological benefit here is that genes from the 'settlers' are incorporated into the gene pool of the 'migrators'. This would have the effect of, fairly literally, slowing down the conquerors – i.e. of reducing the migratory and perhaps the aggressive urges in future generations, if these are genetic. On other occasions the nomads might not wipe out the entire settlement – merely plunder and move on – leaving as mementoes a seeded crop of half-breed babies. On yet other occasions still the meetings and minglings of wanderers and settlers might have been wholly amicable. All such events, in any form, would nevertheless on the one hand have prevented settled groups from degenerating into any kind of pastoral idiocy; and migrating groups from becoming totally, if the phrase is appropriate, 'de-humanized'.

In a limited sense, but only in a limited sense, it is possible to say that settling behaviour eventually in evolutionary terms won out over migrating behaviour. The struggle is by no means wholly played out, however, as a brief consideration of modern society shows.[14]

14. Reverting again to the Old Testament and to the suggestion that it constitutes, among other things, a more or less symbolic account of man's biological evolution, the story of Cain and Abel is perhaps to be interpreted as a portrayal of the clash between the 'wandering' and the 'settling' personality types – a struggle perhaps still echoed within each of us.

Let us first be clear that that story does not accord in every point of detail with the suggestion just made. For example, *both* these men are described as farmers. Abel, however, is certainly the more passive and is described as a keeper of sheep. The punishment which Cain receives for killing Abel appears to be the loss of the 'right' to be a farmer: 'Now you are cursed from the ground which has opened its mouth to receive your brother's blood from your hand . . . it [the ground] shall no longer yield to you its strength.' Then follow the, for present purposes, more significant words: 'You shall be a fugitive *and a wanderer on the earth.*'

In the context of this discussion perhaps one should emphasize that 'settler' in any case does not and cannot mean 'farmer' until about 11,000 B.P. Prior to this period crops were not cultivated. When one speaks here of a settlement one means essentially the establishment of a permanent dwelling site, which is thereafter inhabited by the descendants of the founders, in a perhaps extensive neighbourhood capable of bearing continuous hunting or foraging.

In the terms of this book, I propose the establishment of a permanent settlement as a B1 phenomenon. In our present society, for instance, it is men far more than women who travel, explore and discover, and are in all senses more generally restless.[15] Marriage, significantly enough, is referred to as 'settling down'.

To the migrators one is inclined to assign a more masculine-aggressive and ego-based – that is, belonging to the type A-B2 personality. This is an assumption. However, Cain's reply to God – 'Am I my brother's keeper?' – is typically ego-assertive and truculent. Cain is saying 'Who do you think I am – his *mother*?' Jack Kerouac (if poetic evidence is admissible) said that 'ego is the greatest hobo'.

Let us apply these generalizations to Neanderthal and Cro-Magnon. If the Cro-Magnons were indeed warlike men with a

---

Cain's reply contains perhaps the admission that religion (submission? passivity?) is not his scene: 'Behold, thou has driven me this day away from the ground; *and from thy face I shall be hidden;* and I shall be a wanderer on the earth.' (Genesis iv)

With that Cain departs from his home territory.

15. While, as ever, one would not wish to deny that these behaviour differences are culturally reinforced in society, there would seem to be a sufficiency of biological reasons for the existence of a sex-difference in this case. In later life the caring for young children effectively puts paid to any ideas of extensive travelling a woman might have. What one would be interested to know, however, is whether among animals which in general neither form a marital home nor rear their offspring – fish, crabs, frogs etc. – the male covers more ground in his natural wanderings than does the female.

fairly high level of practical intelligence, this would be more likely to make them the victors in a clash with all Neanderthals. If the *Neanderthaloids* were less fierce, less intelligent in a practical sense, perhaps preoccupied with religious ceremonial, etc., having essentially a B1 b2 a profile, they would not constitute a formidable threat to Cro-Magnon. *Classic* Neanderthal was a rather different matter. One is tempted to assign him a B1 B2 a profile – a somewhat unusual profile, but then, classic Neanderthal *is* somewhat unusual – implying that he was partly a fierce, but not especially well-organized fighter. One can imagine that if classic Neanderthal fought his *own* kind it was as the result of shortage of food, not out of a desire to conquer or a joy in combat. We are here, naturally, wholly in the realm of speculation. There is, however, at least a possibility that classic Neanderthal was cannibalistic. This is strongly suspected of certain paleoanthropics outside Europe.[16] The most likely prima facie use for the curved-arm and thigh-bones would seem to be wrestling or hugging, and the most likely object for this style of attack would be other Neanderthalers. That would even appear a reasonable and possible basis for the evolution of curved limbs – but there is another, far more arresting possibility, to which we shall come later.

Cro-Magnon, essentially modern anatomically, as authorities unite in agreeing, pushes up into western Europe from the east around 35,000 B.P. From that time on, large numbers of 'anatomically pure' Cro-Magnons are found in Europe. There is the question, as mentioned, whether he *originated* in the Near East, or whether he arrived there already fully evolved from some

16. A story originally considered in an early draft of Chapter 1 is as follows. Lycaon (cf. *lycanthropus*, a wolf-man), a mythical king of Arcadia, when once entertaining Zeus, the chief of the Gods, put before him human flesh (some say the flesh of Lycaon's own son) to test his divinity. Zeus was so enraged at this blasphemy that he produced a deluge which devastated the Earth. Among Greeks generally it was held that the eating of human flesh was one way of becoming a werewolf. The vampire (Dracula) in drinking the blood of his victims is likewise practising cannibalism. Our reasons for reviving this issue again here will become clear shortly.

other part of the world – this book siding, as indicated, with those who argue the latter. I envisage several waves of migration, or one rather piecemeal and drawn-out migration, the first, or first part, of which faltered and came to a halt in the Near East. Here these new-style men settled, and in the course of time intermingled with local Neanderthaloid populations. The major, later wave of Cro-Magnon, I suggest, progressed intact through the Near East – no doubt leaving further mixed marriages in its wake – *but nevertheless not taking strangers into its ranks* – to reach western Europe. The significance of the 'uncontaminated' condition of the groups arriving in western Europe appears to have been largely underrated. Those who would argue for the evolution of Cro-Magnon from a mixed but basically Neanderthaloid population in the Near East have the task of explaining not only how the separation into a '*pure*' modern strain took place there (under conditions of not better than relative isolation) in the comparatively brief space of some 30,000 years – but how then these pure individuals *but nobody else* took off for western Europe. Mixed types, moreover, persist in the Near East and eastern Europe very much later than the time in question. (It is on balance much less difficult to imagine the evolution of Cro-Magnon taking place outside this whole arena, in the greater isolation of the Far East or Africa.)

This being as it may, we arrive at the point where a pure Cro-Magnon population appears in western Europe. Relatively pure descendants of these invaders, incidentally, survive today in the Dordogne, the Canary Islands, Scandinavia and elsewhere – a fact universally agreed in the specialist literature. Now Cro-Magnon is confronted by a people of perhaps repellently, even horrifyingly animal appearance – squat, immensely powerful, perhaps excessively haired in evolutionary response to the rigours of the climate, with jutting brows, jutting muzzle and flattened skull.

Did – one speculates – Neanderthal's style of fighting involve the use of his mouth – and were the brows therefore to protect the eyes? Still more, what of the massive head as a whole with its thick bones? Did Neanderthal run at his enemy using the lowered head as a battering ram?[17] We shall have significant occasion to raise this point again.

Was Cro-Magnon's reaction to classic Neanderthal only one of physical horror or disgust at his appearance or the conditions in which he lived? Or are there perhaps grounds for postulating a *superstitious* fear also? We do know, of course, that Neanderthal practised ritual burial of the dead. What further 'religious' ceremonies, one wonders, might he also have practised?

Cro-Magnon had, perhaps, been prepared to acknowledge his kinship, his common humanity, with the Neanderthaloid peoples of the Near East – possibly in a wholly contemptuous, dismissive way. That view is strongly supported by the circumstance that no Neanderthaloids accompanied (were allowed to accompany?) Cro-Magnon into western Europe: perhaps the Cro-Magnon male considered Neanderthaloid woman 'an excellent fuck' and made full use of that facility, while disowning the halfbreed offspring. But *classic* Neanderthal was something else again. Here, possibly for the first time, (Western) man's consciousness would have been confronted by the inescapable

17. One recalls here that the 'devil' of pagan, pre-Christian Europe, that is, Pan, was a goat, and that goats fight by butting with the head. It appears also to be the case that animals which fight by butting evolve a flattened forehead. Could, then, either or both of these observations have any possible relevance to Neanderthal?

In connection with the first point, the chance is taken here to remind the reader of the widespread bull-fighting and bull-worship of ancient Greece, Crete, etc. – the deformed Minotaur, too, at the heart of an underground maze – which survive still in the Spanish bull-fight and similar spectacles. Also very relevant here is the cult of Mithra. Mithraism was a Mystery religion (from which, interestingly, women were excluded) which spread into the Roman Empire from the Near East. The ceremonies took place in small underground crypts, sometimes in caves. The legend, re-enacted in the crypts, tells how Mithra the god of *light* slays the bull, which is the enemy of the *sun* (*n.b.*).

The connection between symbols of light and System A has already been discussed at various points. Here, too, we have apparently a further linking of the bull – a horned creature – with System B.

evidence of his own animal origins.[18] The psychological mechanisms of Denial would ensure his making every effort to destroy that evidence.

It is tempting to allow the imagination full rein in this area. One can picture perhaps a period extending over many generations of continual slaughter of the native, classic Neanderthal population – possibly the first systematic pogroms. Perhaps too a time when internal schisms within Cro-Magnon tribes occurred about how to deal or come to terms with the 'Neanderthal problem', accompanied also by the purging on the part of extremists of any liberal elements in the Cro-Magnon ranks. These phrases and the situations they suggest may seem out of place in this context. Yet why should we imagine that Cro-Magnon differed significantly from ourselves in anything other than in a lack of our technical skills?

The picture of systematic slaughter indicated here might also have seemed less likely prior to the last world war. Now it would seem only too possible. Still, the deliberately political drift of the comments made may yet appear misplaced in respect of our primitive forebears – who perhaps (it can be argued) did not even possess a language. Doubts on *this* score, however, are probably groundless. While, of course, in the absence of writing we cannot have direct evidence of whether any particular variety of early man did or did not possess a language, one may make inferences based on other behaviour. Already 500,000 years ago *homo erectus* had the use of fire, a tool culture, lived in permanent cave sites and practised cooperative hunting. It would seem unlikely that even then some primitive form of communication can have been totally absent. I am personally not in doubt that a

18. When one looks again at classic Neanderthal, in terms of his size, hairiness, mystical connections, possible cannibalism and so on, is one not suddenly reminded of the troll, the dwarf and other such figures of Northern mythology? Nor do we have to stretch the imagination very far to bring in here other figures such as the werewolf – the human being who grows hair and becomes an animal – and many related figures of myth and legend. The parallels seem altogether too frequent to be the result of mere chance. Still further evidence will suggest even more strongly that they are not.

complex language and perhaps a complex social order could have and did exist by 35,000 B.P.

The notion behind the phrase 'purging of the liberals', for instance – of some selective process of survival or non-survival – is necessary from one important point of view. For this book wishes to suggest that some of the fear of Neanderthal or of Neanderthal qualities was *bred into* the ancestors of western Europeans at this time: became constitutionally, that is, genetically, fixed. Many hundreds, even thousands of years of bitter, continuous fighting could have formed an adequate selection mechanism.

The general view of this book, as already proposed, is that legends and stories, no matter how they originate, do not persist unless they contain some psychological truth or satisfy some inner psychic necessity. Though they frequently derive from actual events or the lives of actual people (Abraham, Robin Hood, the Trojan Wars), or arise out of some factual error such as the reporting of 'canals' on Mars, they survive and grow, one suggests, for other reasons. Turning the view around, one argues also that all details of a legend, no matter how trivial they may seem, are each of the greatest importance. That is to say, there are *no* chance inclusions. Freud, of course, took this view of the dream.

Let us look again at the paleoanthropic or Neanderthal skull with its projecting brow, the 'pronounced, bi-arched, supraorbital ridges', as they are called. The Hominid 9 skull offers here the best example of this brow formation. If now we look again – are we not looking at the skull of the *Devil*? *Are not these his horns*?

In proposing this view of the Devil's physical origins one is in no sense going back on the interpretation of the Devil offered in Chapter 8 and the link with the cerebellum.[19] But what we see

19. Hopefully the reader has by now been dissuaded from any expectation of single explanations in human activities. This is an erroneous frame of mind, fostered by the physical sciences and inadvisedly applied to human motivation and behaviour.

in the skulls of the paleoanthropics is perhaps the 'real', the actual *physical* basis of the myth.

There is a good deal more to be said here, but let us first deal with two possible objections.

(1) Even conceding a certain reasonableness in the suggestion that the form of the paleoanthropic skull could give rise to the later much stylized idea of 'men with horns', is not the time-gap between ourselves and Neanderthal (some 35,000 years) too great to allow for the survival and handing down of the legend to modern times?

(2) Even if the concession is made regarding the survival of the legend over the very considerable period in question, why is it necessary to argue for the genetic fixing of the legend – that is, for a biological basis? Will not the handing-on of the legend by word of mouth suffice to account for the facts?

These two aspects are interlocked. If the second, biological basis were established, for example, the first objection would be largely quashed. Let us, however, treat the two objections as separate issues.

First the time-gap. Anthropologists and other involved parties are agreed that Cro-Magnon is *solely* dominant in western Europe by 25,000 B.P. That must mean that others shared the environment with him till then. This is already something of an improvement on our earlier figure of 35,000. May one not ask, however, whether there was any conceivable way in which isolated remnants of classic Neanderthal could have survived even beyond that date? First let us glance as it were sideways at another important find from Africa – Rhodesian Man. This skeleton comes from Broken Hill in Rhodesia, and is dated between 40,000 and 25,000 B.P. (The disagreement among authorities on a precise date rests on technical matters.) While the body-skeleton of Rhodesian man is thoroughly modern, the skull – which is what interests us at the moment – is wholly paleoanthropic. The brow ridges are described as 'simply enormous', the forehead retreats sharply, the vault is low and the skull bones extremely thick (see Figure 22). Thus, whatever the fate of classic Neanderthal in Europe, it appears that 'men with horns' were still perhaps flourishing elsewhere, at a time when Cro-Magnon had become dominant in Europe.

Reverting again now to classic Neanderthal it seems likely that he was no longer at large in Europe proper by 25,000 B.P. However – as we shall see later – until as recently as 10,000 B.P. Scandinavia, like Tibet, was still in the grip of the glaciations of the last Ice-Age, and in this state inaccesible to modern man. Could it be that a few small bands of Neanderthals eked out some kind of existence in such areas? The answer must be a tentative yes. For we note that stories of trolls, dwarfs and giants[20] are 'coincidentally' more common in the *north* of Europe: and that these mythical creatures supposedly lived in the mountains in inaccessible parts. And then, what do we find in Tibet? Stories of an ape-like creature living above the snow-line. One certainly hesitates before mentioning the Abominable Snowman, the Yeti, in a serious context. But the legend is too useful to ignore. We are not suggesting that he is alive today, but that he *may* have been alive almost into historical times – since the legend, logically, *cannot* be more than ten thousand years old, due to the aforementioned glaciation of the areas in question.[21]

Even if the foregoing suggestions in connection with the lingering of Neanderthal are unacceptable, what certainly seems likely is that occasional Neanderthal skulls would have continued to turn up from time to time – as they do very occasionally to the present day. Such finds could once again trigger off the retelling of the old tales.

Viewing these various considerations, and bearing in mind also that by 9,000 B.P. Jericho is a flourishing 'modern' town, is

20. Without wishing to have my cake and eat it, I suggest that 'giant' in this context conveys the idea of 'monstrous' or 'looming large in the imagination' and *not* 'tall'. In the course of time, possibly, the grossness of the 'giant' was expressed as more conventional tallness.

21. An expedition is about to set off for the Himalayas to explore the extensive caves in those mountains. Let us hope it returns with a few skulls to finally settle the matter.

not the time gap between ourselves and classic Neanderthal, though still great, rather less formidable than it first seemed?

But what in any case of our second suggestion – that the response to classic Neanderthal became genetically fixed? The general view of this book as stated, is that oral traditions do not survive indefinitely, much less grow, unless they reflect some real psychological/physiological need or condition. In particular our whole-hearted, many-faceted and persistent acceptance of the Devil in his numerous forms – in other words, both the *intensity* and the *pervasiveness* of his presence – for then argues a constitutional-genetic component working in his favour.

Aside from merely repeating this article of the present book's faith, let us consider the more detailed features of witch-hunting during fairly recent historical times – in particular the behavioural concomitants and the physical marks – the witch-marks – associated with being a witch. These were, in no special order, physical misshapenness (stunting, deformity, ugliness, etc.), excessive hairiness – especially in women – extending (perhaps by stimulus generalization) to warts, moles, birth-marks, and so on – magical practices (often nothing more than an interest in natural medicine), having the second sight, solitariness, hypersexuality, being abroad at night, left-handedness, and so on.

One could almost protest that *any* marked deviance of behaviour or appearance would seem to have been enough to arouse an accusation of witchcraft, but this is not actually true. There is no record, for example, of very tall people or exceptional athletes or warriors being burned as witches.[22] The range of witch qualities, though large, is by no means infinite or all-embracing. In short, do not the main characteristics in question *all seem to be possible features exhibited by classic Neanderthal*?

Briefly, it is possible that 'western Europeans' experience an involuntary urge to destructive action in the presence of certain

22. A near-psychotic or pathological *denial* of sex, for example, or a delight in inflicting pain, would conversely count as *recommendations* for the rank of priest or chief torturer.

stimuli or releasers, in line with the kinds of involuntary responses to stimuli identified and described in animals, by ethologists such as Lorenz, Tinbergen, Baerends and others. A chapter will be explicitly devoted to these issues. For the moment one need mention only that these responses and releasers which elicit them are variety-specific (in some cases species-specific, and so on) – that is, particular releasers and responses characterize certain groups and not others. They are in fact one of the bases by which we distinguish between varieties. These responses are acquired in the course of the evolution – often, obviously, the recent evolution – of the variety. The reason for their acquisition is that the survival of the individuals who possessed them was thereby somehow promoted, while the survival of those who did not was disfavoured. It is proposed now that the so-called witch-marks are releasers (or at least the residue of such), specific to the variety of sapients termed Cro-Magnon and his descendants – while the resulting reaction, in its full-fledged form, is a witch-hunt.

This line of argument may seem more convincing at a later stage. It is now suggested here as a *partial* explanation of the persistent and savage persecution of Jews, gipsies, Negroes and others, which has so long been a feature of Western society.[23] Is it perhaps the 'accidental' (?) resemblance of members of the groups just named to a long extinct variety of man which triggers, in some present-day individuals, a very ancient and destructive response – once essential to survival?

Whatever the precise facts may be, I am personally persuaded that say, the attitudes to the Jews in Germany and parallel events in other countries at other times could not have taken so extreme a form or gained such impetus, or for that matter continue to be so broadly vigorous in our society at large, purely on the basis of a body of anecdote or a catalogue of imaginary

23. 'British settlers in Tasmania in the last century cheerfully organized parties which amused themselves by hunting and shooting the Tasmanian natives – now extinct. The Tasmanians were obviously regarded as some sort of animal.' Douglas MacEwan, *Question 3*, R.P.A., 1970.

crimes – and that they must therefore be ascribed to a deep psychological need, and an innate psychological response.[24] These innate tendencies are then of course susceptible to cultural reinforcement.

There remains a further and crucial aspect of classic Neanderthal's life still to be considered. Hopefully the reader has himself detected the frequent and, to me, remorseless parallelism that exists between the features of the suggested clash between Cro-Magnon and Neanderthal, and various aspects of the myth and legend, both religious and secular, of Europe and the Near East. One final piece of evidence appears to tip the scales definitely in favour of the position that there are several coincidences too many for the parallels noted to be the result of chance.

A well-known feature of the Neanderthaloid and paleoanthropic skull has, deliberately, been left undiscussed till this point. Its significance had long escaped the present author, but in the course of re-reading Romer's *Man and the Vertebrates* (to which I am therefore indebted) its probable meaning became clear. Romer's own words are:

> In Neanderthal man the eyesockets are very large and round.

Among what kind of creatures does one find large, round eyes? The answer, of course, is among nocturnal creatures.

The reader is invited to inspect the skulls reproduced in

24. In no sense, however, is this explanation intended to account in full for the antipathies between Systems A and B. As we have been at pains to show, these, as well as those between the sub-Systems B1 and B2, are far more pervasive and ramified – in evolutionary terms reaching as far back as the emergence of the vertebrates and the bi- or tri-partite brain.

Thus *both* the Nazis and the Jews 'correctly' referred to *each other* as devils. In that general sense, as opposed to the specific sense discussed above, the qualities of the 'Devil' are simply those which are the opposite to those of your own dominant System or sub-System. The 'Devil's' intellectual, aggressive, sexual or whatever qualities are thus variously available for parading to suit any occasion.

Certainly it is here suggested that the alleged events in Europe some 35,000 and onward years ago both sharpened and conferred certain specific features on this general conflict.

Figures 22 and 23. These are the front and side views of various paleoanthropic skulls, and the front and side view of one neanthropic, actually Cro-Magnon, skull. Do not the eye-sockets of the former impress us as the eyes of a nocturnal, or semi-nocturnal animal? The sockets of Cro-Magnon, one notes, are longer than they are higher and, characteristically, downward slanted. It is from the side views of the skulls that one best appreciates how much larger and apparently more bulging were the eyes of the paleoanthropoids. One says 'apparently', for of course we have no actual idea of the appearance of the physical software of any ancient man.

The possibility that Neanderthal was nocturnal in his life-style enables us to understand a great deal of what is otherwise puzzling or simply meaningless (such as – a small example – the association of large-eyed creatures of nocturnal habits, cats and owls, with witches). Again and again throughout our explorations of System B the trail pointed puzzlingly to a nocturnal setting.

Although, as will be shown, 'night-life' reaches perhaps its most marked expression with classic Neanderthal, its influence is nevertheless detectable among the paleoanthropics as a whole.

One can, of course, in general only speculate as to what factors might have led or forced *classic* Neanderthal to adopt a markedly nocturnal existence. A possibility is the avoidance of predators abroad by day, another is snow-blindness. Present-day Eskimoes wear a piece of bone with a fine slit in it over their eyes – without such protection permanent blindness rapidly ensues. Certainly, however, the arrival on the scene of Cro-Magnon would have reinforced any tendency to avoid being seen by day, and perhaps also led to a retreat to the deeper caves. One recalls that in legend dwarfs and similar figures work underground and inside mountains.

That Neanderthal's nocturnalism, however, preceded the arrival of Cro-Magnon can be seen from a fact that is wholly obvious, once one has begun to think of Neanderthal's life in these terms.

Classic Neanderthal has curved arm- and thigh-bones, enlarged ends to the main bones and many other skeletal irregularities. He has in general a somewhat stooped posture. The first Neanderthal skeletons discovered were initially thought to be those of modern individuals suffering from illness or deformity.

An illness widespread in industrial slums in the first part of this century and earlier was rickets. It is, perhaps, not generally appreciated that this disease produces curved arm-bones as well as leg-bones, in addition to other effects. It is a deficiency disease caused by lack of vitamin D, the amount of free vitamin D available in nature being extremely limited. The majority of animals, including human beings, synthesize this vitamin in the skin under the influence of ultra-violet light – hence its name 'the sunshine vitamin'. Lack of the vitamin in childhood produces various types and degrees of curvature and deformity including enlarged joints, the precise effects, however, depending on the amounts of calcium and phosphorus also available. The slum children of our own industrial era were receiving insufficient sunlight – while the deficiency is rarely experienced in the tropics.[25] If the deficiency occurs, or persists, in *adult* life the spine may become deformed. One recalls at this point that the troll and the dwarf are frequently hunch-backed, as is Punch. Once again, it seems, it can be asserted that no detail of legend is unimportant, or included without cause.

If classic Neanderthal rested in the caves by day, and hunted and socialized principally by night, rickets would be endemic in the whole population.

Slightly curved limbs are found among some Neanderthaloids – arguing that a tendency to nocturnalism existed among paleoanthropics generally. Perhaps we have also solved the puzzle of the straight limbs, but Neanderthal head, of Rhodesian man – who lived in the tropics and could therefore hardly have avoided the small amounts of sunshine necessary for normal growth.

25. Or among Eskimoes, since their specialized diet happens to include rare natural sources of this substance.

One further point can be made in connection with nocturnalism, but before making it – in case it is felt that we have been making much of nothing in this discussion of sunshine deprivation – let us consider momentarily the phenomenon of blond hair and fair skin. If a demographic map of Europe is made showing the incidence of fair skin, and a further map showing the incidence of overcast days in the year, the two maps agree rather well. It is clear, then, that fair skin – the ability to make best use of limited amounts of sunshine – has proved a decisive survival factor along classical Darwinian lines. Any tendency towards a lighter skin has been selectively reinforced.[26]

Reverting to classic Neanderthal, what was the form of his religion? We can only speculate about this. But we have noted in earlier chapters a tendency for older religions to be linked with earth-moon-goddesses and later religions to sun-light-gods. It is likely, then, that Neanderthal was a moon – and not a sun-worshipper – and would not such behaviour be in any case more natural and logical for a nocturnal people? Can one go further and say that any people worshipping the moon are likely to have been nocturnal in their habit and life-style? Were, once again, all paleoanthropics night- rather than day-oriented?

One proposes that since their inception (whenever and why-ever that occurred) paleoanthropic and neanthropic man have found darkness and light respectively more congenial. This statement implies some kind of conscious choice or intention. This can be avoided by suggesting simply that paleoanthropic man is the evolutionary product of a somehow nocturnal environment, while neanthropic man was subjected to the evolutionary pressure of daylight conditions, with all that these two sets of circumstances imply.

In thinking along such lines one may as well get one's priorities straight. It is natural for us when considering darkness, the nature of the unconscious, and so on, to imagine neanthropic man banishing his more sinister relatives to the gloom, to the

26. Thus the modern 'sun-worshipping' practices of present-day North Europeans would almost certainly have a genetic component.

edges and margins of the 'real' world. This is probably a serious misconception. Paleoanthropic man was probably after all first on the scene by some millions of years. Is it not far more likely then that *he drove neanthropic man into the daylight*?

This is perhaps yet one of the further meanings of the story of the banishment of Adam and Eve from the Garden of Eden, or of Cain from his homeland, and the forces behind all stories and legends of banishing.

To generalize at this point, it is the conviction of this book that in the two broad groupings of paleoanthropic and neanthropic man we have respectively a System B-dominant and a System A-dominant variant of *homo sapiens*. In the former case nature appears to have capitalized on what we may term the old nervous system, in the latter case on the new nervous system. This claim is somewhat borne out in the next section where we examine, as best we may, the actual brain structures of the two variants. In the grossest possible terms, the difference may be perceived in the contrasting development, in the two cases, of the *rear* of the skull and the *front* of the skull respectively.

For reasons which are by no means entirely clear, System B seems to function optimally under conditions of reduced light or actual darkness, while the exact converse is true of System A.[27]

Representatives of the two major variants of *homo sapiens* did not, however, altogether succeed in evolving into two wholly separate species – the possible exceptions to this statement being Cro-Magnon as compared with classic Neanderthal – by reason, one has suggested, of fairly continuous cross-breeding, either of a forced or of a more amicable nature. One can, perhaps, imagine such cross-breeding taking place at the periphery of contact (perhaps between peoples of not too dissimilar appearance, who then as it were fed back the acquired genes to their

27. One recalls in this connection the tendency of women to seek darkness and men to seek light in sexual matters.

These actually rather firm statements, and others made in the further course of this chapter, give rise to readily testable hypotheses, a few of which are indicated later.

more distinctly paleoanthropic or neanthropic relatives). Such might be the symbolic significance in such expressions as 'the twilight zone' and 'no man's land'.

I have proposed that these suggested crosses between varieties of man at least occasionally resulted in 'hybrid vigour' and the emergence of new qualities or personality characteristics. This, one argues, has resulted in the continuous step-wise development of man from the ranks of the primates. One has also proposed that the cross between Cro-Magnon and the Neanderthaloids of the Near East produced an exceptionally vigorous hybrid – namely ourselves – and resulted directly in the rapid, major advances which thenceforth lead to the emergence of modern civilization.

Cro-Magnon peoples arrived in western Europe initially anatomically free of mixed characteristics. Even today readily recognizable descendants of this physical type exist in peripheral areas such as Scandinavia and the Canary Islands. In subsequent millennia, however, repeated waves of migrants of mixed characteristics fanned out from the Near East across western Europe. Thus, in time, modified Neanderthaloid characteristics *were* incorporated into the European gene pool. A. S. Romer notes:

> To this 'dash' of Neanderthal blood, some suggest, is due the rugged build of the Upper Paleolithic people of Europe and the occasional primitive features noted in their skulls.

Our attention centres on the fact that all major advances towards our present civilized state, as far, of course, as the admittedly piecemeal nature of the finds allows us to conclude, are associated with mixed populations and *not*, in particular, with the Cro-Magnon variety in its purer form. This seems to hold true whether we consider variously the emergence of art and art forms, the establishment of the first permanent towns, the practices of farming and animal husbandry, the commencement of the Bronze Age, the development of handwriting, or the rise of latter-day religion.

Looking back briefly at 35,000 B.P. we find at that time Cro-Magnon and classic Neanderthal in existence as pure varieties, each possessing in their different ways a brain, as far as we can judge, equal if not superior to our own. Yet what had either actually achieved? Neanderthal, we believe, had evolved some kind of religious belief, perhaps a form of 'natural magic'. At any rate, he buried his dead in ritual fashion. Aside from this, all Neanderthal leaves behind him are a few unexciting, poorly executed tools.[28] No buildings, no art – nothing of this kind. The Cro-Magnon legacy – *that is*, of the first 15,000 or so years of his occupation of Europe – is even poorer. We *suspect* that he was a first-class fighter – the skeletal build alone argues this, and in this connection we have noted one or two suggestive scraps of legend. He may have been barbarically cruel in a rather off-hand, 'psychopathic' way – his pleasures and interests centring chiefly in the hunt, the combat and the kill. Apart from

28. Since I wrote this chapter Professor Ralph Solecki's book, *Shanidar: The First Flower People*, already cited, has appeared. The book reports Professor Solecki's recent archaeological finds in Iraq. He has located and excavated a cave occupied by Neanderthaloids over very considerable periods of time. His finds are dated around 45,000 years B.P., and they are remarkable.

One of the skeletons found, for example, was that of an individual severely physically handicapped from birth. He nonetheless lived to the then ripe old age of forty years – was, therefore, accepted and supported by his people till the day of his death. There is every evidence that the dead of these people were buried with ritual celebrations, involving a funeral feast and a funeral fire. Most remarkable of all is the finding that the dead were interred with flowers. There is even the suggestion that among these flowers those today recognized as having medicinal properties preponderate. Finally, the remains of an individual wounded by stabbing with a wooden blade and in process of recuperation were found. This is one of the earliest evidences of a (probable) attack on one human being by another. It is, of course, possible that this act was committed by some Neanderthaloid neighbour in a domestic quarrel. But is it not interesting that we find this particular evidence just as the very time when Cro-Magnon was probably first infiltrating this area? Had, in fact, Cain arrived among the people of Abel?

In general, Professor Solecki's finds appear to go a considerable way towards removing some of the suggestions in this chapter from the realm of pure speculation.

this Cro-Magnon would appear to have had few noteworthy features.

It is important not to romanticize either Cro-Magnon's or Neanderthal's way of life. By our standards they scraped little more than a bare existence from their environments, by whose *charity* they continued to exist. In this they cannot be said to have achieved significantly more than any other species of animal in existence at that time. One is saying, in short, that their *control* over environment and events was, in general, extremely slight.

It is, indeed, very tempting to argue that it was the physical, *genetic* crossing of these two varieties which enabled the enormous potential of each to be realized. Whether or not this is true, what is certainly true, and beyond all argument, is that between the years 30,000 and 20,000 a number of dramatic changes in the sapient way of life occur.

The site of the earliest known art-find is, perhaps significantly, in eastern Europe – at Brno, Czechoslovakia – and dated 27,000 B.P. The find itself is the sculpture of a man, of mixed characteristics – '. . . the brow rather low, but straight, the eyes deep-sunk . . .' The owner of the sculpture is, to us, even more interesting:

> The man who owned this sculpture was of a rather primitive physical type: robust, with unusually heavy brow-ridges very slightly hollowed between the eyes. . . . It seems possible that some of his ancestors belonged to the less specialized Neanderthal-like race known to have lived in the eastern part of Europe. (N. K. Sandars, *Prehistoric Art in Europe*)[29]

Art-finds in the *west* of Europe occur much later – from the early twenty-thousands on. Is it that man there had to wait longer for his booster injection of Neanderthal genes?[30] The

29. Sandars goes on: 'The same robust type has been traced into the late Bronze Age and even the Middle Ages.'

30. It is possible that the belief that 'the seventh son of a seventh son' possesses the second sight is actually a statement of the rarity of the genetic combination that produces a 'Neanderthal' in the ranks of 'Cro-Magnon'.

exquisite flake tools of this area, for which Cro-Magnon is rightly famous, appear even later – from around 19,000 B.P. onward.

One or two points are worth making in connection with the cave paintings of the era, which are also attributed to Cro-Magnon. These are found in southern France and northern Spain. Two brief extracts from the specialist literature follow:

> Prehistoric painting was born in most instances, and always in the great classic sanctuaries, in the entrails of the earth where it found its meaning. . . . Paleolithic painting consists primarily of rich and recurrent ensembles in the darkness of caves. . . . Prehistoric painting of the Upper Paleolithic cannot be separated from a long experience of the subterranean world. (P. M. Grand, *Prehistoric Art*, Studio Vista)

> The fine engravings, like the paintings to which they are often joined, usually lie far away from the places where men lived. . . . The question of situation belongs to a different order of problem from the purely mechanical; it is an aspect of prehistoric art which immediately and powerfully impresses everyone who has visited the caves, *especially the tortuous labyrinths* like Tuc d'Audoubert . . . [my italics] (N. K. Sandars)

These cave paintings, then, are found deep underground (sometimes as much as a *mile* from the cave entrance) in places unusually difficult to approach and in areas far from where man actually dwelt. There is thus something to recommend the suggestion that it was classic Neanderthal who introduced Cro-Magnon to the caves – that it was here that the remnants of that race were perhaps finally trapped and slaughtered.[31] One has, of course, no direct evidence for this – nor for the next suggestion, that classic Neanderthal himself conducted religious ceremonies

31. It is the present author's opinion that the Mystery religion of Mithra, with its ritual, underground slaughter of a bull, the legend of the Minotaur, and the public spectacles in which a bull is slaughtered are all (stylized) reenactments of the triumph of Cro-Magnon over classic Neanderthal. Further mention will be made of this.

in these deep caves and that Cro-Magnon adopted the practice thereafter. In general, however, the specialist literature does *not* contest a link between these paintings and rites of sympathetic magic connected with hunting, the movements of game, and so on.

There would seem, whatever else, little doubt that the Mystery rites and the labyrinths of ancient Greece and the Near East are somehow descended from these ceremonies in the deep caves – a suggestion also made by N. K. Sandars:

> . . . when these sacred places were no longer visited, the image of the caves may not have died, but have remained, no less obsessive for being no longer understood. As well as the physical barrier there arose also a spiritual barrier, and the real journey to the cave sanctuary *became the journey to Hades, the Land of No Return, the 'descent' of mystery religions* . . . and the Underworld journey of Shamanic ecstasy. [my italics]

Here then we have at least *one* practice which would appear to have survived, fairly definitely, for some tens of thousands of years. One should perhaps also recall that the cave paintings and associated ceremonies in question were still being performed in those deep caves at the very least already some 10,000 years after classic Neanderthal had vanished forever from the area. (It is salutary, too, to remember that modern Western civilization has so far lasted a mere 2,000 years.[32])

An extremely interesting feature of the paintings, mentioned previously, is that the majority appear from internal evidence to have been executed by left-handed artists. This apparent fact,

32. In the evolution of religious practice two movements over time are discernible – one already mentioned in Chapter 2. The first is a vertical movement – from the depths of the lowest caves (in prehistoric times) to the slightly subterranean labyrinths and crypts of ancient Greece – thence to ever taller surface buildings and finally cathedrals, built occasionally even on the tops of hills. The second movement is a lateral one – from places far away from human habitation (in prehistoric times) to the outskirts of habitations (in early classical times) and thence to the centre of towns and cities at the present day.

along with other considerations, has led some authorities to propose that left-handedness predominated among the peoples of those times. This may or may not be going too far. There would seem at least to be grounds for believing it to have predominated among cave artists (cave-artist-priests, if such there were) and possibly among the makers of tools also.

The view of the present book, nonetheless, is that classic Neanderthal and the Neanderthaloids *were* predominantly left-handed, while Cro-Magnon was predominantly right-handed. If such were the case left-handedness could come to be associated with witchcraft; perhaps in later generations left-handedness was one of the features associated with a general 'Neanderthal' inheritance. Perhaps actual witches were more often left-handed than not.[33]

33. There is a great deal more one could mention on the subject of left-handedness, but this would unfortunately take us too far – such as the fact that the tribe of Benjamin (which means 'the tribe of the son of the right hand') had a famed battalion of *left*-handed slingsmen, every one a picked man who 'could sling a stone at a hair and not miss' (Judges xx, 16). Apart from the left-right paradox itself here, there is the matter of the number of left-handed slingers, which is stated as 700. The total combined forces of the tribe of Benjamin is given as 27,400 – thus the slingers represent around 2.5% of all armed men. This is not of itself exceptional – but the question is, of how many left-handed men were these the *pick*? We have no way of estimating that number. A further reference to the tribe, moreover, I Chronicles xii, 1–2, tells us that '. . . they were among the mighty men who helped him in war. They were bowmen also who could shoot arrows and sling stones with either the right or the left hand; they were Benjamites, Saul's kinsmen.'

Benjamin is the youngest son of Jacob. Does that mean the most recent in evolutionary terms – or neanthropic? J. Hastings, in his *Dictionary of the Bible*, comments: 'The character of the country [of the tribe of Benjamin] was fitted to breed a race of hardy warriors rather than peaceful agriculturalists.' In Genesis xlix, 27 Benjamin is described as 'a ravenous wolf, in the morning devouring the prey, and at even dividing the spoil'. These are characteristics which the present argument would certainly associate with a tribe described as being 'of the right hand', that is, of System A.

One does not wish to make too many disconnected comments on these issues – they deserve a book to themselves. But is it perhaps possible that the tribe of Benjamin was a group having a somewhat stronger admixture of Cro-Magnon blood in an area of basically Neanderthaloid peoples?

An important feature of the emergence of art, from our present point of view, is the suddenness of its appearance – full-fledged and unrehearsed. For coincidentally or otherwise, this is one feature of the outcome of a hybrid cross: the new characteristic – be it longer fur, greater size, ill-temperedness (as with mules) – is present, and as present as it is going to be, allowing for differences within the normal range of variability of the characteristic, in the very first hybrid offspring produced.

In respect of this aspect of art, P. M. Grand writes: 'We know that art has no infancy; that . . . it either exists or it does not.' N. K. Sandars tells us: '. . . art "flashes up" mature, perfect and puzzling . . . what we [first] find is neither crude nor tentative, but an ivory figure, a female torso and a woman's head which are, I think works of art by any standard.'[34]

Thus art does not *improve* from its beginnings, though it is, of course, ever new.

In this general evaluation of art as a hybrid (?) capacity, and not as a learned function in the normal sense of that term, one notes that it is impossible to teach any other primate, even chimpanzees, to produce any recognizable *representation* of anything. They lack as a species all capacity for pictorial or plastic reproduction.[35]

34. The present writer would wish to go somewhat beyond even that statement by describing that torso (Sandars, plate 5) in question as breathtaking – suggesting as it does so strongly the classical art of later Greece, but something also of a Picasso figure.

35. There *is* another matter involved, and this we may term 'learning the use of the cortex'. Here one does see the gradual acquisition of certain abilities – which is perhaps what makes the apparently abrupt emergence of art so interesting a phenomenon. In the cases discussed below we are not speaking of a *genetically* acquired capacity in that *exact* sense but of *culturally* or *socially* acquired abilities.

In the same way that a person learning a foreign language can understand far more of it than he himself can actually speak, so chimpanzees and other animals can be taught the uses of quite sophisticated pieces of equipment, which they would be unlikely ever to arrive at unaided – even assuming they were made a present of the equipment in the first place. In the case of man no higher organism exists which could demonstrate such unknowns to

As pointed out earlier, the major steps towards what we term civilization seem invariably to be associated with mixed populations, and not with the earlier or purer Cro-Magnon strains of west and north-west Europe. Thus farming and agricultural communities appear in the Near East around 11,000 B.P. The first town of which we have any knowledge is that of Jericho (9,000 B.P.). The discovery of metal is made in Mesopotamia and there the Bronze Age begins. Modern religion as practised today in the West also originates here – the Jewish calendar taking its beginning over 5,700 years B.P. In this general area arises, too, the first great civilization – that of Egypt – and writing is invented. So one could go on.

What, meanwhile, was happening in Europe during these millennia? Of course *something* was going on. But it probably consisted principally of hunting, exploring and killing. What can certainly be said is that nothing of permanence comes down to us (apart from the cave paintings and the excellent flint tools), no monuments, no buildings, no townships. Why, then, does one assume exploring and fighting? Mainly, perhaps, because these figure so largely in the Germanic and Nordic sagas which do come down to us. These, of course, at least in the form in

---

him or otherwise test the upper limits of what he is capable. Man must instead await the birth of the individual from his own ranks (or, in a microcosmic context, the visitor from another culture), who has been able to discover and can show him the next step – who teaches man what man did not know before. The implication here is that man is therefore always operating below his best level.

A historical instance of this is provided by the use of the abacus in counting operations. The cortex of the individuals living at that time was without doubt fully capable of symbolic counting – i.e. counting with figures – but this had not been demonstrated to them. Similarly, when reading was first evolved men read aloud (as a child often does), later silently but with lip-movements (sub-vocal speech) – again as children sometimes do. The individuals who first demonstrated true silent reading without lip-movements were held to be magicians. Yet the people of former times were certainly *themselves* intrinsically capable of true reading.

This 'gradualism' is precisely what is not found in graphic art.

which we have them, are very recent products, perhaps only a few thousand years old. Still, the Vikings, that warlike and inquisitive people, seem to be at least anatomically the descendants of Cro-Magnon – and they exhibit precisely the qualities we are discussing.[36] Throughout the long period in question one pictures, on an ever-increasing scale, the rise and fall of 'empires', trials of strength (both personal and national), the pendulum of alliance and treachery.

While these goings-on may sound extremely romantic, especially when contemplated from an armchair, one should not hesitate to point out that these were in all likelihood eras and areas of extreme barbarity, with very few redeeming features, and that this kind of behaviour does *not* constitute civilization. These people lived and died by the sword, or more accurately by the flint axe. The result is the same. The European *barbarians* (the Cro-Magnonoids) were gradually civilized from the East over a period of many millennia.[37]

At this point we discontinue these sociological speculations in order to resume those of physiology – still, certainly, very much in the hope of demonstrating the hybridization of varieties which, among other things, we have been discussing.

It will by now be clear that this book considers that in paleoanthropic man we see nature in the process of producing a System B-dominant sapient, and in neanthropic man a System A-dominant sapient. As stated, this gives the situation a flavour of intention. We can avoid any such idea by stating, neutrally,

36. Possibly, too, among Cro-Magnon's descendants are the 'blond adventurers' whom Thor Heyerdahl (*South American Indians in the Pacific*) proposes set off from the Canary Islands or North Africa to reach Central America, travelled thence down the west coast of South America, and ultimately took sail for the Pacific Islands, to give rise to the undoubted caucasoid strain found there.

37. In fairness to Cro-Magnon, and in order not to bolster artificially my theories of the virtues of mixed populations, it must also be pointed out that after the dispatch of classic Neanderthal, Europe's climate became once again extremely inhospitable, so that simply staying alive was once more something of a full-time occupation.

that a particular kind of life, such as the nomadic, once undertaken, and for whatever reason, could tend to favour thenceforth the survival of individuals who were – what? – aggressive, tall, good at handling environments and the unexpected? Another kind of life, perhaps one involving a permanent home base, might favour the quality of *living with* as opposed to dominating (raping) an environment; if the site were to be really permanent one would have to be ready neither to over-hunt nor over-crop: qualities, then, of unobtrusiveness or patience might be required, and possibly a tendency to move about at half-light or by darkness.

Given that two such broad ways of life became established, for whatever initial reasons, perhaps equally favourable to survival in their different ways, interaction between tribes of these opposite persuasions could well reinforce these tendencies. 'Night' tribes might deliberately stay out of the way of 'day' tribes (probably at all times) while 'day' tribes might take care to attack their cousins only by day ('Benjamin, a ravenous wolf, in the morning devouring the prey, and at even dividing the spoil'), never by night.[38]

The favouring, then, of what are recognizably System B attributes leads to the paleoanthropic form of *homo sapiens*, the favouring of System A qualities to the neanthropic sapient. In general terms this suggestion accords well with the findings of earlier chapters. It is further borne out, as we shall see, by an inspection of the brain structures of these two forms of man.

Before undertaking that examination, a final comment. A fear of the writer is that having read this section his reader will say: 'Ah. He has made out a case, based on a reasonably impressive amount of parallelism, for the view that the ancient myths and legends of Europe are at least partly a distorted and fantasized

38. Psychologists have established that some individuals have a higher metabolic rate and are more active in the morning, while others function better and have a higher metabolic rate towards evening. This finding may be linked with the present subject.

account of a historical event – the overrunning of Europe by Cro-Magnon. How interesting. Excellent. Well, that's that then. Another stronghold of the fantastic has succumbed to patient, scientific investigation.'

The author feels that if he hears this, he will go out and shoot himself. For while that statement is, as he believes, true, it is the least important facet of the material we have examined. Those historical events were the starting point of the legends. They are not what gives those legends their vitality and longevity, are above all not their *explanation.*

There are two strands here, which we shall try to unravel without letting matters become too complicated. Classic Neanderthal is enshrined in our legends not so much because he was once a formidable, real-life enemy (though there *is* a contribution from that direction) but because he is/was an embodiment of our unconscious fears. He was/is our own, other secret Self made flesh – no longer content merely to haunt our more uneasy dreams, but suddenly physically there before our waking eyes. Symbolically, as has been indicated in detail elsewhere, the dwarf, the troll and his companions represent the cerebellum – our own 'night' personality: the shadow Self; the unwelcome lodger that lives within each of us – relatively passive or relatively active as the case may be.

Classic Neanderthal involves us in our own System B *precisely* because he is the embodiment of many of its actual qualities.

Let us consider one example only of the pointlessness of trying to argue that classic Neanderthal was a historical event *and nothing more.* In Persia, a mere 3,000 years ago, there flourished a Mystery religion, the cult of Mithra. It was an extremely vigorous cult, clearly, with a wholly current appeal, for it spread rapidly through the Roman army. This event was taking place some 30,000 or more years after Cro-Magnon moved up into western Europe and dispatched classic Neanderthal – a considerable time-gap. But what is still more to the point is that classic Neanderthal was never *in* Persia or anywhere near it. It is, perhaps, just possible that one or two bands of stragglers wandered down this way after the (alleged) sack of Europe. But certainly the original inhabitants of Persia could never have fought for their lives against any 'Neanderthal peril'. Yet the cult of Mithra enthusiastically slaughters a bull in an underground crypt, or cave, for millennium after millennium. What must then be the origin of this situation, and whence did it derive its then-and-there psychic energy? Must it not be that stories of battles against classic Neanderthal gradually filtered back east *where they were immediately understood symbolically*?

A further point. It is perhaps possible in some sense to account for *Nordic* legend solely and simply on the reality basis of a probable protracted struggle against Neanderthal. Many details of these legends fit with observed and hypothesized (though probable) attributes of Neanderthal. But look now at those other legends of the Near East – e.g. those of the tall, warlike, sons of the gods, which we have suggested refer to Cro-Magnon as perceived by Neanderthaloid peoples. What of *angels*, what of men ascending to heaven in fiery chariots? Did Cro-Magnon (or anybody else) really possess *those* attributes? Or is one ready to dismiss *these* particular facets of the legends as nonsense – while retaining the 'more reasonable' ones and agreeing that these could well be a slightly distorted or colourful reference to Cro-Magnon? One can do this if one wishes – but one then risks the charge of special pleading. Why is one allowed to reject some aspects of the legends and not others, and on what precise grounds?

By proposing the concept of the *archestructure* the present book is able to accept all aspects of these legends. The concept, and many examples, of archestructures (though not by this name) have already been discussed in these pages. An archestructure can now be defined as a felt or perceived function or structural feature of the nervous system, projected or unconsciously acted out in the life-style or the beliefs, customs and social structures of the individuals concerned, or of whole communities.

The moon and the sun as they appear in legend are therefore

archestructures, as are those religions and philosophies which describe the universe in terms of opposing forces of Good and Evil. The political divisions of left and right, too, are essentially archestructural in origin. The symbol of the Christian cross is yet a further specific example.

The term archestructure is deliberately chosen to align it with the Jungian concept of the *archetype*. A clear difference of emphasis will be seen to exist between the two concepts (though one would hesitate to say that the distinction is absolute). An *archetype* is defined here as the psychological concomitant or accompaniment of a biological response (or a group of related responses) which a species – in this case, man – has acquired in the course of its evolution, in reaction to a quite specific situation or crisis. This response in some way selectively favours survival – is, therefore, thenceforth biologically fixed or bred into the variety or species in question.[39]

One can now see that classic Neanderthal, the dwarf, the witch and all these associated figures are *both* archestructural and archetypal in form. These are the two strands mentioned earlier. The 'survival response' which Cro-Magnon made in the face of Neanderthal (which, one suggests, therefore became biologically fixed) has produced our *archetypal* experience of such figures. The *archestructural* response to those events was already present in the make-up of Cro-Magnon – as indeed of all sapients. The arche*typal* response is involved in, among other things, the real live witch-hunt in the objective everyday world. It is the arche*structural* aspects, however, which are concerned in, and give strength and meaning to, the *symbolic* bull-ritual or bull-worship, to the Mysteries, to the christian Devil and to religion in general.

Now, although the troll, the witch and so on, are at once both archetypes and archestructures, this does not hold true for the demigod – the shining warrior and the angel. These figures are only archestructural. They are not archetypal. We

39. Examples of Jungian archetypes are the Wise Old Man, the Wonderful Boy, the Guardian, along with a great many others.

have no wish and little evidence to argue for any *variety-specific fixing* of that response.

When man experiences the Earth he somehow experiences something of his own nature and origin. When he sees the moon he 'sees' his cerebellum. When he sees the sun he 'sees' his own cerebral cortex. One is anxious to avoid the word *projection* here. Rather it is that certain features of the environment *suggest back* to us certain features of ourselves, of our nervous system. (What *might* have happened, one wonders, if the Earth had had two moons and two suns?)

Thus not only luminosity of a fairly marked variety, but such further aspects as movement upwards, somehow echo symbolically aspects of our self-experience – that is, of our experience of our own nervous system. One recalls that messages from the senses and somatically induced sensation, as well as messages relayed from the cerebellum or the sub-cortical centres *travel upwards* to reach waking awareness in the cerebral cortex. The sun, too, apart from having both the attributes of strong luminosity and movement upward, also conveniently goes down in the evening when waking consciousness (at least in System A-dominants) is reduced. More to the point is that any object or phenomenon which has any one of the 'attributes' of the cortex or of waking consciousness *will tend also to be held to possess other System A qualities*, even if it does not actually do so.

Thus, when the Near-Eastern Neanderthaloids (System B) saw Cro-Magnon they *actually* saw, for example, (1) someone taller than themselves, i.e. above them (2) someone with a high, clear forehead (unlike their own recessive and perhaps hairy foreheads), (3) someone with at least some body areas free of hair – say, the upper arms – whose skin, perhaps, shone in the sunlight, (4) someone extremely strong, ruthless and masculine. They did *not* see someone who soared up into the air; they did not *actually* see flames playing around someone's head – or any of the other features which eventually come to be ascribed to angels and other similar mythical figures. These are fantasies or, more precisely, archestructural accretions added (perhaps long) after the actual events which initially triggered the myths in question. Angels soar into the sky largely because the sun does; that fact in turn has significance – or has prior significance – because, I suggest, some function of the nervous system is therewith echoed.

This rather telescoped account of these features of our general theoretical argument is expanded in Chapter 12.

In making statements about the brainstructure and the related behaviour or personality of early man, a number of points need to be kept firmly in mind.

First, that in many respects the brains of Cro-Magnon and Neanderthal were essentially similar. These two varieties of man, although at the time of their suggested meeting having apparently made tentative steps along different evolutionary paths, nonetheless shared very recent common ancestry, in evolutionary terms. This aside, experimental observation has in any case shown (a) that functions performed by a higher centre can, in the absence of that centre, in many cases be taken over by a lower centre (without, necessarily, dramatic loss of fineness of response), or, more importantly here, simply by an adjacent cortical area; and (b) that localization of function, especially in the cortex, is in any event something of a myth except in very general terms – so that speech for example appears to be, in modern man at least, a combined parietal-temporal-occipital operation, with the assistance of yet another area in the frontal lobes. And finally (c) that despite recent advances very little is in fact known of the functions of many parts of the cortex.

One must be quite clear, therefore, that although for example the frontal lobes of Neanderthal's cerebral cortex were, to judge from the receding forehead, less well-developed than those of Cro-Magnon, there is no question of Neanderthal *not having* those areas – only of their being less well represented; and that even were they wholly absent, which is not true, one still could not even then say with certainty that the functions they subserve in Cro-Magnon were therefore absent in Neanderthal.

With this battery of provisos it may be felt that there is little point in proceeding with the inquiry. However, something can nevertheless still usefully be said.

Some areas of the cortex have been shown, by electrical stimulation and other techniques, to have definite motor or definite sensory functions – are, that is, concerned with particular movements and sensations respectively. The remaining and extensive areas of cortex which do not have these functions are termed 'association areas' – implying that they have associative or interpretive properties in respect of already stored or incoming information.

The frontal lobes of the brain, already mentioned, are generally concerned with volitional motor movements. The forward part of these lobes – termed the pre-frontal area – has, however, no motor function. It is possible that it was in this particular region that Neanderthal was most deficient. Experimental data – including that obtained from psychotic patients who have undergone pre-frontal leucotomy – show that destruction of the pre-frontal areas produces varying – and that word very much means what it says – degrees of: dulling of general awareness, reduced socialization, loss of creativity, and an overall primitivation of behaviour. Responses to the operation, as stated, vary considerably, depending on which part, and what amount, of these areas is put out of action. In monkeys this operation leads to loss of the ability to solve at least some types of complex problem – those which necessitate holding one aspect of a problem in mind while attending to another aspect, as in serial tasks; or which depend on a delayed response. From such experimental evidence psychologists have concluded that the pre-frontal areas contribute to the experience of events over time, to the complex patterning of behaviour and to orderly thinking; and seem linked with conscious learning and conscious thought processes.

While, as already emphasized, there is no question of any of these functions being totally absent either in classic Neanderthal or the Neanderthaloids, it may nevertheless be that these functions were less highly developed in these varieties of man.

When the skulls of classic Neanderthal and Cro-Magnon are viewed from the front, it is seen that, apart from the generally lower vault, the Neanderthal skull is widest just above the ears at the temporal region; and that that of Cro-Magnon is widest *above* this point, in the neighbourhood of the parietal lobes. This argues that the former was better endowed in terms of temporal cortex relative to parietal cortex, while in the latter the position was reversed.

The temporal lobes contain the primary auditory projection areas and the auditory association areas. They appear also to house the individual's memory store. Penfield and others have shown that electrical stimulation of parts of the temporal cortex evokes vivid, complete memories from the subject's past. Dogs with the temporal lobes excised cease permanently to respond to their names. These lobes additionally contain an important speech area (speech recognition). Finally, they seem of great importance in the facilitation and control of emotion, including sexual behaviour and aggression.[40]

The parietal lobes are the seat of *stereognosis* – the sense of tactual recognition of shape. They appear also to be the main projection area for the body-sense(s), touch, the awareness of the position of the body and limbs in space, and these last in relation to each other from within the body. The kinesthetic memory is found here – the memory of learned movements. For the foregoing reasons the parietal lobes are held to be directly involved in manipulation and in skilled movements of the hands, mouth and eyes. (Manipulation in a *general* sense is a property of the whole cortex.) Lastly, in conjunction with the temporal and occipital lobes, the parietal cortex is considered to be involved with language and some higher thought processes.

40. Obviously the role of the temporal lobes is a crucial one. Yet despite the range of functions acknowledged, Grossman comments: 'We do not, in fact, know very much about the functional role of the temporal areas – most of it is therefore classified as an association area.'

Viewed from the side, Neanderthal's skull is characteristically elongated at the rear. Thus the occipital lobes of Neanderthal were probably larger than those of Cro-Magnon. It may also be that the *cerebellum* of Neanderthal was larger – possibly even much larger.

The occipital lobes house the primary visual projection areas and the visual association areas.[41] It will be recalled that it is from the surface of the occipital cortex that the alpha rhythms are emitted when the eyes are closed or in day-dreaming. This part of the cortex is also concerned with *reflex* movements of the eyes.

On the basis of what has gone before, it seems tentatively possible to conclude that Neanderthal would not have been ideally equipped to handle the technical and technological problems in which our society has become so specialist. Equally, it seems possible to conclude that Cro-Magnon *was* so equipped. This is not to imply, however, that Neanderthal did not enjoy a rich mental or emotional life – using the term emotional in its widest possible sense. Such behavioural evidence as we have would also support this view, since Neanderthal had evolved a religion. In terms of brain structure too there are grounds for supposing that he possessed an excellent though perhaps non-practical imagination, and a memory for events possibly in excess of our own. This last, coupled with the possibility of strongly developed powers of vision and hearing may have provided an indispensable survival kit in the environment in which the paleoanthropic was, typically, at home – that is, in darkness?

Is it perhaps permissible also to suggest that Neanderthal was fundamentally inward-turning – preoccupied with the Self (as we have defined it) and valuing his inner life – while Cro-Magnon was outward-turning, the manifestation of Ego in terms of his conquests and journeying; that while Neanderthal built no cities of stone, he perhaps built cities of dreams?

41. Thus the paleoanthropics not only had large eyes – but a large amount of visual cortex.

It may be wondered why one is concerned to venture into this area of speculation at all, and what benefit is thereby derived. One of our reasons is because art (as are many other forms of higher behaviour) is very much a *marriage* or *synthesis* of the personality factors we have assigned to these two kinds of man: that is, of content and form; of imagination and technique; of conception and execution. Freud among many others has expressed the view that art is the unconscious made manifest, the forbidden, made available and acceptable to consciousness. While this next analogy is neither an exact nor a flattering description either of the unconscious or the nature of art, it is as if we see in art the capture and caging of a wild beast, who may then be safely brought into our zoos or our sitting-rooms.

Broadening the discussion very briefly at this point, it will be recalled that in previous chapters mention was made of the destructive or self-protective power of consciousness, due to which unconscious contents arrive in consciousness dead (as plankton, removed from its element, loses its irridescence). Art – that is, the formal qualities of art – permits unconscious materials to arrive in consciousness substantially still in their own (magical) forms and very, very much alive – indeed, immortal.[42]

In classical art, which leans more towards System A, form – the degree of 'control' – is paramount. The rules or conventions, be these the dramatic unities of classical Greek tragedy, the alexandrine of classical French poetry, the formality of classical architecture or whatever, are wholly obligatory. In romantic art, however, which stands closer to System B (that is, the unconscious and the basic instinctual drives) the essential ingredients are inspiration, intensity and passion. Form is a tiresome, wearisome nuisance, a hindrance rather than an aid to creativity. Yet, under the extreme forms of both viewpoints – i.e. too much form or too little – art ceases to exist, degenerating on the one

42. In this general context, the story of Creation in the Old Testament tells of the intellect ('light', consciousness, System A) giving form and permanence to Chaos (unflatteringly, System B). This legend is perhaps a symbolic, generalized account of what is under discussion here.

hand into mere technique or technology, and mere emotion on the other.

The proposal of this book, to revert to the matter immediately in hand, is that at a genetic level, no matter in what other way (cultural, psychological, and so forth), Neanderthal contributed the basic inspiration and Cro-Magnon the basic form not simply to art, but to our civilization.

Modern man, I have suggested, is to be regarded as a hybrid or mongrel descendant of two fairly well-marked varieties of early man – the paleoanthropic and the neanthropic. I have postulated fairly continuous cross-breeding between these two broad varieties over the millennia since their inception. Without this circumstance one may with some legitimacy suppose that the two lines would have steadily diverged past a point of no return, into two quite separate species. The frequency of individuals' showing mixed characteristics throughout the long period in question can be used to support the view of continuous hybridization – though this of course is not the only possible interpretation.

Two main points are made in this section. The first is that the crossing of a well-marked A-dominant variety with a fairly well-marked B-dominant variety in Europe, produced not merely an exceptionally vigorous *but rather unstable* hybrid. The second point is that while, certainly, neanthropic and therefore essentially A-dominant sapients are now found in all parts of the world, it is proposed that some varieties of man are characterized by what may be expressed either as a greater admixture of B-dominance, or a more unified mix – this arising perhaps from the fact that the neanthropic element in those parts did not undergo the particular 'sharpening' which Cro-Magnon is assumed to have undergone in Europe.

Before taking up these points it is as well to remind oneself of the yet wider background to which all of this relates – of the fact, first, that System A and System B are *also* sex-linked. That is, that System B is normally more active and dominant in the female, System A more dominant in the male of the species. This holds true within all other divisions – paleoanthropic, neanthropic or whatever.

Let us also not forget – above all, and at all times – that each and every one of us individually is at once both a System A and a System B: both a striving Ego and a receptive Self: both a male and a female.[43]

Finally, let us not put entirely out of our minds (though its relevance is not crucial at this particular juncture) the fact that the two systems are age-linked – System B being more dominant in the child and System A more dominant in the adult. This, too, applies within all other divisions.

First, the suggested 'instability' of our hypothetical hybrid variety. Reverting briefly to an earlier terminology, it is clear that no real problems are likely to arise in connection with a Ba or an Ab profile – always providing that the psychological dominance coincides with the physical sex of the individual concerned. Problems arise, at least in our society, where the inappropriate system is behaviourally dominant in an individual physically a member of the reverse sex.

Problems of another kind arise, however, when the personality profile AB is present – when, that is, both components are strong or near-equal in strength and are both, therefore, potentially dominant. One proposes that the crossing of Cro-Magnon and Neanderthal produced a relatively high incidence of that very situation, of the personality profile AB.[44]

43. The last is amply demonstrated, at least physiologically. Injections of female hormone into normal males result in loss of beard, a soprano or high-pitched voice, the growth of breasts and many appropriate behavioural changes. The reverse situation obtains when male hormone is administered to normal females.

44. Individuals who appear to have coped more or less successfully with such a combination, without essentially denying either element, are Goethe and Leonardo da Vinci. Goethe's *Faust* would then be an externalization of this personal, inner conflict. (The poet himself, indeed, described all his work as 'fragments of a great confession'.) In addition, both Goethe and da

By way of analogy one may liken the personality of modern (Western) man to a sports car. The 'souped-up' engine of that vehicle results in a high level of performance, but a high rate of breakdown, rapid wear of components, and dangerous and difficult to handle side-effects.

The analogy, a critic might say, is all very well; but where is the precise *evidence* for the alleged instability, and what precise forms does it take? To attempt an answer to that question would involve us in very considerable problems of definition, however, and lead us a long way from the central issue here. Therefore our reply is only a further metaphor. Just as when one used to look at the balance of payment figures for month after month during the late sixties, which, it seemed, were always under the pressure of special factors – strikes at home, strikes abroad, revaluation of someone's currency, the purchase of military aircraft, the holiday season, and so on and so on – one nevertheless ultimately came to the conclusion that the balance of payments situation was basically and somehow unsound. Something like this happens when one looks at the history of mankind in the last few thousand years. One can argue climatic factors, economic factors, population pressures, this precipitating cause, that precipitating cause – yet when all these points have been made, one is left still with the impression that a majority of mankind's difficulties are self-produced and self-discovered – that the constant factor in every equation is man himself, and his somehow flawed constitution.

Perhaps one solid index of man's dis-homogeneity may be suggested, namely, the many forms of polarization in modern society, of which the political parties are one.

Vinci were left-handed. The view of handedness adopted by this book further suggests that Goethe's left-handedness and his preoccupation with the Faust theme are no coincidence. Nietzsche (an extremely gifted individual also) was likewise left-handed, and I propose again that his interest in the alternatives of Apollo and Dionysus, together with his eventual psychotic breakdown, are not unconnected with that circumstance, and with what I have indicated lies behind it.

On the second question, the matter of the relative strengths of System B in various parts of the world, we can be somewhat more forthcoming.

One takes the opportunity, first, of meeting a possible objection. Has one the right to be speaking in a simplex manner of only two varieties of man, when it is so widely agreed, for instance, that the racial spectrum in the world today offers a very confused picture indeed and where the notion of any kind of 'racial purity' is rightly regarded, except by cranks and other disturbed individuals, as complete nonsense?

One can make two points here – the first deriving from Darwin's *Origin of Species.* Chapter 1 of his book – entitled 'Variation under Domestication' – discusses the frequently observed phenomenon that individual animals and plants under domestication vary much more markedly than do those same organisms under nature:

> When we look to the individuals of the same variety or sub-variety of our older cultivated plants and animals, one of the first points which strikes us, is, that they generally differ much more from each other than do the individuals of any one species or variety in a state of nature.

It is easily overlooked that man is himself an animal under domestication. May one then expect this circumstance to have had the same effect on his variability that it has on all other organisms? The domestication of man is not, of course, something which happens overnight. The degree of domestication increases as we approach present times. Might one, therefore, reasonably expect the amount of variation to have similarly increased only gradually? There are two points involved here. Domestication leads (a) to greater variation – that is, a greater range of variability *within* varieties and (b) also to a larger number of varieties.

This then is one aspect: as we move back in time we might expect to find fewer varieties of men and less variation within those varieties, than seen among man today.

A further line of argument is that whereas paleoanthropic and neanthropic man were two clearly marked varieties, a fair way along the road to becoming separate species, what we see in the world today are in general not more than sub-varieties. What is more, they are essentially sub-varieties only of the *latter* variety: for we should be clear that all races of man living today are neanthropic – with the *possible* exception of the Australian aborigines and one or two other groups. But even though aborigines do show some paleoanthropic features, most of their characteristics are essentially modern. What we are discussing is in fact only the varying extent to which the claimed admixture of the paleoanthropic characteristics can be detected in otherwise modern neanthropic man in various parts of the world.

One is not ultimately to stop short at the mere (?) theoretical discussion of possibilities, and in particular of the possibilities suggested by this book. Our final section contains proposals on how the hypothesis involved can be fairly simply investigated by the straightforward statistical treatment of some relatively simple data.

For our present purposes, however, mankind is now considered in terms of three broad areas or, more accurately, three broad distributions. The names given to the three distributions carry here a much wider definition than they usually bear – so these are written throughout within quotation marks. The areas in question are 'Europe', 'Asia' and 'Africa'.

'Europe' is defined as the whole of geographical Europe including Russia, all countries adjacent to the Mediterranean (hence also those of North Africa) and, further to the east still, Iran and India. By extension this term also embraces all people of 'European' origin in the Americas, South Africa, Australia, New Zealand, etc.

'Asia' refers to all of that geographical continent not already mentioned, but principally China and Japan, and additionally all the native (Red) Indian populations of North and South America – who originally migrated to these areas from central Asia.

'Africa' means the whole of Africa south of the Sahara but also includes all peoples of Negro origin in all other parts of the world.

The three 'areas' named constitute three relatively self-contained *gene-pools* – a useful, non-emotive term. Within each of these large reservoirs are, beyond any doubt, further relatively separate, smaller gene-pools, but these are not our direct concern.

There are well-established grounds for believing that gene-flow between the three broad areas or distributions defined was restricted during certain known geological periods, in ways described below. The specialist literature is, it may be added, agreed on these points. The *interpretation* placed on these events here is, however, hypothetical.

Direct access to Asia from the west and north-west of Europe was very largely or entirely impossible throughout the whole of the Pleistocene (that is, during the last two million years), due to extensive glaciation. Ingress from these directions became possible around 10,000 years ago. The peoples of Mongolia came into being at this point. Circuitous routes into 'Asia' through eastern India have, however, always existed.

Africa south of the Sahara, on the contrary, was always *easy* of access *whenever glaciation was advanced in Europe*, and not when it was not. During those times of glaciation the Sahara desert was a grassy plain, with an extensive system of rivers and occasional afforestation. Such was actually the situation in Africa around 25,000 B.P., when renewed glaciation in Europe followed on the warm interval that had *previously* encouraged Cro-Magnon to move up into Europe from the south-east and had brought about, one way or another, the extinction of classic Neanderthal. After an interval of several thousand years the climate of Europe then once again gradually improved, yielding the present mild conditions of the last 10,000 or so years. As the climate of Europe once more improved, however, so conditions in the Sahara again deteriorated. By 10,000 B.P. the ecology of the Sahara was not dissimilar to that prevailing today. At that point this vast locality quite definitely no longer served as a land-

bridge but was again a formidable land-*barrier* to movements between the continents. One may assume that migration across the Sahara tended to decrease after 25,000 B.P. – slowly at first, but more sharply towards 10,000 B.P.

Gene-flow between 'Europe' and 'Africa' is, then, a reducing quantity from 25,000 B.P. onwards. However, the presence of Cro-Magnon on the Canary Islands and other finds on the mainland argues some genuine traffic. Yet, either because the mixing was on too small a scale, or for whatever other reason, no significant *cultural* advance or change is detectable in 'Africa' as a result.

Traffic between 'Asia' and 'Europe', and therefore gene-flow, was confined until recent times to the route passing through the Middle East and northern India and the relatively narrow 'gateway' of north-east India. From 10,000 B.P. on, however, the north-west gate from Europe is open – and mixed Caucasoid-Mongoloid peoples arise in Mongolia.

I consider it extremely significant that there are *no records of prehistoric art either in 'Africa' or 'Asia'*. The earliest art-finds in these regions date from a few thousand years B.C. One can say that art is unknown outside 'Europe' prior to 5,000 B.P.[45]

For such but also many other reasons one is inclined to regard the emergence of Cro-Magnon, followed by his merging, though not swamping, by Neanderthaloid varieties, from some points of view as a purely 'European' phenomenon. One is, of course, not necessarily arguing that *no* neanthropics existed in other parts of the world – after all, Cro-Magnon came into Europe from somewhere. But it is proposed that populations outside 'Europe' retained and retain a rather higher proportion of Neanderthaloid genes; that they are, in short, to be regarded as including rather

45. There is also a case for suggesting that art is culturally introduced into those areas at that time, as opposed to arising independently. There is a considerable difference, as already noted, between being shown a step and evolving it for oneself. However, to accord with the views earlier expressed on art, one is perhaps obliged to concede that some *genetic* capacity for artistic expression, though unutilized, must nevertheless already have existed in these peoples.

more individual B-dominants, or rather more B-dominance generally, throughout the population as a whole, than is the case among 'Western Europeans'.

Whether or not this is so – or, as this book suggests, *because* this is so – the civilizations of those countries outside 'Europe', and in particular their religions, philosophies and literatures, *are based upon and are contributions to Knowledge II*, and not to Knowledge I.

The claim, then, is as follows – that all peoples *outside* 'Europe', and considered as a whole, show greater B-dominance (the dominance of Self, as defined) than do all peoples inside 'Europe', considered as a whole – these being, relatively speaking, A-dominant (and showing the dominance of Ego).

Inside 'Europe', however, the greatest incidence of or admixture of B-dominance is found, in the following descending order, among: (1) Indian, Middle-Eastern and some Mediterranean peoples; (2) the Latin and Celtic varieties; (3) East Europeans – Slavs, and so on; (4) west and north-west Europeans and finally (5) the *migrated* Europeans of America, South Africa, and elsewhere.

Our reasons for these views are numerous. Some have already been mooted, others are now discussed.

It is not seen as accidental, for example, that in 'Europe' religion flourished and flourishes principally in India, the Middle and the Near East, somewhat less strongly in south and east Europe, and least in the north and north-west – where eventually the Protestant Reformation and later still the industrial revolution take place, science and the scientific method finally emerge.[46] The map of religion and anti-religion in 'Europe'

46. Somewhat anomalous here are perhaps Greece and Rome. It is with the former that, among other things, the first stirrings of modern science and the objective method are observed. We have noted already, however, how threatened System A apparently felt itself to be, at that period in that place – judging from the extensive content of Greek legend (notably the unconscious fear of women), the rigid (though of course not unartistic) formalization of classical art, the tendency to (?a-sexual) homosexuality, and so on. The importance assigned to the crime of *hubris* (conceit against the gods = too

is largely – and too coincidentally? – a map of the concentrations of Neanderthaloid and Cro-Magnon elements respectively, of some 30,000 years earlier.

How might one then more generally categorize or describe the proposed similarities between 'African' and 'Asian' behaviour as a whole – together with the differences between it and 'European' behaviour as a whole, and the similar, but less marked difference, which, we are claiming, exists *within* 'Europe' between the East (east-and-south) and the West (west-and-north)?

System B or 'Neanderthal' populations are firstly characterized by what one might term a greater willingness to admit the existence of the unconscious, a greater willingness to evoke it by various methods, and much less fear or aversion (or a 'pleasurable' fear) to its actual manifestations. Thus in 'Asia', 'Africa' and those parts of 'Europe' adjacent to them one has – or has had till recently – complete social acceptance of the use of hallucinogenic drugs, both hard and soft – the opium of China, the marijuana and hemp of the Middle East, the Indians and South American Indians and so on – and the deliberate inducement of advanced states of 'possession' on a whole-tribe basis, e.g. by dancing, accompanied by the intake of 'soft', locally-evolved alcoholic beverages. It is interesting in this connection to note the American Indian's delight in, and great vulnerability to, the whisky and other hard alcohols introduced by white men.

The foregoing does not of course apply so literally to east and south 'Europe' – but compare, for example, the 'whirling Dervishes', Bacchanalian practice, and the use of soft drugs, already mentioned.

---

much Ego), and legends such as that of Icarus, offer additional evidence for this view. The civilization of Rome was still more 'conscious', i.e. technological, and more confidently so. However, the use of augury, for example, persisted. The System A genes of both these civilizations *may* ultimately have been depleted by continuous warfare over many generations, along the lines suggested in Chapter 7, as well perhaps by miscegenation.

The following comments, however, are more literally applicable to all the areas under discussion. In all System B countries and among all these peoples there is a marked willingness to accept what 'is', and not to search for objective causes or the means to control those causes. There is, too, a much greater acceptance of and indulgence in sexual activities. Although these may be surrounded by *specific* taboos, this is quite unlike the *total* taboo that has tended to characterize Western (e.g. Victorian, Puritan) societies.[47] It would, hopefully, prove possible to demonstrate by a study of the legends of 'Asia' and 'Africa' far *less* unconscious fear of women – or actually terror – in those countries than is found, say, in Greek legend. It would indeed be of the greatest interest if extra-'European' legends should in any way show a fear of the masculine.

The kind of generalized categorization of peoples of the preceding paragraphs is one to which Jack Kerouac has given expression at various points – and which, however incoherently, is contained in all versions of the hippie philosophy. The following extract is from Kerouac's *Lonesome Traveller*.[48]

> You just wait patiently like you always do in America among those apparently endless American policemen and their endless laws *against* (no laws *for*)[49] – but the moment you cross the little wire gate and you're in Mexico you feel like you just sneaked out of school when you told the teacher you were sick and she told you you could go home, 2 o'clock in the afternoon . . . you look around and you see happy smiling faces, or the absorbed dark faces of worried lovers and fathers and Mexican policemen. . . . It's a great feeling of entering

47. Even Catholicism – that otherwise enemy of sexuality – has always given its full approval to sexual activity *within* marriage, providing the outcome in children. The Protestant ethic lacks even this limited joy in sex and fecundity.

48. Mayflower Books.

49. Is it of no significance that the one country which has attempted a total prohibition of alcohol (the United States) is one of the countries with one of the highest concentrations of what this book terms System A genes? Puritanism was a manifestation of another country also having a high System A count – England.

the Pure Land, especially because it's so close to dry-faced Arizona and Texas and all over the South-West – but you can find it, this feeling, this fellaheen feeling about life, that timeless gaiety of people not involved in great cultural and civilisation issues – you can find it almost anywhere else, in Morocco, in Latin America entire, in Dakar, in Kurd Land . . .

If there is substance in what this chapter is proposing, certain, and ominous, corollaries follow.

It would follow, for instance, that the view of man proposed by Western philosophers and thinkers, and more latterly by Western psychologists, is a view only of *Western* man. While this view *might* be *wholly* true of Western man – though this book is attempting to show that not even that is true – it is at very best only partially true of races of man outside Europe. What one is implying, in other words, is that Western man is in some sense a special case – and that 'Eastern' and 'African' man are also special cases. And Western man's social order, political systems, attitudes to what is important and absolute are largely the product of local conditions – including, however, in local conditions (and here one parts company with all environmentalists, who might otherwise be in agreement with this last statement), the product of the fairly recent evolutionary history of this particular variety of *homo sapiens – of his genetic make-up*.

It would follow from this, if true, that our attempts to interpret and explain the behaviour of peoples outside 'Europe', and even within 'Europe' – say, attempts to impose upon them a two-party political system, or to prevent their use of certain mild drugs, or to replace their 'obsolete' religious and philosophical world-views with others of a more scientific or rational nature – may be largely misguided, and in any case quite unjustified.[50]

50. Hopefully none of this sounds anything like a hard-core right-wing (or any other wing) view. The present author holds no brief for any political group. To draw attention to possible differences has, as such, nothing whatsoever to do with the labels 'better' and 'worse'. On the other hand, to suggest or act as if varieties of man, separated from each other for periods of time

Naturally, it is not disputed here that there are in all this cultural and environmental components. A society, a particular social environment, may and does encourage the development of System A or System B qualities respectively. While admitting, then, that the end-product personality *will* show the effects of such shaping, the present book is nevertheless not of the opinion that such influences make after all so very much difference. In stating this one is at variance with all those who would, and do, argue that what this text calls the System B way of life of under-developed countries is the product of some kind of backward-ness: of an economic, or social, environment that does not give the individual a chance to become anything else – that is, to become Western man.

Admitting that such situations are complex and that the peoples in question have so far been refused anything even remotely re-sembling a fair opportunity to participate on equal terms with caucasians, nevertheless what is envisaged by the 'progressives' does not seem to have *begun* to occur in the cases, say, of the American Indian or the Australian aborigine. On the contrary, a number of observers, not all of whom are sentimentalists, have expressed the opinion that these peoples are psychologically 'starving to death'.

In general terms one is prepared – to say this now for the last time – to agree that environment and opportunity, of course, make a difference. What one maintains, however, is that neither the 'African' nor the 'Asiatic', even if wholly reared in a situ-ation of bottle-feeding, no-touch mothering, anti-sex fathers, objective evaluations of emotion, etc., would ever demonstrate the degree of Denial – of the body, of feelings – that so notably characterizes Western society.

---

long enough to produce clear physical differences, are not likely also to differ psychologically and behaviourally seems, frankly, naïve.

To say one last word: a colleague to whom I had been outlining these general issues said after a little thought: 'You mean our present troubles are due to 25,000 years of Cro-Magnon mismanagement?'

That one might go along with.

To revert, then, to the particular argument – and taking account now also of the material of earlier and later chapters – the proposition is that the populations of 'Asia', 'Africa' and eastern and southern 'Europe' are *biologically predisposed* both to the evolution and the acceptance of mass 'solutions' – such as religion or Communism – while the peoples of Western 'Europe' are biologically predisposed to notions of individual differences, rank hierarchies and rule by élites. Further distinguishing behaviours of these two broad groups will be considered elsewhere. The two types of communities which arise on these predispositions are, respectively, the System B and the System A society. These two types of society now briefly outlined are, it must be borne in mind, like the individual, a *mixture* of A and B elements. It is, as always, a matter of which element predominates.

The System A society is characterized by strongly demarcated, possibly complex hierarchies within which movement is limited and subject to the fulfilment of precise requirements – the 'examination system' of one kind or another. The System B society is characterized for its part by a relative absence of these features – by a tendency towards levelling out and equalization, a breaking down or reduction of barriers and differences, not a proliferation of these.[51]

System A rules by right – meaning, strictly, right of birth, though the term broadens to take in the right conferred by superior qualities and natural endowment. Thus, in the élite society pure, the blood descendant and only the blood descendant (indeed, only the *senior* blood descendant) of the ruler is the next legitimate ruler. His claim to the title is automatic and undeniable.

This element, certainly, is found in the nevertheless System B society of Christianity. Christ was, for example, claimed to be the lineal descendant of the House of David and the only Son of God. Thus he fulfilled all élitist requirements. He also made such

51. The proliferation of ranks and titles typifies the highly evolved (perhaps degenerated) aristocracies of System A societies.

statements as 'no man cometh unto the Father except by me' (John xiv, 6) – a clearly hierarchical, divine-right attitude. This is later confirmed in the Catholic view of the priest as essential mediator between man and God. Nonetheless, *despite* these elements, the dominant tone of Christianity and the fundament of its mass-appeal is the fact that *all men are equal* in the sight of God. (They are, clearly, *not* equal in the sight of the College of Heralds.)

England of the Victorian era offers an example of an essentially System A society – in which each 'knew his place' and kept it. A much more extreme instance was provided by Nazi Germany. Orthodox armies – not revolutionary armies – are always excellent models of what is under discussion here. Great pains are taken to ensure that differences of position, rank and regiment are on obvious display – and these distinctions are rigidly enforced.

In the System A society B-dominant individuals are found, variously, as an amorphous mass at the bottom – actually, below the bottom – of the hierarchy, or taking refuge in the priesthood, the teaching professions, and so on, as do-gooders, poets, hermits and what have you.

Instances of System B societies (or would-be such) are seen in Buddhism, the Christian movement, Communism and the Kibbutzim or collective settlements of Israel. As noted already Catholicism, for example, has distinct System A overtones – but even so it sees all men as, theoretically, equal. Russian Communism, it may also be objected, has likewise a complicated System-A-type hierarchical structure. Two points can be made here. First, that at least matters such as accidents of birth are not in general a barrier to progress within the Communist hierarchy nor (as far as one can see) is nepotism extensively practised. Second, the use of such terms as 'comrade' and 'brother', plus the total abolition of *hereditary* titles, show at least lip-service to the concept of equality.

It is perhaps the communal settlements of Israel, however, which represent the purest *realization* of System B principles.

There all possessions and clothing, as well as land and produce, are the collective property of the Kibbutz and not of any individual. In these respects the Kibbutz has close parallels with another System B community – the monastery. Additionally, all the work of the settlement, including managerial and administrative functions, is rotated so that all perform all tasks in turn. The present-day hippie commune, again, contains many of these features.

It might well be objected at this point that a great deal of equality (cf. the slogan 'liberty, equality and fraternity') exists in America, allegedly a stronghold of System A – and perhaps more liberty than *actually* exists in Russia. One is, however, discussing here two very different though outwardly similar commodities. In Russia everyone is a follower, in America everybody is boss. Thus at the end of the last century 80% of the American labour force was self-employed – the number since dropping to 37%. The difference can be expressed thus – in the Communist state everybody has no car, or everybody has the same car. In America *everybody has a bigger car than everybody else.*

It is not too sweeping a statement to say that the history of 'Europe', particularly of Western Europe, is the history of the Ego – of the development of the Ego – and of élites. This is seen clearly in the nature of its cultural heroes – Hercules, Siegfried, Beowulf, Alexander, Superman and so on. The chief function of these heroes is the killing of other people, be it indiscriminately or selectively. For, hopefully, obvious reasons, there are very few System B cultural heroes – an outstanding exception being Jesus Christ, together with the lesser figures of the Saints and Martyrs. The function of the System B hero, however, is to get *himself* killed.[52]

52. There are links in these two forms of hero both with Freud's concepts of the life and the death wish, and also on the more biological level (but stil with Freud), where 'the independent individual organism is in opposition with itself in its other capacity as a member of a series of generations', i.e. father versus mother, the assertive versus the submissive or 'sacrificial'.

This last statement is one index of the true nature of the System B hero. The whole concept of the hero is essentially alien to and undesired by the System B ethos. As far as he exists, this hero remains still one of the people. (He is their property, not they his.) He does not actually *wish* to *lead* – leadership is thrust upon him by circumstances or necessity. (So perhaps: 'O my Father, if it be possible, let this cup pass from me: nevertheless, not as I will, but as Thou wilt.' Matthew xxvi, 39.) Under System B there are no individual heroes – no 'cult of personality'. The hero of System B is the People.

For the moment, a final word on the nature of the orthodox standing army, compared with the revolutionary army as seen in the French or the Cuban revolutions. Though there will be more to say in another connection, for the present one would like to suggest that the differences between the two types of armed forces are the differences between classical and romantic art – that is, between Form and Content. The orthodox army has a minimum of passion or emotion – at least, that is to say, it is consciously channelled and held in check by formalization and ritualization, in much the same way that a horse is controlled by bridle and bit. The revolutionary army, on the other hand, has a high charge of emotion or passion, sketchy or ragged discipline and formation, and relatively permeable barriers between leaders and men.

In closing this section, two possibilities may be mentioned. One is that the numbers of B-dominant individuals in a population will probably tend to increase over time. Considering the general record, one finds tentative grounds for assuming that 'Neanderthal' as a variety is more hardy – perhaps more fecund also – than 'Cro-Magnon'. The survival of classic Neanderthal in the inhospitable conditions of the Ice Age, for instance, argues in a general sense for this. If this reasoning is correct one would expect more B-dominant individuals than A-dominant individuals to survive to adulthood. Hence the fact that more females survive birth and childhood than males? This would perhaps hold particularly true under conditions of hardship or deprivation. The peasant or working-class population, as already noted, has frequently – even habitually – lived under just such grim conditions in many societies. It may be that the very conditions which in one sense helped to keep the sheer numbers of peasantry and workers in some sort of bounds also selectively favoured the survival of B-dominants. While one does not wish to seem to be overlooking the many other precipitating and causal factors involved (both social and environmental), it seems possible that the emergence of Communism and Socialism is, in some sense of that term, a 'Neanderthal' backlash.

A question which may or may not have occurred to the reader, but which is in any case a very fair one, is as follows. If System B is, as alleged, dominant in various members of the human community past and present – in the Neanderthaloids, in women, in children and so on – could one not expect to isolate and describe features which are common to all System B-dominants of whatever persuasion? And if not, would that not leave the general theory of this book in a seriously weakened condition?

As agreed, the demand is fair, and the purpose of this section is, hopefully equally fairly, to meet it. Since one of the intentions of the book is not so much to 'win' arguments but either to lay or raise certain ghosts once and for all, the skirmish will be fought over the area least susceptible to opinion and counter-opinion and most susceptible to public, objective measurement – that is, over purely physical and not over behavioural attributes.

However, in passing, let us make one or two comments on the behavioural side. What common *behaviours* possibly characterize all System B-dominants? Dreaming, either expressed as the proportion of dreaming sleep to orthodox sleep or the absolute amount of time spent dreaming, is a probability. It seems that

B-dominants dream more. But how can one know whether the Neanderthaloids and the paleoanthropics dreamt more? There are possible tangential approaches. Do, for example, the Australian aborigines, probably the most paleoanthropic of present-day races, dream more than Western man? Furthermore, we know that Neanderthal had evolved a religion. Do, then, present-day religious or superstitious people dream more than practical or atheistic individuals? Were that established one would have *some* grounds for assuming it to hold for Neanderthal also.

Another possibility is suggested by the word-association data cited in Chapter 5. From that data it seems that children more than adults and women more than men fail (no value-judgement implied here) to come up with a word of 'opposite' meaning when asked to free associate to a stimulus word. In *some* sense, then, children and women seem to perceive the world more wholistically (?), certainly more associatively or Gestalt-wise, than do adult men, who apparently see the universe in terms of (paired) opposites. When considering ancient languages we there found evidence of a tendency to use one and the same word for a concept and its opposite: that is, in a sense, a tendency to perceive as a whole what we today (as far as we see any link at all) perceive as the polarity of opposite qualities – and for which we therefore use two different words. Now, since the practice of using one word both for an object or an attribute and its opposite (a) increases as we go back in time but (b) even in the earliest recorded times is an almost totally vanished process, have we not some grounds for assuming that practice to be of very great antiquity indeed – *reaching back in origin, perhaps, to the paleoanthropics?*

In these tangential, conditional terms, then, one sees again that it *is* possible to argue tentative behavioural parallels among System B-dominants ancient and modern.

However, what I intend to concentrate upon here is the purely *physical* aspect.

In moving into this frame of reference – without wishing to

seem to make excuses – it is hoped that the reader appreciates the risk which this text is voluntarily running. One is putting what is actually an extremely subtle process of psychological interaction at risk to, and on trial by, evidence obtained at the grossest physical level. Nonetheless, a robust hypothesis is rarely wholly destroyed by crude measurement or other weaknesses of experimental design. In the belief that the central hypothesis of this book *is* robust, I am, therefore willing to take the risk.

While it would be interesting to consider *all* the possible correlates of System B-dominance, many of these are far from straightforward, either in their provenance or their actual manifestation – and would require therefore a great deal of prior debate. For such reasons, the list below includes only certain of the more straightforward items. In making statements about the attributes which *are* included, it must also be made quite clear that what is being discussed is *the average or mean result to be obtained from the inspection of very large samples.* One must not expect the proposals necessarily to hold in respect of any given *individual*, both because of the slightness of the effect as such, but more importantly because of the wide ranges of variability which are involved. These result in considerable overlap between samples of what is, effectively, a series of bi-modal distributions. A second very important point is that when one uses the phrase, 'larger than', one means in general *relatively*, and not *absolutely*, larger than. If one says, for example, that a woman's eyes are larger than a man's, one means larger in proportion to other aspects of, or the rest of, the skull in each case.[53] One must sometimes be prepared for appearances to be deceptive.

With this, the following proposals are now offered. The way to read the chart is to link any of the same-numbered items in the extreme right and left-hand columns using any and all of the intervening central statements.

It is proposed that:

---

53. Thus, to labour the point, the head of a cat is smaller than the head of a fly, when considered in terms of head-body ratio.

(1) Women		(1) Men
(2) Children		(2) Adults
(3) South-east 'Europeans'	have larger eyes than	(3) West 'Europeans'
(4) 'Asians'	have larger mouths than	(4) 'Europeans'
(5) 'Africans'	have larger palates than	(5) 'Europeans'
(6) Neanderthaloids	have fuller lips than	(6) Cro-Magnonoids
(7) All paleoanthropics	have less prominent chins and externally less developed man-dibles than	(7) All neanthropics
(8) All and any individuals defined in any way as System B dominants: therefore	have more hair than and are shorter than	(8) All and any individuals defined in any way as System A dominants: therefore
(9) Neurotics		(9) Psychotics

Some of the further items one might have included in this list, but have not by reasons of the various complications involved, will no doubt occur spontaneously to the reader. A few points arising out of the list given urgently need comment.

By larger eyes do we mean larger eye-sockets or simply somehow rather more prominent eyes? Probably both. Probably one means also, for instance, larger irises than, pupils capable of greater expansion than, and still other eye-properties which need not be itemized here.

In that context, however, one is impressed by such expressions as the 'soft-eyed lover' or poet, wide-eyed innocence, 'come-to-bed' eyes, 'eyes to drown in', and so on; and by the fact that women's face make-up seems, whatever else, always to have involved high-lighting the eyes in one way or another – either by painting them directly, or by not doing so in order that they stand out by contrast. The basic intention appears to remain the same even though the method differs. One is impressed, too, by the prominence often given to eyes in tribal totems. On the other (System A) side one encounters phrases like steely-eyed, flinty-eyed, hard-eyed, gimlet-eyed, beady-eyed and so on. In that connection one notes that interest in a subject or attention to an actual object is accompanied by a reduction in the size of the pupils – sometimes, of course, also by a visible narrowing and tightening of the whole eye musculature. It is a fact, too, that the eyes quite literally harden as one grows older – presumably the eye as a result is smaller. These last observations begin to suggest the very complex interactions which seem to be taking place continuously at all levels, within and without the personality.

Larger mouths means larger mouths. Certain other features of the mouth, such as the tongue and teeth, would probably also bear closer inspection. One notes in passing that spinsters are frequently referred to as thin-lipped, as is also disapproval.

The chin and the jaw mandible will be commented upon later in a more appropriate context.

To state that e.g. children have more hair than adult men is a proposition which involves us in some difficulty. After all, adult men have beards, and hair on their chests. Let us first put the matter somewhat out of context and say that the statement in question refers rather to *something* like the fact that a young kitten – or any young mammal – is fluffier than the adult cat. To make a quite specific within-context claim, however, it is very likely that the hair of a group of adolescent or pre-adolescent boys if allowed to grow to its fullest extent would, on average, exceed the average length achievable by a group of adult men, if they similarly allowed their hair to grow to maximum. A further point is that the adult male, even when he does not actually lose his hair, often evidences one version or another of the adult hair line – for example, the 'widow's peak'. A child's hair line, aside from being usually rather less well defined, tends to be straight across and relatively low on the forehead. Hopefully these brief observations on this subject will meet at least *some* of the reader's possible objections.

Straightaway, however, we are faced with further difficulties – now in connection with height. There is no doubt at all that some of the tallest men in the world are African Negroes – not simply on an individual, but on a whole-tribe basis. This difficulty is readily admitted. One can make two points by way of partial defence. First, that *some* occasional traffic between the 'European' and 'African' gene-pools certainly occurred (are the tall tribes Northern rather than Central African?); second, that nonetheless the Negro *average* would still be lower over-all than that of 'Europeans' over-all – among whom are included, for example, white North Americans and Australians.

Is it either dramatic or useful news that children are smaller than adults? Clearly not. The fact that women are shorter than men (all female mammals are smaller than male mammals) is equally self-evident. This, however, is less readily accounted for. In that connection one question is asked. Is the size difference between male and female vertebrates as marked in species where the cerebellum is relatively large – e.g. in birds and fish – as it is in species where the cerebellum is less well-developed? Put another way, does small stature (particularly in males) correlate with a large cerebellum, and vice-versa?

Since, allegedly, all System B groups are shorter than all System A groups, it should be the case that the average height of, for instance, Socialist members of parliament would be less than that of Conservative members of parliament. (However, even if that should be so – and I understand that this has already been demonstrated to be true – there are clearly many other factors affecting the result. The Labour government contains, for instance, considerably more Jews and Welshmen than the Conservative house, the former being, in general shorter than Anglo-Saxons. Equally importantly, many Socialist M.P.s have emerged from economically underprivileged backgrounds. Poor, or merely less adequate, nutrition in the formative years is known to exercise a constraining effect on final height as an adult.)

Rather more free of such considerations are our neurotic and psychotic group.[54] On the argument being here proposed

54. Were the links between these two particular groups and the physical variables in question ultimately to be demonstrated, the central theory of this book would gain enormously in credibility. Or could one imagine mere coincidence reaching this far?

hospitalized psychopaths in particular should prove taller than the national average and, in particular, taller than hospitalized neurotics. (The term 'hospitalized' is important – for while diagnostic methods differ somewhat from hospital to hospital, confining one's attention to the interned population gives at least some focus of standardization to the undertaking.) It would prove an easy matter for researchers to establish whether the average height of all hospitalized male *neurotics* in this country is below that of all hospitalized male *psychotics*, since this information already exists on hospital files. Would, one wonders, the neurotics also show a lower incidence of 'male pattern baldness' than the psychotics? (Doubtless *that* statistic is not recorded!)

A matter not so far considered at all in these pages is that of ears. One recalls, first, that Neanderthal was well-endowed not only with visual, but also with audial cortex. May one assume that above-average powers of hearing would be accompanied by larger ears? It is impossible to be certain, of course, but it seems a reasonable assumption. One notes, with no little satisfaction, that dwarfs, goblins, pixies, and so on are very often drawn with large ears. A well-known Enid Blyton character has a name based on that fact. These are the *kinds* of links we have previously found significant. This is not to suggest that Neanderthal actually had *pointed* ears, but that they may have stood well away from the skull, especially at the top; or, quite simply, they may just have been large. Observing one's fellow-men one notes that large ears appear to give the face a distinctly puckish appearance. Clark Gable provides a fairly good example.[55]

It would *indeed* be interesting if one could show that Labour M.P.s had larger ears than Conservatives, and neurotics larger ears than psychotics. One could hardly explain that away by nutritional or any other such simplistic factors.

55. Is it the case that large ears are more often associated with a short face, and small ears with a long face? The short face is a fully-agreed characteristic of Neanderthal, the long face an equally well-established characteristic of Cro-Magnon.

There is a concluding word to be said, namely about left-handedness, that will bring us to the next chapter. In addition, a renewed word of warning is in order in connection with what has been proposed in this section concerning System B and its manifestations. What we have been discussing, both here and elsewhere, are group generalizations – trends, in some cases very slight trends, detectable in large samples. This trend is not only sometimes entirely undetectable in a given or particular member of the sample – who is nevertheless by general definition a member of that group – but the individual in question may, indeed, show features nearer to the norm of an opposite group. This is due, among other things, to the wide range of variability involved. Neither does the presence of any one characteristic guarantee the presence of any other. An individual might show a bewildering assortment of the anatomical, physiological and personality characteristics that have been discussed. Although the single phenomena have been discussed here in simplex terms, it is clear that their action and interaction must in reality be extremely subtle and extremely complex. Finally, the reader should remember a comment made in an earlier chapter – that System A and System B represent two *universes* of behaviour. Two System B individuals might therefore be totally unlike each other in many respects – perhaps unable even to communicate or to relate to each other at all. There are, of course, certain general underlying principles which underpin all System B behaviour, and others which underpin all System A behaviour. These have been indicated at various points, and will be indicated still further. Were this *not* so, of course, one would be quite unable to define the systems. They would, in short, not exist.

So what of left-handedness? This is an excellent topic to raise at this juncture, in view of what has just been said regarding the confusions and pitfalls which exist in the application of our general theory. For we argue left-handedness as a notable System B phenomenon, but are able to demonstrate it only in a mere 5% of the male, and 3% of the female, population. Worse, much worse, this allegedly *System B* characteristic has a *higher* incidence in a System A group – men – than it has in its appropriate System B group – women. What can be done?

There are suggestions from various sources that the proportion of left-handed individuals in the general population was once considerably higher than it is now. The present book has, additionally, specifically proposed that left-handedness was more associated with the paleoanthropics than with the neanthropics.

The earlier notion that once large-scale or fairly continuous warfare became common, left-handers were at a disadvantage and were thus selectively bred out of the population over a period of many thousands of years, would neatly enough take care of the reduced number of left-handed males surviving. It does not at all account for the still greater drop in the numbers of left-handed women – especially when one expects this characteristic in any case to be sex-linked in favour of women. That fact *alone* should result in a higher incidence of left-handedness in females.

Examining our own young, we find that all our children pass through a stage of ambidexterity, but do not pass through a stage of left-handedness. This establishes quite conclusively that sinistrality was never a *biological* stage passed through by neanthropic man – at least, not by the ancestors of currently surviving neanthropic man.[56] If contrariwise the alleged left-handedness of past ages *was* however biological, and not just a cultural habit

56. At a slight tangent, it can be said here that the fact that our children do not pass through a left-handed stage in their early life, but *do* pass through an ambidextrous stage, taken with the apparent fact that left-handedness was once common, at least among some early peoples, argues strongly that the split between these two types of sapient occurred very, very long ago – that is, at the time when the common ancestor was purely ambidextrous. One has here in fact a piece of persuasive evidence to support the notion of two basic strands of humanity existing side by side over millions of years, as the zoologist suggests. One branch became right-handers, the others became left-handers and some, again, may have remained ambidextrous. This view would give a very real and literal meaning to the mystical or symbolic notion of the 'left-hand path' of the occultists. Once again one would have found an accurate biological statement in what is usually regarded as mere mumbo-jumbo.

of some kind, then where has it gone to now? To argue that this was a purely *paleoanthropic* biological or evolutionary stage, which largely disappears from the scene either by reason of the smallness of the paleoanthropic admixture in the neanthropic ranks, or because right-handedness is genetically dominant when a right-hander mates with a left-hander, or both, would at best only partially answer the question. There remains the matter of the lower incidence of left-handedness among women.

To lose or reduce the incidence of a genetically based attribute in a given population one must literally and actually remove those particular genes from the gene pool. Thus, if one wished to prevent the occurrence of red-headed people, it would be necessary to kill all the red-headed individuals in every generation before they reached the state of sexual maturity (or, after that point, to kill their children also). This is, quite precisely, what the animal-breeder does. If he wishes, for example, to produce a straight-haired variety of a particular dog, he selects from each litter those with the naturally straightest coats, breeding then from these only, and always raising his standard of acceptability as time goes on. Fewer and fewer *curly*-haired animals are produced in each succeeding litter.

How, then, would it be if the last 50,000 years or more have seen various versions of the witch-hunts such as took place in Europe only a few centuries ago? These other witch-hunts might have been organized or casual, thorough or careless, and intermittent or cyclic, rather than continuous. The long-term effects, nonetheless, would be ultimately similar to those achieved by the more thorough-going methods of the breeder. And how if in those ancient witch-hunts, as in our more recent European ones, *more women than men were the victims*? And how, too, if among those so destroyed there was always a higher proportion of left-handers? Would not this suggestion not only account (in part) for the general dearth of left-handers, but in particular for the dearth of left-handed women?

Whatever else, one would imagine also that those so destroyed included a high proportion of the *rather superior* B-dominants –

those, that is, gifted in the 'higher dream-processes' and other forms of Knowledge II.[57] In the diagram which follows, it is the item indicated by the hatched area which one deems to have suffered most from the 'genetic pruning' we have envisaged.

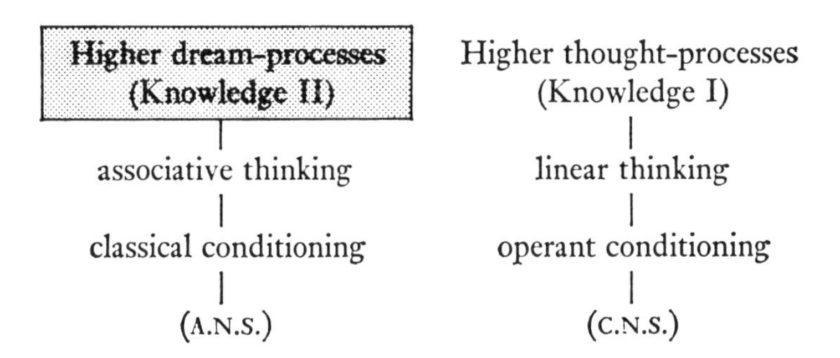

With this we can turn, at last, to a direct examination of the nature of Knowledge II.

57. Here, perhaps, one reason why left-handedness is more common among the mentally retarded.

What one very badly needs is a precise breakdown of the incidence of left-handedness in various social groups and professions – a plaint also voiced by Martin Gardner in *The Ambidextrous Universe*: 'Unfortunately, accurate statistics about the incidence of left-handedness in various professions are hard to come by.'

Part Five:

The Rise to Tyranny of Western Consciousness

The Momentary Universe 11

But how has this reaction come about? Because I threw three small coins in the air and let them fall, roll and come to rest, heads or tails up as the case might be. This odd fact that a reaction that makes sense arises out of a technique seemingly excluding all sense from the outset, is the great achievement of the *I Ching* . . .

The Chinese mind as I see it at work in the *I Ching* seems to be exclusively preoccupied with the chance aspect of events. What we call coincidence seems to be the chief concern of this peculiar mind, and what we worship as causality passes almost unnoticed . . . While the Western mind carefully sifts, weighs, selects, classifies, isolates, the Chinese picture of the moment encompasses everything down to the minutest nonsensical detail, because all of the ingredients make up the observed moment.

This assumption involves a certain curious principle that I have termed synchronicity, a concept that formulates *a point of view diametrically opposed to that of causality* [my italics].

It is a curious fact that such a gifted and intelligent people as the Chinese has never developed what we call science. Our science, however, is based upon the principle of causality, and causality is considered to be an axiomatic truth.

Causal connection is statistically necessary and can therefore be subjected to experiment. Inasmuch as situations are unique and cannot be repeated, experimenting with synchronicity seems to be impossible under ordinary conditions.

C. G. JUNG, Foreword to *I Ching or The Book of Changes.*[1]

What we have called Knowledge II is on view in the Old and New Testaments of the Bible, in the *Upanishads* of ancient

1. Translated from the Chinese into German by Richard Wilhelm, with a Foreword by C. G. Jung and from the German by C. F. Baynes.

India, the Tibetan *Book of the Dead*, the *I Ching* (or *Book of Changes*) of classical China and in many, many other texts. Those mentioned are formalized and codified statements of Knowledge II. Essentially the same principles, however, are also embodied in the utterances of the Delphic Oracle and other auguries; and at the lowest end of the scale in fortune-telling, astrology, crystal-gazing, tea-leaf reading and palmistry. This latter statement will not do very much to warm the critical or sceptical reader to what follows. However, that is a risk we must take.

The predisposition to scepticism may be somewhat tempered by the statements that Schopenhauer, for instance, fully acknowledged his debt to the *Upanishads* and that Leibniz declared himself 'amazed' by the 'purity of structure' of the *I Ching*. C. G. Jung's admiration of this latter work may be discerned from the quotation at the head of this chapter. He said also: 'For more than thirty years I have interested myself in the *I Ching* . . . for it has seemed to me of uncommon significance.'

It is with a clear sense of the difficulties involved that one undertakes to convey anything of the import of even a single one of the works mentioned above in the short space of a part of one chapter – however, one trusts that the limitations of the present book have already been sufficiently admitted. It is the *Book of Changes*, the *I Ching*, which we examine here.

Richard Wilhelm in his Introduction cites four authors of the *Book of Changes*. The first of these, Fu Hsi, is of great interest to us for reasons which follow – as is perhaps the last, Confucius, by reason of his reputation in the West.

Fu Hsi is apparently a legendary figure 'representing the era of hunting and fishing and the invention of cooking'. The suggestion is, in other words, that the first version – that is, some part – of the *Book of Changes* arrives in historical times with an already considerable antiquity behind it. There is in fact good reason to believe that the book existed as an oral tradition long before writing was introduced.[2] Placing the origin of the book in the

2. In general, as already suggested, I am firmly persuaded that we must be prepared to assign a far greater antiquity not merely to the origins of lan-

era of 'hunting and the invention of cooking' would give us the *possibility* of a staggeringly vast history: for, as was shown in the last chapter, man possessed fire, practised cooking and had permanent cave-sites already by 500,000 B.P. (Coincidentally or otherwise, some of the most important sites of *homo erectus* are found in the vicinity of Peking.) However, in fact we have no *particular* justification for placing the origins of the *I Ching* at the beginning of the vast period in question.

Nevertheless, Wilhelm also concurs that

> . . . the linear signs of the *Book of Changes* ascribed to Fu Hsi have been held to be of such antiquity that they antedate historical memory. Moreover, the eight trigrams [of the book] have names that do not occur in any other connection in the Chinese language, and because of this they have even been thought to be of foreign origin. At all events they are not archaic characters . . .

Compare in this connection the unknown origin of the pre-Greek word 'labyrinth'.

Two collections of trigrams[3] are known to have been associated with two dynasties of antiquity, one traditionally dated 2205–1766 B.C. and the other 1766–1150 B.C. (or 4,000 and

---

guage, but also the origins of philosophical thought, than seems currently assumed in the specialist literature.

In terms of *oral* traditions Thor Heyerdahl has described, for example, how certain individuals among the Polynesians are designated as historians. These are taught a detailed history of their people covering several hundred years, which takes days to recite in full. (That material was, of course, subject to steady and continuous growth. The material discussed *here* would grow far more slowly and would not reach these unwieldy proportions. Nor is it suggested that the tradition of the *I Ching* was simply a rote exercise, but rather a living and meaningful body of understanding.) The interminable epic tales of the Germanic, Celtic, and of course, Jewish, and other peoples were also in existence long before writing appeared. For obvious reasons a spoken-word-based culture, however complex or ramified, can leave no trace of itself if the tradition of continuity is once broken.

3. A trigram is any of various combinations of three solid or broken lines, thus ☲ ☳ ☰ ☷. A *hexagram* is any combination of six such lines.

3,500 B.P. approximately). Thus the great age of the *I Ching* is undoubted.

Before attempting to say what the *Book of Changes* is, one further comment may be made concerning its antiquity. The oldest text and interpretations contain a symbolism or ellipsis of so compacted a nature that without the bridging commentaries and elucidations of later sages, such as Confucius, it is unlikely that we today could make much of them. We may liken the old text to the very ancient, collapsed, dwarf stars, where a cubic inch of matter commonly weighs five tons. The items in question here are, as it were, dwarf dreams, (as are, of course, the trigrams and hexagrams themselves) containing a densely compressed mass of observation. Such compacting itself argues a long period of existence. Three short pieces from the text are here given by way of example. One must bear in mind, however, that what one is reading is a translation. Inevitably much is lost in the transfer; nor are these symbolic images, in general, part of our own traditions.

> Return. Success.
> Going out and coming in without error
> Friends come without blame
> To and fro goes the way
> On the seventh day comes return
> It furthers one to have somewhere to go.
> (from 24, Fu/Return)

> If you walk in the middle
> And report to the prince
> He will follow
> It furthers one to be used
> In the renewal of the capital
> (from 42 I/Increase)

> You let your magic tortoise go
> And look at me with the corners of your mouth drooping
> Misfortune.
> (from 27, I/The Corners of the Mouth)

It is easy to dismiss this material as charming nonsense; or as of so obscure or ambiguous a nature as to be valueless.

The *Book of Changes* is in essence a collection of linear signs which, when used as oracles, provide answers to questions asked of the book. The book is, equally, a non-oracular account of the nature of the universe – to this aspect we turn later. In the most ancient times there were probably only two lines – an unbroken line meaning 'yes' and a broken line meaning 'no'.[4] Later the single lines were combined in pairs to yield more differentiation – four possibilities in all instead of two. Later again, a third line was added to each of these combinations to produce the eight basic trigrams that are associated with Fu Hsi.

With the existence of eight trigrams one has reached a degree of complexity that is not only capable of considerable subtlety, but which is becoming hard to conceptualize or hold in the mind in its entirety. It will be realized that, by this time, the basic broken and solid lines no longer had the simple negative and positive qualities of no and yes that they started out with. Indeed, it was probably felt at the time that the eight trigrams encompassed all that was happening in the universe. And as (or because) the trigrammatic signs readily change into each other by the alteration of single lines, so it was felt too that the transitional and constantly changing nature of the universe was reflected.

Each of the eight trigrams in time acquired particular attributes of various kinds. Thus, for instance, they represent a family consisting of a father, mother, three sons and three daughters. These incidentally are in no sense supernatural or godlike beings. They represent *functions* or roles. Further attributes of the trigrams are not of direct interest here – but one will not be surprised (will, perhaps, be pleased) to learn that the image of the father is heaven, and that of the mother earth. The *name* of the father is Creative, and that of the mother Receptive.

4. cf. the use in modern spiritualism of 'one rap for yes, two raps for no'. Whether this is a deliberate borrowing, or a separate manifestation of an archestructural situation is not easy to judge.

The basic system (the eight trigrams) appears for a time to have met the needs of the day-to-day users of the oracle and of the philosophers who were systematically extending man's 'understanding' of the universe with the conceptual tools and conceptual possibilities provided. One can already see clearly how radically such a subjective view of the universe must differ from one that is empirically based.

In the course of further time, nevertheless, the degree of subtlety and complexity offered by the trigrams was felt or found to be inadequate. The eight trigrams were re-combined with one another (at a very early date) to yield the sixty-four hexagrams of the evolved *I Ching*. 'At a very early date' means prior to 4,000 B.P. Not, however, that one needs to imagine the sixty-four hexagrams appearing overnight, least of all as the product of a conscious decision by any one individual. It is likely that the evolution of the full complement of hexagrams was itself an extended process.

The kinds of subtlety and complexity which result or become possible in the fully evolved *I Ching* are now of an order where a man may study his whole life without having grasped more than part of the total concept. Above this there was the practical application of the philosophy in everyday life, its translation into the minutiae of individual existence and its problems and situations – trivial or otherwise, high or low – which was a further universe of study, somewhat resembling the practice of psychoanalysis in our own day.[5]

The foregoing is not in any sense offered as evidence of the *value* or *validity* of the *Book of Changes*. There is nothing to

5. The study and elucidation of a body of concepts evolved over very long periods of time, with the production of commentaries which in further time themselves become added to the permanent mass – leading then to commentaries on the commentaries – is a feature of all religio-philosophies all over the world, equally of Judaism and Christianity – and, yes, in the shorter term, psychoanalysis and Communism. While readily acknowledging once again that this shared feature is not an index of the validity of the *content* of these philosophies and religions, one must insist on the significance of the repeated pattern of activity in these System B products.

prevent one regarding it still as a very complex folly, as a tragic monument perhaps to the wasted energies of a considerable section of humanity over a considerable period.

However, let us revert to the basic concepts of the philosophy of the *I Ching*. The foundation on which the whole ultimately rests and from which it develops is the duality of the broken and the whole line, the 'negative' and the 'positive' – in Chinese, the concepts of *yin* and *yang* respectively.[6]

The primary meaning of '*yin* is 'the cloudy', 'the overcast'. *Yang* means literally 'banners waving in the sun' – that is, something shone upon or bright. How well these terms align themselves with the terms which describe Systems B and A respectively in Chapter 3: dark–light, night–day and so on.[7] Indeed, all these and many other terms are applied to, and inhere in, the concepts of *yin* and *yang*. As pointed out, *yin* and *yang* are also mother and father, female and male.

Moreover, it is certainly no accident that in the habitual phrase *yin* precedes *yang* (the female precedes the male): as we know, System B is older than System A in evolutionary terms, and is the primal part of the personality. In all this we have clear

6. The term 'negative' is better understood in the photographic sense than in the sense of 'no'. The negative of a photograph wholly reverses the shading of the positive or print – it is black where the other is white, white where the former is black. The negative, one must point out, is also the *mirror image* of the positive, which links this concept with the significance and meaning of the mirror, discussed in detail earlier.

7. A glimpse into the subtlety of the *Book of Changes* is contained in the following. When one looks at a mountain, the southern side is the bright side (the one bathed in sunlight) while the north side is dark. When a river is viewed from the bank, however, the *northern* view is bright, by reason of the reflected sunlight, while the southern aspect is the dark one.

One of the things the *I Ching* is saying here is that there are in *human* terms no absolute values for situations: their interpretation, their very initial appearance, depends upon the standpoint from which they are viewed by the subject. In describing subjectivity in this vivid and useful way, the *I Ching* does not then make the mistake of Western psychology of saying that the subjective is therefore valueless (because forever changing). Rather the book says that such change, from the experiential point of view, is what for human beings life and the universe are about.

evidence that we are in System B country. That is, we are observing personality through the eyes, and from the standpoint, of the 'unconscious', and in terms of the intuitive.

In the original text and in the very oldest of the commentaries, however, *yin* and *yang* are not used. Instead the two expressions used are 'the yielding' and 'the firm'. These of course align equally well with our general concepts of System B and System A; for in general the character of the female is yielding and that of the male firm or dominant. One may state the matter still more literally, however. The 'yielding' is also the opening and softening of the vagina in preparation for intercourse, just as the 'firm' is the hardened, erect penis. The text and some commentaries of the book are, I suggest, often most clearly understood in these terms. There is no doubt in my own mind that the oldest layers of the *I Ching* are very close to this primal symbolism.[8] This fact further argues for the great antiquity of the text and philosophy of the *I Ching*.

The *Book of Changes* is both an oracle and a 'Book of Wisdom'. As the latter it attempts a description of life and the universe, and the meaning of the universe. As the former it is a 'visionary power' to be consulted in times of distress or in the face of difficulties of any kind. With this we appear to have entered the area of mumbo-jumbo, ignorant superstition and magical divination. And, indeed, we have.

It is not easy to convey one's own convictions about and experience with the *I Ching* even to an open-minded and neutral observer – let alone to the sceptic. However, for what it may be worth, I offer one example of an experience with the *Book of Changes* which the reader is free to take or leave, as he thinks fit. The situation described by the example will be a familiar one to

8. It is probably less correct to interpret the subsequent development of *yin* and *yang* in this primal way. The later concepts are considerably more evolved. However, it is possible to perceive (somewhat amusingly) a suggestion of the male organ in 'banners waving in the sun'.

The broken — — and solid —— lines are, incidentally, certainly thinkable as stylized renderings of the male and female genitals in a state of readiness.

those who know and have employed the *I Ching* – an instance of the astonishing accuracy which (the devotees of the book agree) is displayed in the *large majority* of applications of the book to a problem.

It is, naturally, permissible to regard both myself as well as many others as misguided even though well-meaning. This is the reader's prerogative.[9]

The original method of consulting the oracle is a somewhat complicated one, involving a series of arithmetical and ritualistic exercises with a large number of yarrow stalks. However, a later and simpler method exists involving the use of only three coins, as described in the opening quotation at the head of this chapter. The purpose of the method in both cases appears to be to make quite certain that the choice of text which is produced is wholly random and completely beyond the control of the subject. One notes, incidentally, that such methods as sticking a pin in a list of numbers or horses' names, or opening a book at random – which are well known methods of producing a chance consultation – are not truly random, since for obvious reasons the beginning and end of the list or book tend to be neglected. The coin method of consulting the *I Ching* involves throwing the coins high in the air and allowing them to bounce and settle at

9. Jung has expressed my own position very clearly in his Foreword to the Wilhelm translation.

'I must confess that I had not been feeling too happy in the course of writing this foreword, for, as a person with a sense of responsibility toward science, I am not in the habit of asserting something I cannot prove or at least present as acceptable to reason. It is a dubious task indeed to try to introduce to a critical modern public a collection of archaic 'magic spells', with the idea of making them more or less acceptable. I have undertaken it because I myself think that there is more to the ancient Chinese way of thinking than meets the eye. But it is embarrassing to me that I must appeal to the goodwill and imagination of the reader, inasmuch as I have to take him into the obscurity of an old-age magic ritual. Unfortunately I am only too well aware of the arguments that can be brought against it. We are not even certain that the ship that is to carry us over the unknown seas has not sprung a leak somewhere. May not the old text be corrupt? Is Wilhelm's translation accurate? Are we not self-deluded in our explanations?'

will. The face side of the coin is *yin* (with a numerical value of 2) and the reverse side is *yang* (with a numerical value of 3).[10] One throws the coins six times in all. The further details of the method may be obtained from the *I Ching* itself.

Now to the example. An acquaintance of the author's had been married for two years. His relationship with his wife was and is a difficult one – both parties being not only very dogmatic, but quite opposite in their views on most matters. Despite their apparently genuine affection, or at least need, for each other, the relationship is in continual danger of foundering, and imposes a great strain on both participants. On the occasion in question the man (John) turned up at the author's flat in an acutely withdrawn and defeated state. Relatively helpless in the face of the situation, as one frequently is, I suggested the use of the *I Ching*. Except by name the book was unknown to John. Without enthusiasm he agreed to the consultation. After I had explained the use of the coins to him, a text reference was obtained. This he read in silence for some time.

Eventually John put the book down. The first words he said tend to be echoed, in one form or another, by most of those who come into contact with the *I Ching* under these conditions. He asked if he could go through the process again, 'to see if the first one was just a fluke'. A second reading on any one occasion, however, is not normally allowed – since moments occur only once, and the reading for that moment had already been issued. This aspect of the matter will be discussed again.

The text and commentary which John afterwards showed me comprised a detailed discussion of the relationship of man and woman, questions of following and leading, the psychological strains imposed on both parties in attempts to manipulate or

10. The reader may be outraged to learn that it probably makes no difference if one treats the 'tails' side as *yin* and the 'head' as *yang* – or if one switches method on different occasions. This is a measure of the alien nature of this way of thinking compared with our Western, scientific approach. One could point out, however, that a truly random process is not made either more or less random by incorporating yet another randomization procedure.

control, and much else of a similar nature. The details are in a sense unimportant. What is important is John's own view that the text, for him, 'exactly described my relationship with Carol'.

No doubt many will find the foregoing singularly unimpressive. Or concede that, while interesting, it proves nothing. I would indeed myself agree that the validity or meaning of the procedure and the experience *can never be finally demonstrated at second hand* – though one *can* recognize someone else's experience as being an example of what one has already experienced oneself. In the first instance conviction can be achieved only by means of one's own personal experiences with the phenomenon. Here one sees that the process involved is wholly and diametrically opposite to the criterion, demanded by modern science, of objective and repeated demonstration of findings under controlled and circumscribed conditions: that these phenomena are, by definition, untestable by such means. Again we revert to Jung's words –

> . . . experimenting with synchronicity seems to be impossible under ordinary conditions. In the *I Ching* the only criterion of the validity of synchronicity is the observer's opinion that the text of the hexagram amounts to a true rendering of his psychic condition.

With this we may again broaden the discussion and undertake a consideration of the meaning of coincidence – or, to use Jung's term – synchronicity.

Jung's Principle of Synchronicity is adopted by the present book to function as what has this far been referred to as the Alternative Principle – that principle which complements the Reality Principle, and stands in the same relation to the unconscious (or sleeping consciousness) as the latter does to waking consciousness.[11]

11. It will be recalled that a self-imposed demand of this book was that the new principle must be able to incorporate Freud's Pleasure Principle as a

Jung has expressed his views on synchronicity a number of times besides in the Foreword to the *I Ching*. His most explicit exposition is found in Volume 8 of his collected works, in a long essay entitled *Synchronicity: An A-Causal Connecting Principle.* One cannot do justice here to the full argument; those interested are referred to Jung's own text. However, one or two brief extracts will convey something of his direction, and will serve to show how Jung's (of course, prior) thought tends to resemble the view taken by this book towards System B and of sleeping consciousness.

I hope it will not be construed as presumption on my part if I make uncommon demands on the open-mindedness and goodwill of the reader. Not only is he expected to plunge into regions of human experience which are dark, dubious and hedged about with prejudice, but the intellectual difficulties are such as the treatment and elucidation of so abstract a subject must nevertheless entail . . . there can be

---

special or subsidiary case. It is in the close link between pleasure and the 'moment' that this proves to be possible. Thus, as we should expect, we find many expressions current both in popular and poetic speech which attest to that link – and which often emphasize the momentary, transient and even contrived quality of, certainly, the high points of pleasure. 'Moment' in the present context must, incidentally, be understood to incorporate not only time, but equally space. Particular configurations of objects or events or a specified environment constitute 'moments'.

And so: 'never the time, the place and the loved one all together'; the 'fleeting pleasures' or the 'magic moments' (of a love affair); the 'moment when my heart stood still'; 'for a moment it was as if time had stopped'; 'it was the moment I had waited for all my life'; 'such moments are all too rare', *et al.*

In a rather more explicit or expanded form the moment (in time and space) is a central concept and image of all romantic literature, and of much poetry. We find it readily in Wordsworth or Dylan Thomas or, for instance, in Masefield's well-known: 'all I ask is a windy day and the white clouds flying, and the flung spray, and the blown spume, and the sea-gulls crying'. It is from a *configuration* of events that the sought for experience emerges and is – subsequently – seen to inhere. While one lacks space to examine the relationship of pleasure and time in detail, two of many possible underlying *physiological* elements may be mentioned in passing.

First, it has been proposed that the human organism may be subject to a

no question of a complete description and explanation of these complicated phenomena, but only an attempt to broach the problem in such a way as to reveal some of its manifold aspects and connections and open up a very obscure field which is philosophically of the greatest importance.

[Synchronicity] is certainly not a knowledge that could be connected with the ego, and hence not a conscious knowledge as we know it, but rather a self-subsistent unconscious knowledge which I would prefer to call 'absolute knowledge'. It is not cognition but, as Leibniz so excellently calls it, a 'perceiving' . . .

If we are correct in this assumption then we must ask ourselves whether there is some other nervous substrate in us, apart from the cerebrum, that can think or perceive . . .

. . . It turns out that bees not only tell their comrades, by means of a peculiar sort of dance, that they have found a feeding-place, but that they also indicate its direction and distance, thus enabling the beginners to fly to it directly. This kind of message is no different in principle from information conveyed by a human being. In the latter case

---

complex pattern of (physiological and psychological) cycles. One obvious instance of such a cycle is the monthly menstruation of women. Many women find this an occasion when they feel particularly depressed or anxious – and in fact more suicides occur among women around this time as compared with the intervals between periods – when women, presumably, are happier. If one imagines that we are all subject to such cycles of various kinds (for which there is some evidence), it will happen that we will function better at the high points and less well at the low points. It will happen too, perhaps, that because these cycles may vary in duration, once in a while the high points (or low points) of several cycles will coincide by chance at the same time – with a corresponding 'unaccountable' feeling of great elation or depression.

A second though related point concerns the question of 'orgasms' of various kinds – in particular the sexual orgasm. Here again one has a (momentary) high point in a, this time, usually self-promoted cycle. On the less literally physical level, a love-affair may go through a series of psychological 'orgasms' – perhaps, however, resulting in permanent, and not transient, changes of feeling – which yield moments or periods of great happiness.

These are some of the possibilities.

The special relationship of pleasure, and its opposite unhappiness, to particular moments in time should therefore be kept in mind as a kind of subsidiary theme – if one may call happiness subsidiary – during the wider discussion of the remainder of this chapter.

we would certainly regard such behaviour as a conscious and intentional act. . . . Nevertheless it would be possible to suppose that in bees the process is unconscious. But that would not help solve the problem, because we are still faced with the fact that the ganglionic system apparently achieves exactly the same result as the cerebral cortex . . .
Thus we are driven to the conclusion that a nervous substrate like the sympathetic system, which is absolutely different from the cerebrospinal system in point of origin and function, can evidently produce thoughts and perceptions just as easily as the latter. What then are we to think of the sympathetic system in vertebrates?

Synchronicity is what is normally termed coincidence. Actually, that is not wholly true. It is likely, I suggest, that there not only is but probably always will be a body of synchronous events that are accurately and fairly described by the word coincidental – which are, that is, completely fortuitous. Within that body of events, however, one proposes that there are other events at first sight perhaps in no way dissimilar but which are nonetheless *not fortuitous*. One may call them meaningful coincidences. To request an accurate definition of meaningful coincidence, and to ask for unimpeachable instances, is precisely to state the problem which now faces us.

I am myself of the opinion that Jung's attempts to demonstrate synchronicity in the essay cited by the application of normal statistical techniques may well not be an advisable approach. To concede at any rate the *normal* use of statistics is to allow the 'enemy' – that is, the sceptic – the choice of weapons. This is an advantage to one's opponents which one should not be overready to concede.[12] Quite apart from that strategic consideration, however, bearing in mind what this text has reiterated over and over again – i.e. that the attributes and functions of System A and System B are diametrically opposed – it is likely that the statistical method is *wholly irrelevant* to the matters in hand. Having said

12. How clever of you (one imagines the sceptic to say) to rule out of court that weapon which most effectively reveals the non-validity of your arguments. To which one feels like replying: how clever of *you* to *rule in* that which gives you the victory before the case ever comes to court.

that, let us then however concede that some *modified* statistical approach might perhaps be of use in these areas. It should certainly not be thought, in any case, that one is refusing to admit the need for validation of one kind or another.

Jung's approach is essentially that of demonstrating that the odds against a particular event happening by chance are so astronomical that an opponent is thereby obliged to concede that something other than chance is at work. This is also essentially the approach of J. B. Rhine and others in their investigations of extra-sensory perception (the acquisition of information other than by means of the known senses). The method is, indeed, the standard practice in most psychological research.

At least in the kinds of uncontrolled situations described by Jung, however, one has, in the first place, no reliable means of determining what constitutes a chance result. As a few examples taken from the behaviour of a roulette wheel will later show, odds of a very high order against their happening do not prevent the occurrence of certain remarkable incidents.

Jung cites various estimates by mathematicians and philosophers of the against-chance odds in certain cases of pre-cognition (the knowing that an event is going to happen before it happens) and of various allegedly meaningful coincidences. Xavier Dariex estimated odds of 4,114,545 to one against precognitions of death being due to chance. C. Flamarion reckoned the odds for a particular instance of a similar phenomenon as being 804,622,222 to one. Jung himself, in the course of the essay on synchronicity, describes a detailed investigation of his own into the relationship of personal zodiac signs (as prescribed by astrology) and certain outcomes. He produces results with an against-chance outcome of 10,000,000 to one.

Finally Jung reports (without giving a precise textual reference) that in one of J. B. Rhine's published experiments in extra-sensory perception, which involved guessing ahead of time the order of a shuffled pack of twenty-five cards (made up of five sets of four designs), one subject correctly guessed the order of all twenty-five cards. The odds against this happening by chance are 298,023,223,876,953,125 to one.

The present author's opinion is that in the face of this last dumbfounding result (assuming that it was indeed obtained under the experimental conditions specified) the psychological establishment must, on its own terms, strike its colours and concede that extra-sensory perception, whatever that may be, occurs. In other areas of psychological research it should be pointed out, the psychologist accepts odds of twenty to one against chance as indicating a significant result. That many psychologists have not conceded the victory is, in my own as well as many other people's view, because they have some personal need to reject these findings.[13] In the areas under discussion psychologists seem in general to be looking for negative, not positive, results. It would appear that Western scientists generally are as resistant to (as prejudiced against?) the idea of a non-causal relationship of events as the East has traditionally been to the idea of causality. The evidence suggests that it is as hard for a scientist to accept non-causality (synchronicity) as it is, say, for a Papuan native to accept that illness is caused by organisms in the bloodstream. Our language really requires two words for 'prejudice' – one for System A rigidity, and one for System B's intransigence.

Let us now turn to a consideration of certain behaviours of the roulette wheel. In the course of this consideration the question of the *meaning* of results is introduced.

A roulette wheel, when a zero is included, has on its circumference thirty-seven divisions numbered randomly from nought to thirty-six inclusive. The odds against any specified number

13. My personal view, in any case, is that the phenomenon of synchronicity *cannot* be demonstrated objectively. It can only be experienced *subjectively*. Only a convincing personal experience – whatever that may entail in the individual case, and certainly it will differ from individual to individual – will 'convert' the unbeliever to a condition of belief or acceptance. The religious terms used here are used advisedly. Whereas in (objective) science things must be seen to be believed, in the realm of System B things, perhaps, must be believed to be seen.

being produced on any one spin of the wheel are thirty-six to one.

One evening I was present in a casino when in the course of normal play the *croupière* spun the number '4' four times in succession. I was astonished by the occurrence and asked the *croupière* if she was not also surprised. She replied, indifferently, that she had spun one number four times on a previous occasion, and was now waiting for the occurrence of the same number five times. The odds against a particular number being produced are, as we know, thirty-six to one. The odds against the same number occurring twice in succession are 36^2 to one; and of the same number turning up four times 36^4 to one – that is, 1,679,616 to one. That is, the odds against the event which I saw were over one and a half million to one.[14]

If the *croupière* ever realizes her ambition of spinning the same number five times in succession – not apparently an impossible task in view of her two spins of four, and in fact only thirty-six times more unlikely than either of those – she will have overcome odds of 60,466,176 to one, more than sixty million to one.[15]

Odds of these magnitudes are well into the range of the odds discussed by Jung. Moreover, we have been dealing here with a precisely circumscribed set of events, where we *know* the odds; whereas in some at least of the cases cited by Jung one is drawing on a virtual infinity of events and possibilities of combination. In such a continual flood of possibilities may one not expect the apparently miraculous to occur once in a while?[16]

14. A point worth making is that we actually have very little conception of how large a number a million really is. For example, if one had spent £1 a day every day since the birth of Christ, one would not yet have spent three quarters of a million pounds.

15. I have since been informed that the correct odds against a number occurring five times in succession in roulette are $(37^5)-1$ to one. This raises the figure quoted to the seventy millions. I am also informed that Monte Carlo has a record of two occasions of a number occurring *six* times – odds against of a quarter of a million million to one.

16. J. B. Rhine's work has the merit of being able to state the possibilities involved with great accuracy. However, he loses very considerably in other directions.

Let us revert to the example produced by the roulette wheel. As it happened it was the number *four* which I observed to come up *four* times in succession. Already the situation begins to assume a 'meaningful' aspect. One is inevitably more impressed by the number four occurring four times than one would be by the number thirty-six occurring four times. Statistically, however, the two events have equal value. Now let us suppose an additional and by no means unlikely circumstance. Suppose that on the morning of the events at the casino I had received a letter from my car insurance company stating that as I had made four claims in the last four years, the company felt reluctantly compelled to decline to offer me further cover – might one not have retired instantly to a life of prayer in the face of such cosmic karma? I have given the example a frivolous content. It could easily be given a more serious one – say four friends killed in an avalanche on the same day, which one heard about subsequently.

My feeling about many of the reported 'miracles' of coincidence is that they are, alas, nothing of the kind – that is, that they are indeed coincidences but not in any sense miraculous.[17] We look now at one more, certainly non-meaningful coincidence, and then compare this with further of the possibly meaningful examples cited by Jung.

While waiting for a friend, I observe that a leaf falls from a tree above and floats in at the window of a stationary bus standing at a bus-stop. There are two questions (1) What are the odds against that particular leaf having entered that particular window of that particular bus? (2) What does it mean? The answer to the second question is easy: nothing. The answer to the first is difficult. It is actually impossible to give any real answer. What are the parameters of our estimate to be? The number of leaves

17. One might add finally, incidentally, that the odds against *any* four numbers whatsoever occurring in sequence on a roulette wheel (7,3,0,19/5,17,31,18, etc.) are likewise (36^4) —1 to one – and that this 'miracle' occurs continuously. One could stand by the wheel for the rest of one's life without ever seeing the same sequence of any four numbers twice.

at present on the tree – or the number of leaves the tree has had in its lifetime? What is our time-scale? The life of the bus, or the fraction of its life that it has spent on this route? Must we take account, too, of all the other buses on the route, or all the trees next to all the bus-stops? We flounder in a morass of pointless speculations and arbitrary decisions.

Two further examples recorded by Jung follow.

A woman took some photographs of her son in 1914 and left the roll of film to be developed in Strasbourg. In the interim the First World War broke out and she never collected the film. In 1916, however, she bought a roll of film in Frankfurt to take pictures of her newly born daughter. When these pictures were developed they were found to be double-exposed – under the pictures of her daughter were the pictures taken of her son in 1914. Somehow the old film had not been developed and had got back into circulation with a batch of new films.

In the second of these examples Jung tells how at a critical point in the psychoanalysis of one of his patients, the woman concerned was telling him of a dream in which she was given a golden scarab. As she spoke Jung heard a tapping noise at the window behind him. A flying insect was knocking itself against the pane. He opened the window to admit the creature which proved to be a scarabaeid beetle (a rose-chafer). Contrary to its usual habits it had apparently at that moment felt an urge to get into a darkened room. For Jung, one of the important features of this event was that in ancient Egypt the scarab was one of the symbols of rebirth.

What is one to make of these extraordinary occurrences? For myself a criterion involving meaning, involving a dynamic, involving some kind of outcome, is required. Having said that, it becomes necessary to distinguish between *real* meaning and seeming but actually meaningless meaning (as in the case of the number four occurring four times). This is no easy task.

In my own opinion the case of the photographic film is astonishing but meaningless. That is to say, there was no 'outcome' to the events, no confirmation of anything. Such was the case with the four fours on the roulette wheel; such is the case with any sequence of any four numbers on the roulette wheel; such was the case with the leaf falling from the tree into the window of the bus. Granted the odds against the film turning up as it did were phenomenal – but quite how phenomenal we have in any case no real means of knowing. (Once again, what would be one's parameters?) But the *odds as such* are not really the point.

Jung's second example is less easy to dismiss; for there is an interpretive element in it and a kind of self-justification. There *appears* to be something more to the event than a mere juxtaposition of circumstances which have no dynamic relationship and are joined by no necessity. Jung and the patient were after all *engaged* in something – the analysis – and were *seeking* an outcome. This outcome happened to be precisely a psychological rebirth. In short the situation had a certain structure or purposiveness – not the case in any of the examples discussed in the last few pages.[18]

Precisely this purposive and structural element is what we find in the *I Ching*. When consulted in the manner prescribed, it apparently yields an answer which is appropriate to the question and to the condition of the inquirer. The 'coincidence' has not here arisen *fortuitously* – it has been *produced*. Nor has it arisen in the context of an infinity of possibilities. The scope of the text of the *I Ching* is large, but by no means infinite.

The *I Ching* is probably as near as we will ever get in this area to the controlled experiment of the physical sciences. To produce a 'miraculous' – that is, an extremely unusual – outcome *on demand* is of course precisely what the 'believer' would so much

18. A hypothetical instance of a wholly meaningful or dynamically integrated coincidence, for this book, might be as follows. A man who has lived ten years in a house is suddenly and unaccountably dissatisfied with it. He makes hurried arrangements to sell it (even at a very poor price), cancelling his planned holiday to do so. He moves into a much less pleasant house elsewhere in the town. A few days later an aeroplane crashes into his former house, destroying it completely and killing several people. This, certainly, is a hypothetical example. The Society for Psychical Research, however, has many actual, substantiated events of this type on its files.

like to be able to offer; and this is what the sceptic and the scientist insist on. One is, of course, not denying anyone the right to ask for some kind of evidence. But one does request the scientist to approach the matter with far less than his usual rigidity – or, if that statement is pejorative, with other than his normally reasonable criteria. For it seems that the very intention to 'test out' such claims as these effectively destroys the circumstances, the essential ambience, in which these phenomena occur.

Once again the sceptic may exclaim: 'How convenient! Whatever I specify as a test or criterion you claim disrupts the phenomena. You agree that it is not unreasonable to ask for evidence, but refuse to accept the conditions or the outcome of a test case.'

The matter is actually worse than that. In place of objective evidence one is even proposing to reintroduce the criterion of *subjective conviction* – the very criterion that, for example, sent so many unfortunate wretches unjustifiably to the stake.

To enter these questions in any further detail, however, would lead us too far from the central argument of this book. One can reiterate this much. First, by reason of the concept of the 'moment' that never occurs twice, whatever else one may ask, one can hardly in this area insist on the repeatability in the sense understood in the physical sciences. ('A man can never step into the same stream twice.') Second, that there may well be a fundamental difference between the artificially contrived moment (for experimental purposes) and the moment which arises from the 'chance', and perhaps once and for all, combination of a possibly astronomically large number of circumstances – and involves in particular a state of mind.[19]

19. Thus the 'scientifically respectable' experiments of J. B. Rhine on extra-sensory phenomena are, in my opinion, fundamentally misconceived. Astonishingly, however, some evidence is produced (and gains then in significance from the fact that it is the outcome of an attempt to obtain such evidence). It seems, for example, to have been established that individuals tend to perform better on their first testing than on subsequent occasions; and that those who believe in the existence of the phenomena tend to score

With this very brief examination of a System B text and its rationale, or irrationale, in some detail, we again widen the discussion to embrace the whole of this type of literature and the knowledge which it incorporates. One has made the briefest of attacks on the question of the *validity* of the claims to value of these writings. This attempt is now abandoned in favour of the further description simply of the general *nature* of Knowledge II.

Scientific knowledge (Knowledge I) is a product of the Ego. It is the most evolved of the ego's attempts to understand, control and dominate its environment. These attempts have their origins in the wish of the Ego to stay alive – in pursuit of which aim it will readily take life from others. The basic function of the Ego is to ensure *the survival of the individual.* All Ego functions are, in the last analysis, directed to this end.

Knowledge II is the product of the Self. As we have repeatedly observed, the functioning and the attributes of the Self always take an opposite form from those of the Ego. Thus the Self submits itself to the environment. It is not essentially concerned with staying alive (much less with taking life from others) but with giving life – that is, among other things, of reproducing itself. The basic role of the Self is to ensure *the survival of the species.* For this reason the role of the Self is frequently a sacrificial one.

---

higher than those who do not. (One also has the impression that women score higher than men.) These differences are usually described – particularly by the opposition – as statistically insignificant or statistically unreliable.

In the present author's view, a modification is required of the statistical method applied to these results (one which, for example, ignored magnitude in favour purely of the direction of results, or which established empirical criteria of magnitude independent of those in general use among psychologists). Above all one must break away from the concept of narrow repeatability.

For reasons which have been sufficiently discussed elsewhere, the Ego is concerned primarily with the external universe, and mirrors many attributes of that universe in its own structure and modes of thought. Thus the Ego is concerned with cause and effect – with serial or sequential events. Therefore, too, its thought is essentially linear and its models and schemas of all kinds show (cannot help but show) progression, development and continuity. Our conscious concept of time is derived wholly from the serial nature of events – from the observation that events take place one after the other. Without some form of change indeed we would have not only no measure of time, but no concept of it at all.

The Self is concerned with the 'inner universe' and the internal environment. It incorporates and reveals many features of that environment in the structure and mode of its thought. Change, certainly, occurs in the inner as well as in the outer world. There are, however, significant differences. First, the rate of change in the inner world is a variable, not a fixed quantity; second, there are periods when the rate of change is slow enough to be considered or felt as stasis (during deep sleep, perhaps). Moreover, of those changes which do occur, many are not linear but cyclic. Such a sequence of events returns continually to its point of origin.[20]

These considerations enable us in particular to comprehend the nature of time as understood and expressed by System B. The concept of the moment (i.e. a stasis, of varying duration) has already been mooted earlier and will be examined further. There is additionally the notion of timelessness – of being outside time – which is a notable feature of many religio-philosophies

20. At one time homeostatic theories of psychology were popular. These posited that organisms constantly sought to attain or re-attain equilibrium. The notion worked reasonably well when autonomic functions alone were considered. It did not, however, very happily accommodate for instance the exploratory and stimulus-seeking behaviour of organisms, and such theories are no longer common. Such theories, however, are very relevant to much System B behaviour.

(the eternal truths, and so on). Timelessness may perhaps best be thought of as the 'moment which lasts for ever' – the condition of *permanent* stasis. It is in their conception and handling of time that System A and System B constructs differ perhaps most widely, and by means of which they may most satisfactorily be identified.

Before pursuing these notions of time, there are other less specific points to make in connection with the general nature of Knowledge II. Unlike Knowledge I, it takes what are tantamount to two major forms. It can exist in more or less its own right (as in dreams) and in the absence of normal consciousness – or it may be found more (or less) integrally combined with normal consciousness. In other words, Knowledge II can be genuinely unconscious or it may be manifest in waking consciousness – and then either as a help or a hindrance to the affairs of consciousness – a partner, as it were, or an opponent – a guest or an intruder. Thus there arises the question both of the style and of the extent of the 'intrusion'.[21]

21. Several points may be settled here concerning the unconscious as an intruder. The medieval concept of 'raising the Devil' is defined now as the wilful calling of the unconscious into consciousness, to the point where conscious (System A) control is lost, or is in grave danger of being lost. A state of having lost conscious control while still awake is a state of 'possession'. Apart from magical and ritualistic methods, the 'Devil' is also raised by the use of alcohol, drugs, self-induced or other-induced trance (as respectively in mediumship and hypnosis), and so on. We need to be clear that there are many, many different manifestations of the unconscious in consciousness, which fact no doubt accounts for the wide variety of demons and devils which the occult literature describes. William Burroughs and Allen Ginsberg in the book *The Yage Letters*, Timothy Leary in *The Politics of Ecstasy* and many other writers have detailed some of the differing states of possession which different mixtures and strengths of hallucinogenic drugs produce. In Scotland, too, it is popularly claimed that whisky makes one more aggressive drunk – 'fighting drunk' – than does, say, gin – and this is presumably one reason why the former is preferred by men.

While the varying states of possession largely still await description and examination, one can distinguish at the macro-level (without exhausting the possibilities) aggressive, sexual, and 'Knowledge II' or 'visionary' arousal. However, these states are unstable and readily pass one into another. An

That System B can *productively* predominate in waking consciousness we see in the stable Ba personality; and what is produced is Knowledge II. When System A predominates, one has a stable Ab personality and the product is Knowledge I.

Knowledge II can be further conveniently considered under four headings:

(1) Self-awareness.
(2) The control of the autonomic nervous system.
(3) Accounts of the universe.
(4) 'Real' understanding of, and possible control over, objective phenomena.

(1) *Self-awareness.* It would seem that System B, by means which are far from clear to us, is able to perceive both itself and the nervous system as a whole. In terms of symbolic imagery it describes itself – the unconscious – variously as a woman, a dark cave, the moon, an animal, a misshapen dwarf, and so on. This book has, of course, suggested that some of these descriptions are

---

uncontrolled variable in every instance appears to be the state of the mind of the subject at the moment prior to entering the 'possessed' condition.

In all ages both legend and factual reportage have warned of the dangers of having anything to do with System B (while agreeing on the treasures guarded by it). It is a playing with fire. At any moment the raised 'Devil' may turn the tables, even when you thought you had him safely within bounds. This is true even in the context of the B1 state of Christianity.

Thus one is never *sure* what type of experience one will have or of its intensity; nor of one's degree of control either in the short or the long run. Here now one is not *simply* talking of the advanced alcoholic or the heroin main-liner, nor of the occasional bad (or very bad) 'trip' of the acid user. One is referring also to the soft-drug user, the religious zealot and even the simple seeker after philosophical truth.

In general and more objective terms it would appear that the barrier, never in any case a total barrier, between the two major Systems A and B tends to become more permeable, or lower, the more often it is crossed. This notion would take in the gradual deterioration of control over the manifestations of the possessed state with time. Although this is perhaps a description rather than an explanation, the widespread symbolism of the Wall and the Barrier, which one cannot discuss here, suggests that there may be in reality some tangible, physiological or neurological obstruction between the systems. That

a literal approximation to the size, location and mode of functioning of the cerebellum. Consciousness is described in masculine images – a God, a King, the sun, a mighty wizard, the vault of heaven filled with angels. The imagery of consciousness – that is, System B's imagery of consciousness – is perhaps less varied and more repetitious than the imagery of the unconscious. In general, System B's view of consciousness suggests approval – or perhaps awe is a better word. However, the negative properties of System A *are* also mentioned – the Midas touch which kills the living, the power to shatter the mirror or to enslave with a name. (It is interesting that both Midas and the sun turn things to gold.[22])

One has already also suggested that certain features of legend are further descriptions of aspects of the nervous system and its functioning – the Journey, the mirror-image, the Cross, and so on.

---

nuns, priests and holy men apart from their contact with the spirit of God receive also visits from 'devils' is well documented. Spiritualist mediums, too, are always on their guard against being possessed by an evil spirit, and sometimes refuse to 'sit' for this reason. That soft-drug users develop a sense of paranoia in excess of the real and justified fear of arrest is the opinion of a number of 'smokers' I have encountered professionally and of several psychologist colleagues. That, additionally, to take a final example, habitual concentration on the 'eternal truths' and the mysteries of life renders one – progressively more? – unfitted for dealing with the real, everyday, objective world – as well as more indifferent to it – is perhaps reflected in the way of life and the standards of living in the East and the Far East.

However, the total *denial* of System B is no better as a solution and in fact is almost certainly worse. This view will be justified in Chapter 12 in a discussion of the 'neurotic' and the 'psychotic' society. A situation whereby the two systems have equal, but not more than equal, expression seems the only viable solution.

22. It seems that the very large majority, if not all, of symbolic representations of consciousness are masculine – e.g. Icarus, Midas, Cerberus, gods, angels, wizards. To an extent, of course, one is perhaps begging the question that is, one may be choosing as representative of consciousness precisely those symbols which happen to be masculine. In fairness to this book, however, it must be admitted that one *has* suggested at various points the reasons which lead to those choices.

(2) *The control of the autonomic system.* This might seem a mere tautology. Obviously the control of the autonomic system is under the control of the autonomic system. However, this is not what is meant. What is meant is the *purposive* control of that physiological system which argues for an 'intelligence' – the appreciation of outcomes and results. One is thinking of such matters as hysterical paralysis, hysterical pregnancy, hysterical amnesia and so on; of the various forms and motivated nature of psychosomatic illness in general; of the production of, for instance, bleeding stigmata, and so on, on the hands of religious 'neurotics'; and, on the positive side, of faith-healing in its various forms, and fringe medicine as a whole.

(3) *Description of the universe.* Most religio-philosophies contain a detailed (and non-scientific) description and 'explanation' of the universe – of its purpose and meaning, the principles under which it operates, its origins and ultimate destination, and so on. System B sees and is much taken, for obvious reasons, with the cycles it observes in the universe – that of the seasons, of the sun rising and setting each day and so forth – and with moments of various kinds (of which more is said again subsequently). The question of whether and to what extent these cycles and moments are 'really there' is partly dealt with under the next heading.

(4) *Understanding and control of objective phenomena.* It is here that we come to the very vexed question of the reality of the System B view. Many might be prepared to concede that there *is* an unconscious, that it is autonomous and self-aware, that it does in some strange way perceive and describe not only itself, but also the external universe. They might even go on to agree that in terms of human relationships and the nature and structure of *society*, System B has a genuine contribution to make – one perhaps that System A cannot provide. Such agreement one might obtain. There is nothing wrong with this as far as it goes.

However, the real crux is whether or not System B has anything meaningful or useful to say about the *non-human*, the physical, the objective universe. And perhaps in addition, or instead, whether it has some appreciation of, say, the forces in the universe which cause life to come into being in the first place, and which may by their very nature be partly or wholly inaccessible to System A. To repeat the first suggestion, however – has System B any direct, *objectively* useful understanding of the external universe? Can it even *affect that universe* in some way?

This latter question is of course one of the banes and bugbears of Western science. One of the ghostly possibilities which still skirt (or surround?) the academic-scientific establishment, and so far refuse to finally lie down and die, once and for all.

The *I Ching* is a powerful weapon in the armoury of the dissidents. It alone must give us cause to refuse, for the moment, to consider these issues in any way settled. On the score of the practical and tangible, it is right to mention also the Chinese method of acupuncture in the treatment of disease. This is not faith-healing, and is moreover now practised in a number of clinics in Western Europe. Its philosophical rationale is that of the *I Ching*.

Finally, the phenomenon of psycho-kinesis (the movement of objects without physical agency) – which is allegedly behind poltergeist happenings, which can apparently be produced by the so-called physical mediums of the spiritualists and has been given a statistical clearance certificate by J. B. Rhine – offers further possibilities. The discussion of this matter is not a direct concern of this book. However, quite why a religious movement dedicated to *spiritualism* should have such a direct and apparently paradoxical connection with physical matter is a question that does not usually seem to be asked.[23]

At this point we can revert to the question of time, first to the moment then to the cycle, as displayed in expressions of Knowledge II.

Astrology is centrally founded on the concept of the moment – in particular, on the moment of birth. Although for general

23. Another instance, perhaps, of the gods with feet of clay, of the link between the spiritual and the physical noted elsewhere.

purposes the day of one's birth will give the astrologer something to go on, for a really accurate horoscope he requires not only the hour but the very minute of that occurrence. According to astrology it is in the constellation of planets pertaining at the moment of birth that one's character and fate may be discerned.[24] Thereafter, in later life, particular groupings and regroupings of the planets will provide moments favourable and unfavourable to various kinds of action on the part of someone.

In the general context of astrology one need hardly recall the role of the bright star in the birth of Christ. Christianity, however, is not so much concerned with the moment of birth as with the moment of *re*birth – which is both the moment of conversion (or dedication) to Christ; and the moment of death – that is, earthly death – and the condition of the soul at that time. In one sense it matters little what has preceded these moments. All past crimes and failings are wiped out.

The moment is an extremely important aspect also of magic. The whole of the magic ritual builds up to and is designed to procure the moment, when the devil appears or the magic works. One is also speaking here very much of the physical moment in space. The pentangle must be correctly drawn, the items of ritual correctly positioned and correctly compounded. Normal time and geographical place (e.g. midnight, the graveyard, the crossroads) may additionally play an important part. Similar insistence, certainly on ritual and to an extent on ingredients, is seen for example in Catholicism. For a dying man to repent while alone is no guarantee of salvation. Without a consecrated priest present at that moment to shrive him and give him absolution, his soul is in peril. In the *magic* ritual there are also, however, strong elements both of the cycle and the orgasm.

24. It is interesting that many words in common use originate from the study of the stars and planets. Constellation itself of course means a collection of stars (from Latin-Greek *stella*, a star). Consider and consideration mean originally 'to observe the stars' (Latin *sidus*, a star). Disaster means literally 'against the stars' or 'ill-starred' (Latin-Greek *aster*, a star).

Outside black magic, though of course linked to it, are the moments of sympathetic or natural magic. The compelling moments of nature must have been considerably more impressive for primitive man than they are for us. The middle of the night – actual midnight – has already been mentioned, and the middle of the day also seems special to us, the highest point of the sun. So do midsummer and midwinter day, and the midpoints in the year between these – the equinoxes, the points where day and night are of equal length. Certain unusual – and occasionally disastrous – movements of the tide and runnings of fish and so forth are also observable on these occasions.

The role of the moment in augury and fortune-telling has already been discussed. What is of additional interest is that in connection with the moment of consultation many different peoples in many parts of the world have independently evolved, in essence, the same method – that of a-causal randomization. Thus one consults, in various times and places, the wheeling of a flight of birds, the pattern of an animal's entrails in the dust, or that of a collection of bones, the random fall of tarot or playing cards, the tea-leaves in a cup. With the *I Ching* we use yarrow stalks or coins. Further, the pattern of the stars and planets in astrology and the lines on the palm of a hand in palmistry may, too, in one sense be regarded as examples of the same procedure – though here non-random factors, conceivably, also play a part. As already mooted above, one of the purposes of the randomized consulting procedure appears to be to remove the element of normal causality (and/or of simple human influence). A further purpose may be to focus the attention without giving the conscious – and interfering – mind anything to work on. Is it not as if consciousness is given something to occupy it uselessly while the real work is going on? Compare here the legend of appeasing Cerberus, the guardian of the Underworld, with music or with drugged cakes so that one may pass. Cerberus would here represent consciousness.

Finally, there is the question of moments in human relationships, not in the sense discussed earlier, but in the sense of knowing when to say what to whom. This is an intuitive form of knowledge sadly decayed in the West. The fact remains, however, that a word 'in the right place at the right time' is very much a reality. A remark made to a person at a crux in his affairs can have an effect it would not have had before or ever will again. This position is widely appreciated in psychoanalysis, and some of the art of being a good analyst is to know precisely when to give a patient a particular insight. In the majority of cases such moments cannot be forced and if they do not arise that, in general, must be that. Outside psychoanalysis the concept of the moment is a living, day-to-day reality, paradoxically, perhaps only for the salesman – the good salesman – who knows when to edge his customer further towards the moment of purchase and when to refrain. *Is* this possibly one reason why Easterners make better merchants – if they do?

The notion of endless time (as suggested, perhaps the moment which lasts forever – the wish often made by lovers) is similarly widespread in System B literatures and practices. Sometimes one has the variant of a vast cycle, usually repeated endlessly.[25]

In Christianity we have world without end, eternal life, 'for ever and ever amen', 'from everlasting to everlasting', immortal, unchanging, 'a thousand ages in thy sight are like an evening gone' and so on – the concepts of eternity and infinity. As in the case of the moment, eternal time is here also translated into other psychological modalities – 'boundless grace', 'infinite wisdom' and 'undying love'.

The element of eternity is also found widely distributed in legend and, in particular, romantic literature. Thus, as cited above, one always swears to love 'for ever'. This concept has brought forth some of the most superb romantic poems ('but thy eternal summer shall not fade'). In the fairy-story the survivors, usually the lovers, live happily *ever after*. The general setting of fairy-stories is equally timeless – 'once upon a time'.

25. So, in the view of astrology, the full life of the world is a Great Year of some 36,000 years, divided into twelve periods each related to a sign of the zodiac.

Vampires and such beings, unless interfered with, also live forever.

In another area, it is of considerable interest that Adolf Hitler felt motivated to set a *finite* time limit to his (System A) empire – the Thousand Year Reich. Perhaps a long time by conscious standards, but nothing at all to System B. On the other side we have an extremely popular film in Russia (which the Russian Government has made some half-a-dozen times, presumably to accommodate changes in its attitudes to certain political figures) called *Lenin Lives Forever*. Communism, as is typical of the System B product, is as eternal as it is inevitably correct. Marx viewed the *capitalist* system as excessively finite, replete with the seeds of its own destruction.

The foregoing comments concerning time and the formulations of System B are, it will be appreciated, not in any sense offered as a *validation* of these formulations. That is another matter again. It is a question here only of 'by their fruits ye shall know them'. One has principally been offering further evidence of the postulated unity of System B views – and, conversely, of System A views. To repeat an earlier analogy, just as all varieties of Negroes, despite their many within-group differences, have that in common which makes them Negroes and not caucasoids – so the within-system similarities of the products of each of the two major systems are greater than any within-system differences; and these within-system differences are, in any case, of a lesser order than the between-system divergence.

Some further comments are in order here on the relationship of women to time; and on the functions of which women are capable in their own right – that is, in terms other than as pale, ill-functioning copies of men.

The stories of woman's inability to be ready on time and propensity to be late for appointments (and, worse, not to appear at all) – their general difficulties with time-schedules of all kinds

– are legion. These stories may or may not be true – or if true, exaggerated. The interest, perhaps, centres on the fact that such legends exist at all.

If women *do* have some kind of difficulty with normal time – that is, time-on-the-clock, System A time – this would agree well with the view that women are more governed by System B – which has, as we have shown, a very different sense of, and approach to, time. If the cycles-and-moments view of time derives from the functioning and structure of the autonomic system, and since women are more ruled by their autonomic systems (from the physiological level upward), as seems to be the case, then one would expect a woman's view of time to have to do with the moment and the cycle – to be in other words of the System B variety.

A woman late for an appointment is frequently angry at being told she is late. This may be because she knows only too well that she is late. At other times, however, she seems not so much angry as puzzled or bewildered at the charge. It is almost as if she is saying something like 'how can I be late when I am here at the time I should be (am ready to be) here?' or 'I am sure I left in plenty of time'. In these sentences *late* means 'according to the clock' while *time* means 'subjective or System B time'. Subjectively she is on time, objectively she is late.[26]

26. Does this difficulty in coping with objective time, if the case, have any link with the alleged unpunctuality of the Negro, the Indian, the Mexican and other System B dominants?

The suggested view of woman's time-block gains tangential support from other considerations. It can happen, for example, that a man discovers that a woman visitor who has called unexpectedly – who has, perhaps, in the more normal course of events refused to sleep with him – is carrying a toothbrush or other overnight things in her handbag. When confronted with this evidence she may well deny having had any intention of sleeping with him. An obvious explanation for her conduct would be to suggest that for reasons of social conditioning she does not wish to appear to have been taking the initiative. A less obvious explanation – which one nevertheless sometimes hears – runs as follows: 'I didn't know if I was going to sleep with you, but in case I did, I wanted to be ready'. To the male mind this appears to be either double-think, or simple stupidity.

Turning to the second matter, the question of what women can undertake in their own right, we consider briefly the nature of cultural heroes, and in particular the nature of the female of that species.

The vast majority of cultural heroes are men, as we saw earlier. Can one, indeed, speak at all of a cultural 'heroine' and if so who is she?

First one needs to exclude all women who are obviously substitute men, such as Brunhild, Boedicea or the Amazons. These leave one with very few other notable women. Joan of Arc? Nurse Cavell? The female Pharaohs of ancient Egypt? – never an administrative post, incidentally. The Earth-Mother? the priestess of the Mysteries? The witch?

Are not all these women and all these functions actually most *nearly* summarized in the concept of the Witch?

An answer to such illogical and rationally indefensible stands can be found perhaps in the concept of the moment as discussed in this chapter. That is to say, a girl will make love to a man if the moment is right, but not if it is not, and will do so – or not – largely regardless of what she may have said before. A woman may not necessarily be able to verbalize what constitutes for her the right or 'psychological' moment.

While in such moments there may be constant ingredients, they are probably not predictable in the sense that the scientific method understands prediction. Men, consciously or unconsciously, often attempt to engineer the right moment. Hence the 'soft lights and sweet music' syndrome, or the ride in the sports car, or the meal in the expensive restaurant, or whatever. These formulae may appear to work on a number of occasions. One says 'appear to' because such evidence as one obtains suggests that the reasons for success are often other than those one imagines. Thus when the male, enraged by the refusal of a girl to sleep with him this week, demands to know why she slept with him last week, she may reply variously (and perhaps infinitely) because you looked so sad, because I didn't know you, because I had a letter that morning from a former lover, because I didn't have a letter that morning from a former lover, and so on. Attempts to chart these reefs – by the male – are and have been usually such signal failures that he has taken refuge in global statements about the unpredictability, unreliability, cussedness, stupidity, coquettishness, and the many other unflattering and supposed attributes of woman. Like most System A statements about System B these are unfair and unjustified, to the extent that they arise out of inappropriate terms of reference.

It is undeniable that there are certain apparent contradictions involved. Some of this ground has already in fact been covered elsewhere in these pages. One has there shown that the differences between these various females are not as great as at first appears to be the case. Witches, though spinsters who destroy babies and hate marriage, are not in fact enemies of sex – and are in a sense wives to the Devil. The priestess of the Mysteries was no stranger to sex – nuns are brides of Christ – Mary bore God's baby – and so on. There *is* admittedly a range of behaviour here – revealing various differences at various points – but the spectrum is not really discontinuous.

Turning to the present day, we find that it is women far more than men who read horoscopes, have fortunes told and are superstitious. One notes, too, that most palmists and crystal gazers are women (though perhaps more men are astrologers). In respect of spiritualism, a phenomenon of the ninteenth and twentieth centuries, my impression (confirmed also by officials of the movement whose opinion I sought) is that (1) more women than men are mediums, and (2) more women than men are *good* mediums.

The trend seems to be inescapable. Women appear to be 'naturally' drawn to these areas. The present book has argued and does argue that the vast majority of women are fighting a wholly lost cause in attempting to join the System A bandwagon – and should instead set about creating its opposite number, in terms of System B.

One is not proposing, though no doubt I will be misquoted in certain quarters as having proposed it, that women should spend their time organizing Black Masses in Cornwall or dancing naked round a desecrated altar at midnight. I am, however, proposing that they should devote themselves to the areas and activities in which their true interests and their talents seem to lie. In some instances, as we know, these areas overlap or are adjacent to certain of the areas in which men are occupied – writing, acting and so on.

In general I suggest that women should stop trying to justify themselves to men (and most of all stop trying to act like men) and set about justifying themselves to themselves – a task they have hitherto almost wholly neglected.

To the extent that the exponents of astrology, palmistry, clairvoyance, and to some extent religion, *have* attempted to justify themselves to System A they have become trivialized and are a laughing-stock. Women as such have suffered precisely the same trivialization by accepting the terms of justification laid down by System A (i.e. by men). We do not as yet *know* what real possibilities, if any, exist in Knowledge II – there are, however, what one might call promising indications. Similarly, we do not *know* what women are really capable of.[27]

One is not at all trying to say that, for instance, religion is necessarily right about certain matters, or even any matters. What one is saying is that System A is *necessarily wrong* in the approach it has made to such areas of human activity. One is, certainly, only too anxious to separate the 'real' (whatever that turns out to be) from the false in these matters, the essential from the accidental, to introduce whatever stands up to examination into the accepted body of human knowledge. But this cannot be done in the terms laid down by the supporters of System A – that is, by well-qualified, but literal, half-wits.

27. Often one glimpses behind the charade of the so-called emancipated career woman the lonely, dispossessed female eunuch of the male harem. As when a woman, who is an Honours graduate of London University and is employed as a full-time technical translator by a very large industrial engineering concern, recently asked me: 'What *is* the speed of heat?'

The Genetic Prisoner 12

This is the manner, then, in which the hunting ape took on the role of a lethal carnivore and changed his primate ways accordingly. I have suggested that they were basic biological changes rather than mere cultural ones, and that the new species changed genetically in this way. You may consider this an unjustified assumption. You may feel – such is the power of cultural indoctrination – that the modifications could easily have been made by training and the development of new traditions. I doubt this.
DESMOND MORRIS, *The Naked Ape*

Perhaps the correct way of viewing the whole subject would be to look at the inheritance of every character whatever as the rule, and non-inheritance as the anomaly.
CHARLES DARWIN, *The Origin of Species*

Zoologists have recently come under heavy fire, notably from psychologists, for their attempts to explain (complex) human behaviour in terms of relatively simple patterns of inherited reactions – for, in other words, their *reductionist* approach to this subject. The psychologists who mount this criticism are themselves the first to protest when the reductionist nature of their *own* investigations is pointed out. Perhaps we have here an example of projection.

Aside from the reductionist aspect, another element seems involved. That is, everyone, psychologists included, pays ready, even eager, lip-service to the notion that man is descended from and actually is an animal. And yet whenever anyone (say, Freud) actually attempts to indicate specific components of man's animal nature, or to speak more precisely of the relationship of our day-to-day life to it, as in the case of Desmond Morris, outrage or scorn is still the usual reaction. Why is that?

Why did Zeus, in the myth of Lycaon already quoted, fly into an insane rage when his divinity was questioned? Surely it can only have been because he was so desperately unsure of that divinity. Desymbolized, the Zeus of the myth is man, and his 'divinity' our humanity. Lycaon is what – Neanderthal? Certainly, in whatever specific shape or form, our own not-so-distant ancestry.[1]

It can be appreciated that it is not especially flattering for us to have it shown that much of our nature is 'animal' – or, alternatively, mindless. It is not pleasant to have it demonstrated that a good deal of what we like to think of as the exercise of free choice among a range of possibilities is merely some pre-determined response, automatically called forth by a particular, perhaps wholly fortuitous, situation. That, in short, this response is in essence no better than, and little different from, the unthinking actions of ants, barn-yard hens, and sheep.

Yet on the other hand there is no doubt that we do not escape from such 'behavioural prisons', if they exist, by pretending that they do not exist. On the contrary, by recognizing and only by recognizing which parts of our behaviour *are* merely automatic (meaning deriving from the past history of our species) can we begin to exercise any control over what must otherwise control us. Even with this, the most 'insulting' aspect of the affair has not been indicated. I will be suggesting that a good deal even of what we prize as noble or 'divine' in our natures is in fact mindless and reflex in character – though I have, hopefully, already shown myself to be a rabid 'anti-reductionist'. In the context of this particular chapter I would specifically like to go on record as being amongst those who believe that zoologists are often guilty of much reductionism in respect of human behaviour. Nevertheless, when the reductionist element has been so to speak subtracted from the zoologist's contribution, a great deal of value remains.

With these brief remarks by way of prologue, we turn to a consideration of the contributions of zoology to the understanding of human behaviour and human personality.

A branch of that science termed ethology, meaning the study

1. One suggests in fact the myth of Lycaon to be yet another, compacted account of Cro-Magnon's meeting with classic Neanderthal. One suspects the latter of cannibalism, as it happens.

of organisms with particular reference to their environment, has in recent years produced a wealth of data which, from the standpoint of the present book, throws much useful light on the origins of System A – on a primal area in fact, apparently to some extent shared with System B. This area is nevertheless demonstrably the domain of the former.

Much of the subject matter of ethology is what formerly went under the title of 'instinct'. There is actually no harm in retaining this somewhat fallen term, provided one is clear about what it does and does not mean. With the word 'instinct' one refers to acts, the execution of which does not involve or only partly involves, learning and experience: to acts, that is, with an inherited, predetermined or genetic component. The present chapter is not, however, concerned with all of such behaviour. Excluded, for instance, is the single, reflex action such as the involuntary eye-blink or the 'startle' response. What we are here principally concerned with is continuous, patterned behaviour or chains of behaviour – such as the unlearned ability of birds to build nests, of bees to take up a precise role in their community, or of parent animals to look after their young. Already remarkable though these behaviours are, when one considers that they are not learned, they are eclipsed by still others yet more remarkable. Some migrating birds, for example, have hereditary mechanisms for orienting themselves by the stars, coupled with a time sense for relating the various constellations to the geography of the earth at every time and season.

Instinctual behaviours may be recognized in several ways. First, they are shown by the whole species, not simply by individual members of the species (though this is, of course, not to say that the strength of the response cannot vary from individual to individual). The particular behaviours are usually species-specific – that is, they are not found in other species, unless in those obviously closely related or ancestral to the species in question. Thus a few behaviours are genus-specific, order-specific or even class-specific. So the instinctive sucking response of the infant is a behaviour feature of the young of all mammals. The clinging behaviour of the new-born ape, however, though universal through the order of primates, is not common to other mammals.

The fact that much instinctive behaviour is nevertheless *species*-specific means not only that it must have been acquired relatively recently – during the life-time of the species in fact – but means that *we are not obliged to show that behaviours proposed as instinctive in man are also present in other animals* – a quite incorrect assumption frequently made in this area.

By no means all instinctual behaviours are evident at birth – though they are of course genetically encoded somewhere within the organism, awaiting their moment. Obviously, instinctive maternal behaviour will not in general appear until the organism has reached adult status and produced offspring. Again, any migratory instinct a bird possesses cannot be manifested until it has learned to fly properly. And so on.

A central hallmark of instinctual behaviour is its unthought-out, unthinking nature – its unmodifiability in the face of changed circumstances or repeated failure – its essentially 'ritualistic' character, its rigidity and, if one will, stupidity. Let us be clear precisely what this means. The behaviour itself is frequently complex, subtle and highly evolved – to us sometimes even of considerable aesthetic appeal. It is in the application of the behaviour, the value-in-the-circumstances, where the derogatory epithets just used are seen to be fully justified. Instances are given below.

It is convenient to divide instinctive behaviour into two component elements – the factors which produce the behaviour, to use the technical term, the *releasers* of the behaviour; and the behaviour as such.

A great deal of research has shown that instinctive responses and response-chains appear under quite precise conditions – not randomly or *in vacuo*. The term 'releaser' embraces all forms of stimulus which produce the instinctual response. Thus the releaser may be a particular colour, a particular shape or size, a noise, a movement, a non-movement, a territorial area – or a combination of any of these. It may be, then, animate or inanimate, including weather conditions, temperature, and so on; or it may be a hormone or other internal state of the organism itself.

So (to take some examples in no particular order) newly hatched chicks instinctively peck at small objects or spots – something around the size of a small seed. This object is the releaser for the pecking response. Experience *then* rapidly teaches the chicks which objects are edible. (While the behavioural response as such is undoubtedly inherited, it is, subsequently as we see, often additionally shaped and refined by learning and individual experience.)

A robin will instinctively attack the red breast of another robin which appears in his territory during the nesting season. Other birds, including young robins which have not yet developed a red breast, are ignored. Newly hatched fledglings instinctively open their mouths to be fed whenever the mother arrives back at the nest. The open mouths of the young birds are themselves releasers which cause the mother to begin the feeding operation. Young primates, as indicated, instinctively cling to their mothers' fur.

The young of most game birds will instinctively run for cover or respond in other self-protective ways if the shape shown in Figure 25 (which resembles the silhouette of a hawk) is moved from left to right above their heads. If it is turned round, or moved the other way (the shape then is no longer seen or experienced as a hawk – if anything it is a long-necked goose), or held stationary, the game chicks do not respond.

A female virgin rat injected with the hormone prolactin begins to carry out nest-building and other maternal activities. The hormone is normally produced by the rat's autonomic system during pregnancy.

The catalogue is endless.

Many of these behaviours in their appropriate context bear the stamp of reasoned action, of apparent wisdom, of the shrewd appraisal of situations; in short, of intelligence. Such a view could hardly be further from the truth.

If the breast of a black crow is painted red, the robin described will attack this bird furiously, although in reality it represents no threat of any kind. Similarly, even when hatched in an incubator, the fledglings described earlier will open their mouths *whenever, and as often as, their nest or the surface on which they are placed is vibrated.* These youngsters have never even seen their mother, let alone been fed by her.

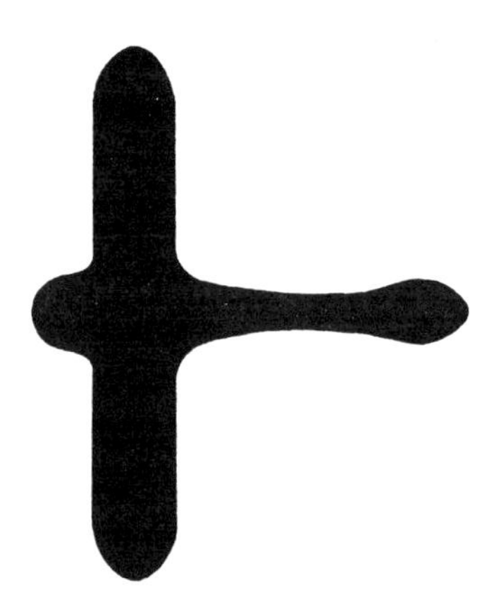

Figure 25.[2]

In species of animals which are incapable of identifying their own young, but which nevertheless 'retrieve' youngsters which have wandered from the nest, a mother will retrieve fifty and more of 'her' babies if these are provided by the experimenter, adding them to the struggling heap under which the nest is now buried. Certain caterpillars, again, will instinctively follow a fellow caterpillar if that second caterpillar should start moving. If a chain of such followers is built up and led round in a circle until the leader is following the last in line, the circular march continues until the caterpillars are completely exhausted. In a colony of mice kept by myself a cage containing a new mother and her litter was accidentally invaded by the mother of that

2. *The Study of Instinct*, O.U.P.

mother. When discovered, the 'grandmother' had eaten her way half through the litter, to the complete indifference of the daughter. The daughter had, apparently, no instinctive objections to the presence of her mother or to what she was doing, nor the other mouse any instinctive inhibitions concerning her 'grandchildren'.

From these very few examples one sees how thoughtless, how lacking in comprehension, how unmodifiable in the light of altered circumstances such inherited behaviour always is. One sees also that it is as inexorably geared to life under *normal* circumstances, as in altered circumstances it is inevitably geared to death. Konrad Lorenz's book, mentioned earlier, gives many further striking instances of this death-process.

A majority of instinctive responses are linked to quite specific releasers. Thus for the robin, as also for the male stickleback fish, the colour red by itself (on a piece of paper, say) is a sufficient releaser for the behaviour of attack. So too any object with a swollen 'belly' will trigger off the male stickleback's courtship ritual. There is here, of course, a response-strength gradient connected with the gradient of stimulus generalization. Nonetheless, the releaser as such is in these particular cases clearly specified. Some other responses, however, are still more open-ended. So a bird or rat will use an enormous variety of materials for nest-building, if traditional materials are not available.

There are yet other cases, however, where the releaser is not specified in advance at all.[3] It is something like a credit-card. The showing of it – even once – establishes credit-worthiness permanently in a variety of buying situations. Slightly oversimplifying the position, the organism for its part has a particular (again inherited) tendency to respond – and whatever appears on cue releases the response. But, much more import-

3. Actually, this is a slight mis-statement. It would perhaps be more accurate to say, in the cases we are to discuss, that the releaser itself is so specific a stimulus that its influence cannot be cancelled out or masked by the presence of no matter how many other even *hopelessly* inappropriate stimuli.

antly, from *then on only* that particular stimulus-object will release the response. A special term, imprinting, has been coined to describe this phenomenon.

Imprinting was first experimentally demonstrated with Mallard ducklings – although instances of it had been observed and reported over many centuries in various animals, without its precise nature being understood. Maximally between thirteen and sixteen hours after hatching, and less strongly for several hours on either side of this period, a Mallard duckling will follow any *moving* object in its range of vision. The *movement* is the releaser. Thereafter the youngster acts as if this object or animal were its mother. Once formed, the attachment is almost impossible to break. In addition, *when the duckling becomes adult*, it will court and attempt to mate with these objects, or animals, and only with them.

Under normal conditions, of course, the moving object is the female parent of the duckling and all goes well. As exhaustive experiment has shown, however, a duckling will imprint as readily also on human beings, hens, footballs, cardboard-boxes on wheels and so on, as on its mother. The later sex-life of the duck is, of course, a disaster.

This extremely precise form of imprinting is found in only a few species. Somewhat less precise forms are, however, extremely common, so that imprinting is found to some extent also in primates.

Two further aspects of instinctive behaviour are of interest for our present purposes: (1) so-called pecking orders and (2) the ritualization accompanying such activities as aggression and courtship.

The term 'pecking order' has been adopted for that particular behaviour because it is most clearly displayed in the domestic hen. It is, however, widespread in other species.

The hens of a flock establish among themselves a hierarchy of dominance. The top hen of this hierarchy has the right to peck all other females, who are not allowed to peck her in return. The second-to-top hen may peck all others below her, to whom she however is sacrosanct – and so on down the line. The top hen has the right to feed ahead of the rest as well as having other privileges; these are enjoyed in turn down the line. The instinctive nature of this arrangement is indicated, for instance, by the fact that it rarely has to be enforced by trial of strength; and by the fact that injections of hormone into a particular hen (leading to altered appearance and behaviour) are able to procure an adjustment in the hierarchy. Social structures of this same kind are observed in lions, primates and other animals.

Any sort of fight to a finish between members of a species would tend to be inimical to the survival of that species. Many animals therefore have in the course of evolutionary time substituted forms of ritual combat for actual combat. The two opponents go through the motions of attack, displaying to each other the calibre of their weapons (by e.g. baring their teeth) and offering demonstrations of how they would deal with an opponent (by stamping on or pawing the ground, tearing at grass and so on). After a little of this the two protagonists can form a reasonable idea of how an actual contest might turn out, and the weaker usually and sensibly retires. To express his surrender and obtain a pass out, *this* one now goes through certain other rituals – perhaps offers his unprotected stomach or flank to the teeth of the stronger animal, or lowers his head, and sometimes generally acts in the manner of the young of the species, signalling in effect that he is only a baby. The victor is usually unable to take advantage of these opportunities for attack – he is prevented instinctively from doing so, by 'negative releasers' or inhibitors. So the vanquished animal may withdraw unharmed.

Courtship rituals have a different purpose, yet, quite often, modified aggression rituals are in evidence in them. On reflection this is perhaps not so surprising. Any approach by another animal of whatever sex is clearly best interpreted as a potential attack, just to be on the safe side.

When a female approaches a male her task is to ensure that she does not get attacked. The male approaching the female must convey for his part that his intentions are not aggressive. Thus both animals may make some of the submission gestures of the vanquished (lowering the head[4], acting in a babyish manner): sometimes mating birds feed each other in the manner of a parent feeding the young – or by turning aside the head (usually the main weapon of attack) show that there is no intention of using it. In addition both partners display sexual releasers to the intended – glorious plumage, coloured patches of skin, swollen parts of the body, etc., and also certain ritualistic movements – the kind of 'dances' which the Walt Disney nature films have frequently presented. Occasionally during the courtship ritual, however, whether because the female is producing too low a volume of submissive behaviour, or whatever, the male will suddenly perceive the female as an intruder and enemy. At this point he will attack her viciously, with a view to driving her away.

In closing this short review, with its emphasis on the extent to which these behaviours are not *thought out*, one would like to make the point here that these actions may be nevertheless thought *of* or *felt*, that is, may be *psychologically experienced* by the organisms concerned. Though in one sense we can only speculate on the precise nature of this mental experience, that a psychological state accompanies the responses would seem at least likely. In human beings, in any case, we have some direct access to such states.

Before moving on to consider what human behaviours might qualify in any sense as instinctive behaviour, some general observations by C. T. Morgan and E. Stellar are of particular interest. But this is to understate: their comments are more by way of a major revelation, from our present point of view. The extracts which follow are found in their *Physiological Psychology* in an, as usual, lucid chapter, on mating behaviour. The authors

4. Compare the human practices of offering one's sword to the victor, of removing the helmet and bowing the head, and so on.

are discussing the relationship of these instinctual behaviours to particular areas of the brain – both cortical and subcortical:

> Since most aspects of sexual behaviour survive decerebrations [removal of the cerebrum] above the level of the hypothalamus one would not expect the cortex to contribute very much to the integration of behaviour. That is how the experiments on *females* turn out, e.g. the estrual [on-heat] pattern of the cat – which is quite striking – survives complete decortication. Such is not the case in *male* animals, however. In most male animals that have been studied, extensive cortical lesions definitely interfere with arousal of sexual behaviour . . .
>
> Despite complete decortication, the *female* rat, rabbit, cat and dog readily copulate. In most gross respects the behaviour pattern is quite normal but the organization of the pattern is impaired in some of its finer aspects. On this score we know most about female rats and there are two points of interest. . . . Second, any tendency the female has to take the initiative in instigating copulation is eliminated. Decorticate female rats fail, furthermore, to show any of the masculine responses such as mounting, which occasionally appear in the intact female. Apparently, then, the cortex in infraprimate [below-primate] females is not essential for the arousal and satisfactory execution of the mating pattern. It does, however, play some role . . . in the extent to which the female will take an active part in initiating sexual activity.
>
> In male animals the position is somewhat different. . . . F. A. Beach found that removal of small portions of the cortex up to about 20 per cent does not abolish copulatory behaviour no matter where the lesion is . . . Cortical lesions involving more than about 60 to 75 per cent of the cortex, however, entirely eliminated sexual behaviour, and even large doses of sex hormones could not bring it back . . .
>
> Cortical lesions have more pronounced effects upon male animals higher than the rat in the phylogenetic scale.

It will be seen how tellingly the above supports the general position that this text has taken up – namely that males are more influenced by the cortex (System A, the central nervous system) and females by subcortical and extra-cortical structures (System B, the autonomic nervous system).[5]

5. That many philosophers, psychoanalysts and other 'verbal psychologists' are so unaware of the support offered them in standard physiological

Having said this, one must add certain riders. First, Morgan and Stellar are speaking specifically of mating and sexual behaviours. One does not know in respect of what other behaviours, and to what extent, it may also be true to say that the male is more cortically, the female more subcortically, directed. Second, the writers are speaking principally of infraprimates – that is, of animals phylogenetically below the primates in evolutionary terms. They specifically caution that: 'no systematic work has been done on the effect of cortical lesions on mating behaviour in primates'. Since no data is available even for primates we must be doubly careful of extrapolating, without reservations, findings obtained in respect of infraprimates on to man. Third, the writers are discussing observed behaviour only. One cannot unhesitatingly assume that what is happening behaviourally is reflected in mental or psychological terms as a state of mind in the animals' heads.[6]

These provisos, nonetheless, do not entirely remove our right to take this evidence as supportive of the general position of this book.

There do exist relatively well-agreed examples of instinctive behaviours in human beings. It has been shown, for example, that the smiling of babies is a reflex, automatic behaviour released by certain features of an adult human face. Not only will a parody of a human face painted on a mask elicit as much smiling from a baby as does its mother's (and anybody else's) face, but

---

text-books is an indictment both of our educational system and other social structures which encourage such compartmentalization of activity.

6. As already mentioned, we cannot absolutely tell by direct means whether any psychological events are taking place in the minds of these animals at all, let alone of what kind, during the behaviours in question. However, it is a far more unwarranted assumption to believe that *nothing* is happening in psychological or mental terms than it is to assume that *something* is happening.

experiments have established that the actual releaser is simply the eyes of the face. Thus a baby will smile readily at two black circles drawn on a sheet of white paper. In this respect the baby does not differ from the robin or the caterpillars, and its smiling behaviour (which the parent interprets as fond recognition) is quite mindless, in that sense. The baby's smile, in turn, releases in us instinctive feelings of affection and protectiveness.

Some aspects of patriotism (though this is purely my own personal view) appear also to be a form of imprinting. There are, undoubtedly, many levels to this phenomenon. But certainly the behaviour shows the typical mindlessness which is associated with instinctive responses. Thus the very large majority of French youngsters come to think that France, the Egyptian boy that Egypt, the English boy that England, and so on, is the best and finest country in the world. They conveniently and *en masse* come to love *its* fields, mountains and rivers above all others. We can hypothesize that at some point in our evolutionary history, perhaps 'when men first began to multiply on the face of the earth', that those who clung most tenaciously to their homelands, say, in times of famine or in the face of would-be invaders, survived best: just as in the case of the ducklings, where over time it so happened that those who followed the parent most diligently most frequently escaped death, so that this characteristic became genetically fixed.

Turning again to less controversial instances of instinctive behaviour in human beings, one can observe frequently, both as an onlooker and in one's own relationships with the opposite sex, that a woman will behave babyishly (that is, talk in a silly voice, or a tiny voice, use baby words, pout, look upward from lowered eyes etc.) towards her boyfriend or husband in order, apparently, to get her own way or to calm his anger. Sometimes, however, the behaviour merely 'takes place' without any idea of an immediate advantage being involved. One can argue, then, that some instinctive mechanism, such as we see in lower organisms, is at work here. One proposes that the woman is instinctively and unconsciously, rather than consciously, reverting to an earlier child-like style of behaviour in order either to release or to inhibit certain behaviours in her partner.

A number of general points made by ethologists and psychologists can make it further possible for us to accept as instinctive in origin still other behaviours, not usually thought of in these terms.

Tinbergen has drawn attention to the fact that in the course of evolution the 'fixed action patterns' we are discussing often become motivationally independent of the situations in which they were originally evolved. This process he terms 'emancipation'. In addition to migrating from the situation which gave them birth, these actions and responses tend to become progressively more exaggerated and progressively more ritualized – the function of the exaggeration, at least, being to reduce the risk of 'misunderstandings'. For the following example of emancipation I am indebted to Desmond Morris's *The Naked Ape*.

When chimpanzees groom each other, a procedure which both parties find enjoyable, the groomer eats any small organisms found on the groomee. In eating them he or she makes lip-smacking noises. Out of this original situation has arisen the further situation that whenever one chimp wishes to approach another, for whatever reason, he will make the lip-smacking noises of the grooming ceremony in an ostentatious manner. This signals to the chimp who is being approached that the visitor does not have aggressive intentions – but on the contrary is the bearer of pleasantness. Thus lip-smacking is the equivalent of the human smile. Indeed smiling among humans is *itself* an excellent example of such migrated behaviour – which has moved from the original baby-and-mother situation into society at large.

A second point is that as we ascend the phylogenetic scale – that is, the more highly evolved the organism – the greater becomes the variety, particularly of external stimuli, which can elicit the instinctive behaviour in question. This is to say both that the situations in which it is used tend to become *more*, not less, numerous, and the stimulus-qualities of the releaser less and less specific, more and more open-ended.

One further aspect of this general broadening of the obviously instinctual situation into what we can begin perhaps to call the instinctually derived or instinctually based situation, is seen in the fact that, again as we ascend the phylogenetic scale, the hormonal or autonomic aspects of arousal – which as we partly know and partly assume are, nevertheless, always stronger in the female than the male at all stages of the evolutionary ladder – become less crucial and eventually virtually expendable. That is to say, the hormonal arousers can and still do perform their functions but they are not now *sine qua non*s; they are sufficient, but not necessary, cause. Thus the mere *presence* of young *can* in time arouse maternal behaviour in a virgin female rat or other animal, and in the absence of appropriate hormones in the blood stream. Although such a virgin animal at first shows little or no interest, after some time spent together with the young, the female will often begin to perform normal maternal functions.

It is generally agreed by psychologists that man is the animal *least* affected by hormonal activity (which is by no means to say, however, that this influence is non-existent) and most affected by stimuli – and by the widest range of stimuli – in the external environment.

It is seen from this general exposition that we are paradoxically, but legitimately, free to impute an instinctual basis to a *wider* spectrum of human actions than would be possible if we were considering lower organisms! Whereas, too, in lower animals instinctive behaviour implies *narrowly stereotyped* behaviour, this implication is *less* justified in respect of instinctively based human responses. In human beings we can accept a wider scatter of what arouses the behaviour, and a wider divergence of the form in which it is expressed, without foregoing the possibility of an underlying unthinking and instinctive basis to that behaviour.

The mistake of ethologists and zoologists, such as Desmond Morris – with whose views this book has in principle no quarrel –

has been, in my opinion, to examine human behaviour on too granular or fine-grain a basis. Thus it is, again in my own view, a mistake to consider the actual hand-shake of Western man, or the raised open palm of the Red Indian and other such details, as being *in themselves* instinctive. The precise form of these actions is, one ventures to suggest, not instinctive. It is not accidental, certainly – having been chosen or having evolved for ascertainable, and probably largely conscious, though not necessarily wholly conscious, reasons. What one has to do here is to widen the spectrum of one's considerations – to, as it were, open the viewfinder to maximum. What is then noted is that *all peoples have some form of greeting ceremony*. And this, really, is the clue. The (ungrammatical) question to ask is 'what behaviours do all peoples have some form of?'

Some of the answers, apart from a greeting ceremony, are: a farewell ceremony; a courtship ritual or period; a marrying ceremony; a burying ceremony; a visiting ceremony. This list is of course by no means exhaustive. What one then further notes here is that all these procedures can be further grouped under two headings: (1) that of approaching someone (of entering somebody's presence); (2) that of going away from someone (leaving somebody's presence). These two occasions, as we know from our brief consideration of animal behaviour, are the two danger periods or danger situations. How to get close to another animal without being killed or damaged; how to go away from another animal without being killed or damaged.

To go into these matters over deeply is, unfortunately, not the purpose of the present book. We have actually other reasons for examining instinctive behaviour. However, one could suggest that a great many, if not all, human social manners have their very real origin in a very real desire not to be killed. Having said that, let us then also say that, in respect of their instinctive content, these are *residual* behaviours – just as the amount of *real* danger present in this area is also residual. As a species and from the evolutionary standpoint we appear to be on the point of abandoning such behaviours altogether. For the moment, however, we have not done this, and the instinctive urges though diminished are still vestigially present. They *still* exercise a considerable shaping force on the nature of our society and our conduct. The new book by the ethologist Irenäus Eibl-Eibesfeldt, *Love and Hate*, has recently given impressive support to this view.

We will consider shortly what our feelings and the feelings of others are when we do and do not yield to these ancient urgings. First let us point out that the response and counter response have, for the most part, long ceased to be *actions* and have become wholly verbalized. Thus when we meet a friend we do not necessarily shake hands or even nod: we instead tend to *say* one or other of a dozen meaningless phrases – hello, how are you, fancy meeting you, and so on. These are the tiny but apparently still not wholly dispensable remnants of once lengthy and highly ritualized manoeuvres. One says 'apparently indispensable' because they do, after all, persist. Without them we still feel somehow naked or undefended. Consider, for example, that when we meet a *stranger* as opposed to a friend – even through an introduction – the ritual is suddenly more important. The vigorous handshake, the warm smile, the momentary burst of heartiness from nowhere, the sudden wish to be liked – are these not the psychological accompaniments of some involuntary push from within?

If these views still seem somewhat debatable, let us consider further situations stripped of these apparent trivia.

If a stranger approaches one in the street with the formula words 'Excuse me, can you tell me . . .', or 'I wonder if you could direct me . . .', the chances are one will not be startled – depending, however, also to an extent on other factors such as the angle and speed of approach. These various circumlocutions and introductory phrases give one just fractionally time to organize oneself, just time to signal to the autonomic system that this is not an attack. If however, the stranger approaches brusquely and says 'Where's the bus station?' we very often experience at least the stirrings of anxiety.

Or what if one entered a coffee-party, ignored the hostess and began to speak without preamble to a group of people one had never met? What might be their reaction? We *know* in fact what their reaction would be, as our language itself shows, and as we see in a moment. Although in reality one is precisely the same person in the two situations, one gets a totally different reception if one merely says a few phrases (which can be *entirely* insincere) such as 'It was so good of you to invite me' or (to the group) 'I couldn't help overhearing what you were discussing', and so on, than if one does not. Without such placatory noises, what of the reaction? People say 'What a rude young man he is.' The word 'rude' conveys – or *used* to convey (for the *reaction* of the others is also residual in nature) – what a 'rough, harsh, unformed, wild, violent, barbarous, marked by unkind or severe treatment of persons' young man he is.[7] In other words, 'what an uncontrolled, *aggressive* young man he is.'

Importantly, not only do they feel bad about what we have done, but we ourselves normally feel bad about doing it. The various psychological sensations one experiences in these situations are not so residual that they escape one's attention.

A final word now in connection with conditioning and the effects of learning and experience. Zoologists are probably at fault in neglecting the influence of these to the extent that they sometimes do when considering or estimating the persistence of instinctive behaviours in human beings – and in particular therefore for not appreciating that *the strength or observed prevalence of a given ritual or ceremony is no reliable indication of the strength of instinctual urges or drives underlying it*. This *displayed* strength will be almost entirely a function of the amount of classical or operant conditioning with which the, actually residual, instinctive prompting has been reinforced and overlaid.

This is to say, the instinctive tendency certainly exists, and although residual it is neither non-existent nor wholly capable of being ignored. The human being is definitely 'happier' going along with his instincts – with the urgings towards 'fixed formal patterns' which remain in his inherited make-up. These urgings

7. *Shorter Oxford English Dictionary*.

one suggests, give an initial impetus and direction to many (to most?) of his actions and many of the forms of his society. (Once again I would refer the reader to Eibl-Eibesfeldt's trenchant discussion of these issues.) They exert fairly gentle but also fairly continuous pressure, especially in the absence of counter-resistance, towards certain actions and certain types of action.

The *precise forms* of rituals and ceremonies are shaped and moulded by a variety of *reality factors* (in which sense we now begin to perceive a conceptual link with the ego) by – what? – the materials to hand, by subsistence practices and amounts of leisure time available – in short, by all the realities which impinge or come to impinge on a particular tribe or people. Parents teach *these* forms, once established, to children, reinforcing their acquisition with, in effect, classical and operant conditioning. These conditioning procedures by themselves I insist would certainly not alone suffice to maintain cultural and social traditions to the extent to which they are in fact maintained.

As a single example of the continuous, uphill battle which has to be fought to condition, in this case, children to express something which at that time they feel *no instinctual urge to express* (and probably never experience) – namely, gratitude – one points to every parent's struggle to teach his or her children to say 'thank you'.

The question which more concerns the present argument is the further psychological, the inner mental, states which accompany the release or expression of instinctual behaviour in the individual human being.

When a baby smiles at its mother what does *it* feel or experience? Perhaps it feels nothing, perhaps its consciousness is a blank at that moment; perhaps it is actually preoccupied with something else – the feelings in its stomach, for example, so that it may smile without even being mentally aware that it has smiled. This is perhaps rather unlikely. Nonetheless, it is probably more profitable to consider situations where we do know, from our own introspection, what mental sensations accompany instinctual responses.

Most men can recall occasions in their lives when they were introduced to an attractive girl and found themselves blushing, perhaps even trembling slightly, or perspiring, stammering or speaking in a strained voice. The inner sensations which accompany this state are a confusion or disturbance of thought, something like disorientation, perhaps even a feeling of panic.

One could easily multiply examples of similar situations, but let us be content with this one. There is very good reason to believe that the presence or sight (the shape) of a girl whom we find attractive acts as a releaser – a releaser, in us, of sexual desire. Various (quite understandable) circumstances, however, frequently block the free expression of the response which the girl (often unintentionally) is calling forth, or 'trying' to call forth: the fact that some cognitive task is in hand (say, an oral examination in French); or the influence of social conditioning, which has taught us that one may not indulge one's sexual feelings indiscriminately; or a *specific* taboo (such as the fact that she is someone else's wife); or, of course, some counter-instinct.

It would seem that whenever an instinctive response is blocked in the way just described and for whatever reason, the resulting state of mind is unpleasant, ranging from mildly to extremely unpleasant. Conversely, *allowing* the response to occur results in a pleasant psychological state – ranging from mildly to extremely pleasant.

With regard to animal psychology, as we saw earlier, when a male stickleback is confronted with a red object (normally the red flanks of a rival) his attack response is evoked. When he is confronted with the swollen – and not red – belly of a female, the courting response is released. If, however, the experimenter introduces a model of a swollen female painted red, the male stickleback is unable to follow any course of action. He demonstrates a condition of extreme confusion, torn as he is between equal but incompatible urges. In this example the block to the instinctive response happens to be another instinctive response. Such really quite desperate confusion can be readily induced in a variety of organisms.

One proposes on the human plane that, for example, the state of being in love – in particular, the great pleasure which results from being in love – arises in large measure from the free expression, and the psychological experiencing, of the many kinds of instinctive response which the partner arouses (and which most women will to some extent arouse, but, of course, usually frustrate in the particular circumstances). Some part of the joy of being in love may be a psychological bonus which comes with doing the right and natural (i.e. instinctive) thing with the partner – touching and holding and caressing and being held, touched and caressed in return.

If, as I am claiming, the 'indulgence' of an instinctive response has mild or strong pleasurable accompaniments (and if conversely the blocking of an instinct produces discomfort or distress), it would follow that when an instinctive response is triggered, all other things being equal – that is, when no blocks to the response are present – the organism will indulge the instinctive urge, and indeed, *actively seek* to indulge it.

We are, however, still more interested in yet another aspect of the mental states which may accompany instinctive responses. As we have shown elsewhere, the human nervous system appears able – in ways that we do not understand – to experience itself sensorily; primarily, perhaps, visually. This is to say, the nervous system experiences *itself* as pictures or shapes. This is a difficult notion for us to accept, and yet the evidence for the view (which we have exemplified briefly from a consideration of enduring legend) seems, to the present writer, extremely impressive. One now suggests further that in addition to consciously remembered actual *examples* of releasers – i.e. some pretty girl we have seen, a particularly venerable old man, and so on – the nervous system also has a mental 'picture' or record of the *physiological sub-structure* – the basic model or archetype –

involved in the arousing-releasing process. These 'pictures' *are* in fact – one proposes – the Jungian archetypes.

Jung's own precise view of the archetype will be mentioned later. For present purposes – and this does not violate the Jungian position – an archetype is, say, the 'essence of woman' as far as a particular individual is concerned – his 'type' as in the phrase 'she's my type' – the picture of woman which most precisely fits the releaser-pattern lodged somewhere in that individual's cortex or subcortex.

There is a large number of these archetypes. A precise number would be difficult if not impossible to arrive at, because it is not always easy to discern whether a *particular* figure or stereotype in a society is merely an accidental figure – meaning, one that has arisen purely out of external, and not inherited, reality – or is conversely a somewhat modernized version of an actually very ancient and genetically-physiologically based releaser, or inhibitor, as under discussion at the moment. In other words, one would probably not be prepared to argue that bus-conductors or postmen are archetypes – although they just *might* be. But what of the policeman – the guardian? Is *he* simply a product or factotum of our modern society, or does he have marked archetypal undertones? Primates and other animals, of course, do set up sentinels to keep watch while the others are feeding.

Other figures, such as kings, mighty warriors, (Earth-)mothers, golden youths, fairest of fair maidens, witches, and so on, are more readily identified as archetypes. What are the indicators? One possible index is the frequency of occurrence and the role played by these figures in legends – *and later as the modern stereotypes of literature.*[8] This latter point is taken up again below.

Jung has written extensively on archetypes and the concept is indeed his original formulation as is the term itself.[9] They were the 'universal images' he found in the legends, literature,

8. And probably also as the sociological concept of social roles.

9. It is permissible to speak as well of archetypal symbols and archetypal situations.

painting and dreams of many different races. While gratefully adopting the term and much that goes with it, my own views on the archetype are not wholly concordant with those of Jung. The notion of the archetype is also linked to Jung's much misunderstood idea of the collective unconscious. His proposal, however, is simply that certain structures and predispositions of the unconscious are common to all of us. In other words that there is an inherited, species-specific, genetic basis to some or all unconscious, and in the end conscious, contents. One could as easily speak, for example, of the 'collective arm' – meaning the basic pattern of bones and muscle which all human arms share in common.

Archetype is to be taken here to refer principally to the 'mental image' of a releaser. This image is 'idealized' in the sense that it will be the closest approximation possible to the 'suggestion' of the physiological infrastructure. One says 'closest possible to that infrastructure' with some reservation, however, for undoubtedly these figures are affected to an unstateable extent by learning and experience, and by the models available in the outside world. Jung certainly held this view. Reality, however, is probably as much modified by the predisposition as the predisposition by reality. This is to say, in part, that while one may invest a man – say, Adolf Hitler – with godlike or king-like qualities (that is, while he may act as a releaser for this response in oneself), he need neither have too much of these qualities in the first place *or* himself change to any great degree in that direction as a result of the perception. One may perhaps be surprised at the notion that such perception, flattering or derogatory, can affect the perceived person at all in reality. Yet it does. Our viewing of him as e.g. a king appears to release what one might call a king-like response from him – and aside even from any attempt to act the part consciously. This notion helps in understanding why a man on taking office may genuinely change – so, apparently, Thomas à Becket when made Archbishop of Canterbury. Thus too, perhaps, the change in the rebel or outsider once he is accepted by the establishment; or the dog given a bad name; or the corruption which allegedly comes from power, and so on.

A table was presented in Chapter 2 (Table I) showing what in fact are certain of the archetypes which exist for the personality at various stages. That table is repeated here (Table 11) to enable comparison to be easily made with Table 12 which follows it. Several points are to be noted. First, that the power of the figures of the earlier table as *living* archetypes is very much diminished in our society. They are in a sense historical figures (evolutionary figures?) which we at one time lived out (that is, experienced with *immediacy*) – and which, certainly, we live out again as youngsters in the course of growing up. The decay of the idea of monarchy, even in Britain, or of the importance of being a virgin, for example, are partly attributable to the development of the average personality beyond this point.[10] These symbolic figures, however, in a certain sense live on in legend, and to a lesser extent in story.

The current living 'descendants' of these earlier archetypes of Table 11, their ranks now thinned, are holding out temporarily in certain kinds of literature, in (failing) social institutions and fictions of various kinds. They are the literary stereotypes of the formula novel or screenplay and the social stereotypes of gossip and prejudice. They are also enjoying a vogue as the social roles of sociology.

By virtue of the fact that these archetypes are in the process of breaking up, one cannot expect in particular the second table (Table 12) to be easy to complete. The table is not intended to exhaust the possibilities. As with fossil skeletons, enough survives to give one the idea. One could say that as these figures emerge into (or, put rather differently, are left behind by)

10. Quite what 'development beyond' may mean is a matter for speculation. It will be recalled, however, that one may *not* equate the strength of a response as seen in a given society with the actual, unreinforced strength of that instinct. The rapid weakening of the virgin archetype in our society, as compared with its status in Victorian society, results *rather* from reduced conditioning during the formative years and after, and to related significant changes in external reality.

TABLE II

As Perceived By		MALE: Non-Sexual Male	MALE: Sexual Male	FEMALE: Non-Sexual Female	FEMALE: Sexual Female
Pre-Genital Female Personality	Pre-Adult	MALE FAIRY	DWARF	FAIRY	NYMPH
	Adult	FAIRY KING	GIANT	FAIRY QUEEN	WITCH
Pre-Genital Male Personality	Pre-Adult	YOUNG PRINCE	? URCHIN	(FAIRY) PRINCESS	NYMPH
	Adult	KING: HERO I*	ROBBER-CHIEF: HERO II*	QUEEN	WITCH
Post-Genital Female Personality	Pre-Adult	CHERUB	IMP/URCHIN	LITTLE GIRL: VIRGIN	'LITTLE MINX'
	Adult	GOD: ANGEL	DEVIL	VIRGIN	WITCH
Post-Genital Male Personality	Pre-Adult	WONDERFUL BOY†	IMP/URCHIN	LITTLE GIRL: VIRGIN	'NYMPHET'
	Adult	GOD: WIZARD WISE MAN	DEVIL	GODDESS: VIRGIN WISE WOMAN	WITCH: SORCERESS

Note: In the pairings each item is the opposite in a certain sense of the one next to it.
The items immediately above/below each other are the young and old versions of the same archetype (stereotype).
* I and II meaning roughly good/bad, a-sexual/sexual.
† Jesus Christ, Hercules, etc.

TABLE 12

As Perceived By		MALE: Non-Sexual Male	MALE: Sexual Male	FEMALE: Non-Sexual Female	FEMALE: Sexual Female
Pre-Genital Female Personality	Pre-Adult	PETER PAN*	RAGAMUFFIN	SHIRLEY TEMPLE	NAUGHTY LITTLE GIRL
	Adult	KING: FATHER I†	STRONG MAN: FATHER II:† OLDER BROTHER	QUEEN: 'STAR'	OLDER SISTER: WOMAN: MOTHER II†
Pre-Genital Male Personality	Pre-Adult	SCHOOL CAPTAIN	TRUANT: 'HUCK FINN'	VIOLET ELIZABETH BOTT**	NAUGHTY (as opposed to UNWANTED) LITTLE GIRL
	Adult	FOOTBALL HERO	GANGSTER	MOTHER I	NASTY OLD WOMAN/HAG

Post-Genital Female Personality	Pre-Adult	LITTLE LORD FAUNTLEROY‡	JUST WILLIAM**	GOOD LITTLE GIRL	BAD LITTLE GIRL
	Adult	SAINT	VILLAIN/ ROTTER	NUN: MOTHER SUPERIOR	PROSTITUTE: COURTESAN
Post-Genital Male Personality	Pre-Adult	YOUNG GENIUS:§ SCHOLARSHIP BOY	ARTFUL DODGER¶	LITTLE GIRL	NYMPHET°
	Adult	PROFESSOR: PRIEST	PLAYBOY: SWINE	'SHE'S TOO FAR ABOVE ME'	PROSTITUTE: GOOD TIME GIRL

Note: In the pairings each item is the opposite in a certain sense of the one next to it.
The items immediately above/below each other are the young and old versions of the same stereotype (archetype).

* From J. M. Barrie's story.

† I and II meaning roughly good/bad, a-sexual/sexual.

** From Richmal Crompton's stories.

‡ From F. H. Burr-tt's story.

§ Mozart, etc.

¶ From Charles Dickens's *Oliver Twist*.

° From V. Nabokov's *Lolita*.

modern consciousness, their glorious trappings fade and fall away, until in the light of present reality their threadbare quality is seen. Certainly they have lost most of their magical, compelling attributes.[11] They appear, in evolutionary terms, to be in the process of vanishing completely.

Using another frame of reference, one can say that the human nervous system as a whole appears to be evolving beyond these methods of regulating life and society, which nature has hitherto used on a fairly large scale. Quite how these behaviours are currently being *bred out* in genetic or evolutionary terms is not at all easy to say. The rapid weakening of the 'releasers' or archetypes shown in the two tables, in recent years let alone in recent centuries, is, to repeat this, almost certainly the result of drastically reduced reinforcement at the conditioning level. The *real* power of these archetypes may have decayed even thousands of years ago. On the other hand, one must not forget that many of these responses were likewise *acquired* within fairly recent *evolutionary* time – since in fact the advent of *homo sapiens* as a species.

Two matters generally allied to the foregoing – the army and the nature of armies; and women's fashion – deserve mention.

One has a good many reasons for attributing an instinctual basis to these areas of human behaviour. One may, I suggest, always *suspect* such a basis whenever and wherever ritual or display plays a central role, particularly where one finds similar practices widespread among the varieties of man.[12]

11. So perhaps Wordsworth's

> 'And by the vision splendid
> Is on his way attended;
> At length the man perceives it die away,
> And fade into the light of common day.'
> (*Intimations of Immortality*)

12. Certainly, of course, one must here also always keep an eye open for the possibility of simple cultural dissemination.

When initially considering these two phenomena I was anxious to associate military organizations, and in particular the structure of military organizations, with the Ego and System A – and equally concerned to associate women's involvement in display and dressing up with the Self. But in another sense, the second also has links with the Ego. There are two principal difficulties. The first is the establishment of a reasonable distinction between uniform on the one side and fashion on the other – for at first sight the two would appear to have more common elements than differences. The second lies in accommodating what one establishes *en route* in our model of the personality.

Taking the second point first – but without wishing to go into the many difficulties involved – one's conclusion is that the structure of the *female Ego* differs from that of the *male* Ego: that the female Ego is much nearer to the Self (that the female personality is more nearly One and not Two?). One can best say perhaps that the division in the female between Self and Ego is less marked. Whatever else, it would seem, in short, that while the archetypes of the Ego are located within the cerebral cortex, *the archetypes of the Self* (if we may now introduce that distinction) are probably located outside the cerebrum in the sub- or extra-cortical centres. This might almost seem a piece of expediency, a sleight of hand to accommodate the phenomena we are about to discuss, but for two factors. In the first place we *can* distinguish very adequately between fashion and uniform (as will be seen) in our now standard terms, without reference of any kind of physiology. But second and more importantly, the experimental surgical work summarized by Morgan and Stellar and quoted earlier *does give us every justification*, as far as it goes, for believing that instinctual behaviour in the male is primarily triggered in the cortex, but in the female is sited primarily elsewhere.

The situation described – with the archetypal sites of the female personality located more within System B – is shown schematically towards the end of this section in Figure 26.

The similarities between fashions in clothes and the wearing of uniforms are many and obvious. Both are for purposes of display, in both situations the functional use of the clothing is secondary or only equal to its non-functional aspects, and so on and so on. What one notes first about fashion, however, is that it is ever-changing, while uniforms, although they do change over long periods, nevertheless *resist* change. There is almost always indignation and lamenting when the decision *is* taken to alter the form of a hat or other features of a uniform. How much worse when it is decided to reduce the appointment of, or actually disband, a regiment. Indignation turns to rage and nostalgia to grief.[13] One might say that with fashion the Form changes continually, although the Content (the motivation) is constant. With uniform, as the word itself strongly suggests, the Form is constant. As for the Content here, we may seriously doubt that there is any.[14] While, too, it is agreed that both fashion and uniform have display functions, it would not be difficult to establish that the main purpose of fashion is to attract (is in origin sexual), while one of the purposes of the military uniform is to overawe and to frighten (is essentially *aggressive*).

Lastly, although with both uniforms and fashion some kind of group identification is established, in the case of the uniform one has essentially *difference-within-unity*, in the case of fashion *unity-within-difference*. That is to say, taking these in reverse order, in fashion one has the *submergence of the individual* (a typical System B phenomenon), while with the uniform (despite that term) individual differences are accentuated in many ways – the more so as one looks away from the basic soldier. Even then, of course, the basic soldier of any regiment has features which distinguish him from the basic soldier of any other regiment.

13. A parallel, I suggest here, to the excesses of rage or grief which are occasioned when the rituals of the obsessional psychotic are interrupted.

14. The meaning of this oracular statement will become clearer. For the moment one repeats merely that a major function of ritual is to *rule out* emotion. Compare again the bed-time rituals of children. The revolutionary army, as noted, is weak on uniform and drill and protocol. Precisely *it*, however, is highly charged emotionally.

Among the means by which individuality is asserted within the encompassing unity of the armed forces there are: actual designated ranks and titles (a great many); regimental insignia, plus further insignia or flashes for smaller special units within these larger units; regimental mascots and specific traditions; service stripes (as opposed to rank stripes); medals for individual acts of bravery and campaign ribbons; and, for officers, the dress uniform. This is the army of *today*. In the medieval and pre-medieval army, however, one saw a still wider range of crests, shield designs, personal trappings and individual armour.[15]

Other links of the army with System A (and in particular with the instinctual, archetypal behaviours which characterize so many animal varieties) are the rituals of parade-ground and battle, the forming and unforming, the giving and obeying of (frequently pointless) orders – whose delivery and acceptance are accompanied by further ritualized movements (saluting, stamping, turning and so on). This mindless charade – certainly meaningless and wholly dispensable in times of peace, whatever its uses may be in war – survives peace because it is an end in itself. It is its own reward.

There is in this notion of an action being its own reward here a further clear parallel with the ritually instinctive behaviours of animals. Acts which in earlier evolutionary stages were a single,

15. A very interesting facet of the armed forces is the division into commissioned and non-commissioned ranks. It is possible, but rare, for a man to rise from the non-commissioned ranks to the commissioned ranks of lieutenant and over. Effectively, however, the rank of sergeant is the ceiling of ascent within the non-commissioned sector. Evidence of the non-permeability of the ceiling (so that any further movement at this point must be a sideways movement) is evidenced by the veritable proliferation of types of sergeant – company sergeant, regimental sergeant-major, quarter-master sergeant, sergeant-at-arms, staff-sergeant and so on.

Although the normal army as such and as a whole is a System A phenomenon, within it the non-commissioned ranks represent System B. The non-commissioned barrier is erected by System A against too-close encroachment by System B. Of the two the non-commissioned ranks (System B) have a more 'uniform' style of dress than the relatively more differentiated commissioned ranks (System A).

simple action became – for a variety of reasons which we need not examine – drawn-out, composite acts. One instance of this is the stalking, and eventual killing of a living prey, and the carrying of the prey to a place of safety before actually eating it – contrasted, say, with the gathering and eating of berries in one continuous movement. Sexual and other activities provide further instances of such behavioural chains. When this lengthening process occurred, parts of the composite act became to some extent, and sometimes completely, independent of the preceding or following section. The performance of the *sub-action* became then its own reward – a goal in itself – regardless of its original connecting purpose in the chain of behaviour leading to the end-act (eating, sexual intercourse, or whatever). Therefore a cat 'enjoys' playing with a mouse, as such. This activity is self-rewarding. Quite often the cat – especially a well-fed one – does not finish by eating the mouse, or even by killing it. (Eating, nonetheless, was the original purpose of the exercise and the original reason why the cat 'plays with' – actually hunts over and over again – the mouse.)

The parallels between the behaviour of the 'playing' cat and the drill and manoeuvring of the armed forces are difficult to ignore. In both cases the end-act has been, at least to some extent, lost sight of. Armies, although they do get a little restless at not being allowed to kill something occasionally, seem in large measure content with parading, manoeuvring, getting ready over and over again for a non-existent battle – e.g. by the daily re-cleaning of weapons which are not dirty.

A similar 'emancipation' from its original goal is seen with fashion – a further *link* between these behaviours. Whereas fashion would appear to have its *origins* in a desire to attract a male and cause him to mate with the female, this is no longer, in general, its main purpose. Unfortunately for women, the male continues to react to the sight of cleavage, emphasized buttocks or exposed thighs as if they *were* invitations to go to bed!

When one says that military behaviour is instinctive or somehow inherited, what precisely does one mean? Clearly, one is not referring to the *detailed* form of actual actions, e.g. the coming to attention with a stamp of the foot, or dressing by the right in groups, nor is one suggesting that these were *ever* as such movements made by primitive man. While, certainly, a few of the actual forms – bowing the head, perhaps – might once have been instinctive signals, such as we see in animals, it is nevertheless obvious that the vast majority of the outward forms and rituals of the army at present have been (on whatever basis) very largely *invented*. It is, then, the *urge* towards a ritual, the *urge* towards a hierarchy (or pecking-order), the *urge* towards giving and obeying orders – *this* is the inherited component. As suggested earlier, as long as no particular objection to them exists, *such residual instinctive urges will be given expression and this expression will give varying degrees of satisfaction or pleasure to the expressor*.

To conclude, Figure 26 offers a schematic representation of the male and female personalities, incorporating the suggestions of this chapter. The significant difference between the two personalities is seen to lie in the positioning of the archetypal contents – closer to System A in the male, and to System B in the female. In the male personality at least they may be thought of as a kind of shield (or no-man's-land) between waking consciousness and the unconscious. So, perhaps, Jung's psychology *is* a psychology of consciousness.

Female fashion display is linked (despite, but not instead of, what was said earlier) to the female *Ego* – which is however in some sense itself situated in physiological areas which, in the male, are the province of the Self. The essential *female* Self – the mother, the nun – is much less concerned with this behaviour – so perhaps the plain, figure-hiding nun's habit, the black dress of the married woman in the Mediterranean, and so on.

The uniformless revolutionary army is rather an expression of the *male Self*. Revolutionary armies do, of course, have more relatively *undistinguished* uniformity. Interestingly, though, this in turn is subject to the dictates of something which looks remarkably like fashion. Once it was a red cockade; now it is an unkempt full beard and an old army shirt open at the neck, with

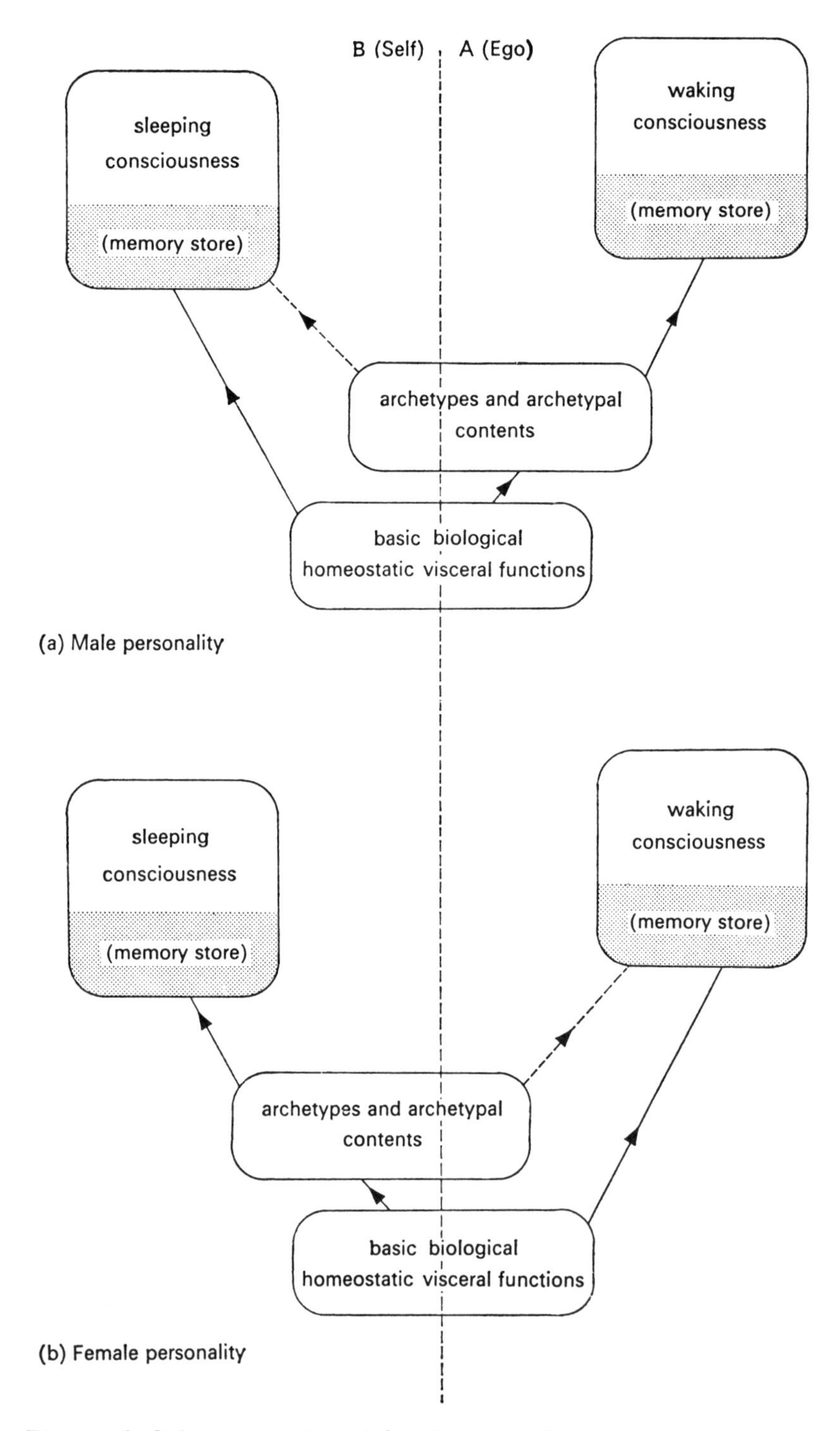

Figure 26. Schematic male and female personality

the sleeves rolled up. (Unless, of course, you happen to be Chinese.)

In psychological terms the archetypal cerebral cortex, or subcortex as the case may be, is a kind of twilight zone between the (new) conscious and (old) unconscious parts of the personality. To an extent it seems available to both major systems. Jung saw these areas and their contents as belonging to the unconscious, and looked for archetypal figures and archetypal situations in dreams. Freud saw the area as one of consciousness and wanted nothing of it in psychoanalysis. One suggests in passing that Jung's psychology is unconscious when applied to women, Freud's when applied to men. One would look, perhaps, by way of verification of this opinion, for more recognizably Jungian figures in the dreams of women than in the dreams of men.

Nonetheless, I must in any case reserve the right to speak both of the archetypes of the Ego and the archetypes of the Self – both of which are present in *both* the male and the female – but with a respective, sex-linked dominance.[16]

16. Somewhat reluctantly I have refrained from raising the matters of fetishism and reification. These are both seen to be connected in their differing ways with System A, the archetypal cerebral cortex and the male Ego.

The links between military practice (say, fascism) and fetishism may not seem all that obvious. Their connections are well known to the pornographer, however, who cannot afford to be wrong!

By way of brief instance of the drift here, a hint comes to us from studies of the Herring Gull. The adult bird has an orange spot on the lower mandible of its otherwise yellow bill. The chicks peck at this and the action causes the parent to regurgitate food. It has been experimentally established, however, that chicks will peck even more vigorously at a *red* spot. Could these chicks be said to be red-spot fetishists?

Reification – the regarding of ideas or systems as if they were actual objects – is a complex matter which, unfortunately, we cannot discuss here. Has it, nevertheless, anything to do with the fact that the psychotic (in his extreme condition) actually *sees* (that is, hallucinates) objects and people who are not present in reality?

Aside from these matters, a word must be said also on one further issue. The term 'romantic' is frequently applied to stories of epic military bravery, to the battles of ancient times and to adventure stories generally. These are,

It was promised in an earlier chapter that a comment would be made in a more appropriate setting on the significance of the enlarged lower mandible and jutting chin of Cro-Magnon and the neanthropics.

One's first thought is perhaps that the larger jaw of the neanthropic somehow came along with the switch from the vegetarianism of the primates to the carnivorous habits of man. On reflection this view is seen to be mistaken, at least in the literal sense implied. For the *last* thing a true carnivore needs, if one thinks about it, is a projecting chin. A comparison of an ape and a human jaw (Figure 27) shows just how 'un-carnivorous' modern man's lower face in fact is. The actual mouth has become smaller as the lower jaw has grown bigger.

A more promising avenue of inquiry is the relation of the jaw to the mode of eating. Man chews his food. *True* carnivores do not chew, but bolt their food. Any chewing they do is to detach pieces of meat from a larger piece and to some extent to render these swallowable. Man on the other hand thoroughly chews (and cooks?) *meat* in order to help his stomach cope with it. *His* saliva is a powerful digesting agent, whereas that of e.g. a dog, is not.[17]

While accepting the foregoing, I have a further suggestion.

First, I am impressed by the ocurrence – especially in the 'western', the detective and the modern adventure story – of such phrases as: 'I like the set of your jaw, stranger'; 'his jaw hardened'; 'he stuck out his chin aggressively', and so on. The

---

however, tales of the Ego, having many archetypal features and containing large amounts of 'day-dreaming'. The present writer would prefer not to call these tales romantic, but to reserve the term strictly for, say, the poems of Keats, Wordsworth, Shelley, for *Romeo and Juliet*, the writings of E. T. A. Hoffman, Edgar Allen Poe, Bram Stoker and such authors as these – that is, for the literature of the Self involving the archetypes of the Self.

17. *Neanderthal*, perhaps, instead evolved stronger stomach juices. Should we look for that in System B dominants?

hero (sheriff or outlaw, it matters not which) is frequently 'lantern-jawed'. Out of that general context we find in common use such phrases as 'keep your chin up'; or 'I can hold my head high', 'a place where a man can hold his head up', and so on,

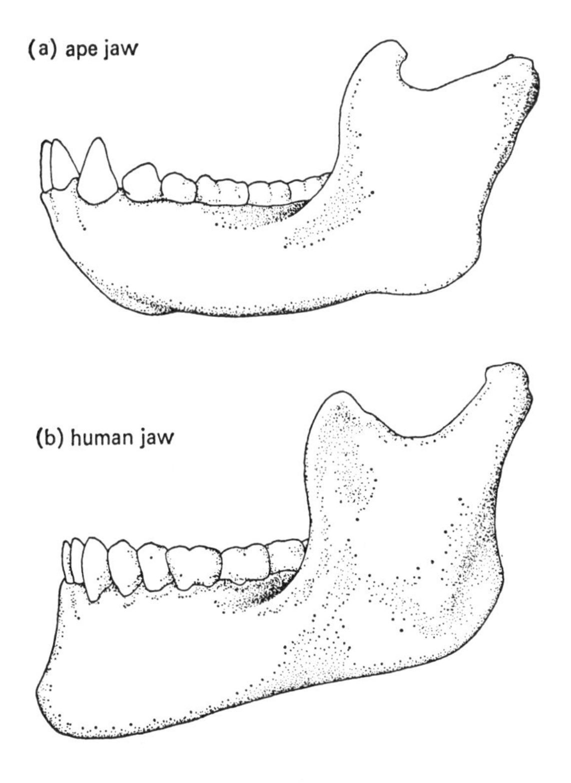

Figure 27

which effectively refer to a similar process. Then, on the other side, among our most contemptible terms of reference character-wise are 'chinless' and 'weak-chinned'. A number of further phrases attest our contempt for the lowered head generally: 'running your head against a brick wall', 'to go at it like a bull at a gate' – even something of phrases like 'a hangdog expression', though here the dimension of despair intrudes.[18]

18. I am inclined to regard the importance of the chin in *boxing* and its non-importance in *wrestling* – together with the greater popularity of these

Why should there be all this fuss about what is essentially a rather unimportant organ?

Not to make too long a story of this, I believe that the enlarged jaw of the neanthropics was (whatever else) a variety-specific sign-stimulus having signal – i.e. releaser and inhibitor – functions in ritualized combat between members of that variety. I believe that the paleoanthropics on the other hand employed the *lowered* head – the symbolic charge – if necessary followed by an actual one.[19]

When man first ceased to be just another primate and took among other things to hunting, numerous changes followed in other of his behaviours. Desmond Morris has cogently argued some of the resultant changes in his sexual activities. We are more concerned here, however, with the evolution of aggression and aggressive behaviours. For a large number of reasons, which we need not examine, evolving man – at any rate *some* evolving men – began to fight on a scale and with a determination lacking in other primates. We are at the moment interested particularly in the period between the initial separation of man from the original ape-stock and his development of reasonable powers of speech – for with the development of speech (and the parallel, taking conscious charge of one's own destiny) the sign-stimulus behaviour so common and so successful in vertebrates begins to become redundant. We can assume in general that *new* sign-stimulus behaviour does not arise in man after the acquisition of speech. An exception to this general rule, however, would need to be the suggested heightened instinctive revulsion which the variety Cro-Magnon acquired to the variety classic Neanderthal some 35,000 years ago.

During the period *after* becoming a hunting-fighting primate,

---

sports in the West and the East respectively – as related factors. In England at least there are appropriate sex and class associations. More of the aristocracy (hence 'my lords, ladies and gentlemen') are interested in boxing than wrestling: and more *women* in wrestling than in boxing.

19. Again one is reminded of this attribute of the bull, and so on.

and *before* the emergence of speech, the neanthropic jaw, perhaps, developed.

What we see by way of its effects today, and what we are discussing now, is the smallest residue only of a once – possibly – extremely specific and powerful instinctive behaviour.

Of course, throughout this section one is merely guessing, although not wholly without justification. Variety-specific behaviour differences, such as we are now proposing, are, however, *exactly* the kind of additional evidence which would support the zoological view – based solely on anatomy – of *two main varieties of man*, the paleoanthropic and the neanthropic, branching away from each other a considerable time ago (some millions of years ago) from a common parent stock.

Let us at this point resort to fancy, and imagine the following complaint on the part of the neanthropic to another – probably before the evolution of speech! – concerning paleoanthropics. The tone and vocabulary of this hypothetical complaint deliberately reflect the picture of, say, a public schoolboy describing a fight with the son of a docker (or any youngster from some underprivileged background). In this actual situation one sees perhaps the very palest reflection of a once rather precise clash involving the instinctive combat rituals of *two* different varieties of sapient.

'I had a set-to with one of those paleoanthropic chaps the other day. You know, everything they say about them is true. They haven't the slightest notion of fairness or decency. I was going to let this particular chap off – because he lowered his head, you see – but do you know what the treacherous blighter did? He ran at me just as I was turning away and gave me an almighty butt in the ribs with his head. Almost had the better of me for a moment. Of course, I gave him a damn good thrashing. It's the only language that sort understands. And what happened last week? One of them *bit* me. Look – nearly took a great lump out of my arm. I tell you, they're bloody-well subhuman.'

This, of course, *proves* nothing whatsoever. Nevertheless there are useful points. Neanderthal behaviour is here described by the fictitious Cro-Magnon as treacherous. We, as neanthropics and System A-dominants, would in general agree with that description. From the *Neanderthal* point of view, of course, Cro-Magnon's behaviour was equally 'treacherous'. But Neanderthal's case is not heard. Cro-Magnon's point of view survives, because it resides within us.

Yet if one examines the facts on a non-partisan basis, in what *logical* sense can it be said to be 'right' to punch someone with your fist, but wrong to bite him in the ear? How is it all right to 'run your enemy through with cold steel' (the word cold is interesting), but impermissible to gnaw through his jugular vein with your teeth? It is all complete rubbish, a transparent rationalization, from a sane and detached point of view. Yet our society persists, in this context, in calling 'civilized' what is merely ritualized or stylized. It is high time this particular double-think was advertised.

I have said that wherever ritual or display plays a central role, instinct-based behaviour may be suspected. This holds, then, also for religion.

Bowing the head – whether it be before a superior opponent, before the king, or before God Almighty – is an act of submission from the Ego. (One still remains *facing* the other party.) In the Black Mass, however, the congregant presents his buttocks to the priest or Devil. This is an act of submission by the *Self.* It is, both symbolically and literally, an invitation to sexual intercourse. One plays here the female to the male. This interpretation is confirmed by the fact that among some primates the loser of a contest between two males will present his buttocks to the victor. The victor will sometimes then mount the vanquished and go through the motions of (symbolic) sexual intercourse.

A final word on the subject of cannibalism. One is indebted to whoever it was who first made the suggestion that the Catholic mass is an act of symbolic cannibalism. We note further that Dionysian and Bacchic orgies frequently included the actual eating of human flesh. Witches are said to eat children, as also are giants.

Can we doubt, then, that these are residues of a *formerly common paleoanthropic practice*, whatever its meaning?

This concludes the review of the contribution of the ethologist to the general notions of this book.

The Psychotic Society

13

Many psychotherapists have pointed out that more and more patients exhibit schizoid features and the 'typical' kind of psychic problem in our day is not hysteria, as it was in Freud's time, but the schizoid type – that is to say, the problem of persons who are detached, unrelated, lacking in affect, tending towards depersonalization, and covering up their problems by means of intellectualizations and technical formulations . . .

There is also plenty of evidence that the sense of isolation, the alienation of one's self from the world is suffered not only by people in pathological conditions but by countless 'normal' persons as well in our day.

ROLLO MAY[1], quoted by Alexander Lowen in *The Betrayal of the Body.*[2]

Most people ask themselves at one time or another, what is the meaning of life. A great many presumably conclude that they don't know for certain – but they manage to live with their uncertainty. It seems, though, that this uncertainty may be contributing to a spreading modern illness.

A psychiatrist from the University of Vienna, Professor Victor Frankl, says there is a 'world-wide phenomenon' which represents 'a major challenge to psychiatry'. He calls it the 'existential vacuum' and claims that growing numbers of patients are crowding clinics and consulting rooms complaining, not of classical neurosis, but of 'a sense of total meaninglessness in their lives'. Other psychologists have encountered the same phenomenon and it seems that one fifth of all neuroses may now be of this kind.

JOHN DAVY, 'The Evolution of Evolution', *Observer* Colour Supplement, 8 February 1970.

Or take the news commentator himself and his manner of reporting the varied happenings of the day. Does he ever break down with grief, does he ever become paralysed by the horror and gravity of his reports? He goes from horrendous, nauseating, shocking incidents to the trivia of everyday life with scarcely a change of intonation in his voice. He avoids comment of his own, even though the news is shattering. He likens himself as much as possible to a machine, a tape recorder, a tickertape.

HENRY MILLER, article in *Penthouse* magazine, Vol. 4, No. 7, 1969.

1. *Existence: A New Dimension in Psychiatry and Psychology*, Basic Books, 1958.
2. Macmillan, Canada, 1967.

These are three faces of what we shall call the psychotic society. No doubt the words of these extracts will recall both the tone and content of R. D. Laing's views on the nature of psychosis reported in Chapter 5.

What then is the psychotic society? Firstly, I must emphasize that the term as such is of my own coinage. It is not, as far as I am aware, used by other writers. What, however, *is* common is the phrase 'the neurotic society', an inaccurate phrase, I suggest, used to cover some of the phenomena which will be under discussion. As is, hopefully, clear I have very profound objections to the interchangeable use of the terms neurotic and psychotic at any level. Most importantly, the words 'psychotic' and 'neurotic' as used in this chapter will *not* carry their full clinical meanings. Whenever I wish to indicate the diagnosed, hospitalizable mental illness to which those labels usually refer, I shall expressly speak of *clinical* neurosis and *clinical* psychosis.

The difference between the term psychosis as used here and clinical psychosis, and between the term neurosis and clinical neurosis, should nevertheless be understood to be principally one of degree, not one of kind. Neurotic and psychotic refer to the psychological-social conditions or preconditions out of which, or within which, the full-fledged clinical condition can readily develop. These may be visualized, then, as a continuum – or rather *two* continua – at one end of which is a condition or state acceptable and accepted as normality; and at the other the neurotic or psychotic breakdown respectively.[3]

Thus, when this chapter speaks of the psychotic society it

3. It would seem that one is saying that mental illness is 'only' an extreme form of normality. Though the view is somewhat over-simplified, and requires certain riders, this is nonetheless essentially what one *is* saying. A point to be borne in mind, however, is that what we may call the 'deterioration growth-curve' is exponential: that is, it rises more rapidly, even logarithmically, as the extreme of the condition is approached – and very rapidly in the final stages. This concept will be seen to be a very useful one from a number of points of view. It incorporates, for example, the observed phenomena of the breakdown – that is, the very rapid deterioration over a short period (at least sometimes after a lengthy period of gradual deterioration).

does not quite mean the insane society. It does mean, however, the society which has moved and is moving perceptibly further along the upward growth curve to that condition.

The styles of the neurotic versus the psychotic society may be considered in the first instance historically. My proposal is that societies of earlier times were in the main neurotic, while our latter-day society of the twentieth century is psychotic. Let it be emphasized for the last time that 'neurotic' here does not mean clinically neurotic. It does not mean neurotically *incapacitated*, except in the senses subsequently defined.[4]

As instances of neurotic societies let us consider those of historical South America – that of the Aztecs, for example. The life of the Aztec was entirely governed by religion and religious considerations. The temporal ruler (as in Egypt) was at the same time the spiritual ruler – himself a divinity or semi-divinity. The economy of the country (again as in Egypt) was essentially geared not to the production of consumer goods, but to the production of vast monuments to the god or gods – pyramids, tombs, temples of expansive and expensive design. So any citizen might at any point find himself nominated for human sacrifice – say, thrown down a well to solicit rain. In this the society acquiesced, not necessarily cheerfully, but by and large unprotestingly. But we, looking back in time, are horrified at a society which regularly and without pity arbitrarily sacrifices and tortures numbers of its citizens to placate the vagaries of climate or the imagined wrath of a mythical god. We have little hesitation in describing such a society as irrational – even, perhaps, insane.

An Aztec Indian, however, could he be brought forward in time and shown the present state of Lake Erie, would have for his part also no doubt that our present society is insane.

4. As Karen Horney and others have pointed out, it is incorrect to speak, for example, of a Red Indian who believes that some god or the spirits of his ancestors inhabit a grove of trees, as clinically neurotic. If anything, indeed, a refusal on his part to accept this belief might rather be regarded as aberrant. Such views are normal both for him and his culture.

Lake Erie is one of the Great Lakes (the fourth largest) on the borders of the United States and Canada. It has a surface area of 10,000 square miles. This lake can no longer support organic life. This lake can no longer support organic life. (The statement has been made twice because we are often in the habit of reading statements without becoming aware of their content.) In common with many other aspects of our natural environment, the pollution of this lake by industrial effluent has been a cause of concern, to a few individuals, for some time. The end-process has now apparently been reached. In a recent television documentary biologists and others confirmed that the condition of Lake Erie is now irreversible.[5] It not only does not support life, but it cannot be made to do so. It is now a chemical soup, to which it is possible only to add further chemicals. The effect of this would be, as it were, to change the flavour, without making it more edible.

To repeat our earlier comment, the Aztec Indian would be quite clear in *his* mind that we are insane. We are destroying, in his view, the very basis of life – the very context in which life, meaningful or otherwise, exists.[6]

5. *And On the Eighth Day*, 'Horizon', B.B.C. TV, January 1970.
Other reports which I have since read suggest that the situation is less extreme than this programme appeared to, and my own paragraph does, argue. Thus one reads that fish taken from this lake are now *poisonous* if eaten, by reason of the amounts of chemicals they have incorporated physiologically. The point, then, is that there are still apparently fish alive in Lake Erie. The nature of this statement, however, seems to me in no way to relieve the nightmarish quality of the situation, and I have therefore left the above paragraph unchanged.

6. It happens that even at this moment we are in a position to make the food we need synthetically. While some of the techniques involved require an organic raw material such as grass, others apparently can make do with fossil organic products such as oil and coal. Quite aside from this, it is even only a matter of time before we shall be able to manufacture our own organic raw materials from inorganic chemicals – of which this planet possesses an inexhaustible supply. These facts the Aztec Indian could not be expected, and many of the anti-pollution lobbyists do not choose, to grasp. Of course, anti-pollution is, quite rightly, concerned with more than just food produc-

The proposed concept of the neurotic and psychotic societies may be further considered from the standpoint of the development of the individual human being. The maximum incidence of clinical psychosis, in terms of admissions to hospital, as stated elsewhere, is between the ages of twenty to twenty-four for men and thirty to thirty-four for women.[7] Thus clinical psychosis may with some justification be considered to be an illness primarily of the *adult* organism. Optimal age at admission to hospital in respect of clinical neurosis is a rather harder statistic to come by, in the literature. One must be content with the implications of the generally agreed tendency (generally agreed among psychologists, not psychoanalysts!) for individuals to grow out of neurosis, even without treatment, as they get older. Thus people (in general) seem to grow *out of* neurosis (as children) and *into* psychosis (as adults).

---

tion. It remains a fact, however, that future life probably *could* go on under some fairly appalling conditions, involving the sacrifice of much of our present environment.

However, this refers to *physical* life only. Whether life could go on *psychologically* under such conditions, or can go on psychologically even under *present* conditions, is another question entirely; and one which, incidentally, has little to do either with sentimentality or aesthetics.

Already in the examples so far quoted above one sees clearly, in the neurotic and psychotic societies, the respective play of the Self and the Ego. The *neurotic* society is dominated by, and is the servant of, the environment. It is aware that its own survival *depends* on the environment, and is fully prepared to sacrifice some of its members in the cause of ensuring that the environment (nature) continues to support the Self (makes possible the survival of the species). The *psychotic* society on the other hand exploits and dominates (or rapes) the environment – as well as all other societies and those members of its own society which it does not consider as direct extensions of itself (those members with which it does not identify). Fundamentally, the psychotic society is unconcerned if the environment utterly perishes or even if the large mass of human beings perishes – for the Ego, as we know, is concerned only with the survival of the individual. This viewpoint is, of course, quite insane.

7. This fact alone, perhaps, suggests that System A matures more slowly in women than in men.

In the non-clinical sense of this chapter, it is legitimate to consider all children and youngsters as neurotic. In early childhood they believe in magic, dream more than adults, are terrified of physical isolation, and so on. As adolescents they are often deeply religious, believe in love, read and write poetry, and so forth.

To what extent adults, on the other hand, may be considered naturally *psychotic* is precisely the nub of our present discussion.

These general considerations appear nevertheless to fit conceptually rather well with the notion that early or primitive societies are neurotic, and recent, more evolved societies psychotic in this sense. Early communities were youngsters; present-day society is grown up.

Unfortunately this description is a little too simple. It is, moreover, by implication unnecessarily derogatory to early societies and certainly too flattering to those of the twentieth century.

The position is rather as follows. First, although one takes the constitutional predisposition of the individual (his inherited tendency to be either a System A or System B dominant) to be the overriding factor, nevertheless – particularly where the inherited predisposition is not of a marked nature – a taught way of life and/or a given type of society together undoubtedly exercise some shaping and disposing influences.

Thus, when System B functions are encouraged during the course of the individual's development to the detriment, or with the neglect, of System A functions, an adult, and hence a society, will tend to be produced biased in favour of System B. When System A is encouraged during the individual's development, a more System A-biased adult, and way of life, tend to be the outcome.

Now, as it further happens, the *child* is naturally predisposed to be swayed by and towards System B; while the adult is naturally predisposed to react in the same way towards System A and System A situations. Therefore it tends somewhat to follow, in the context of *these* predispositions, that the effects of a conditioning or an environment favouring System A will not show up until later in life – because during youth those influences are to some extent counter-balanced, or counter-acted, by the natural, ontogenetic resistance of the organism at that time. This account would among other things accommodate the relatively low incidence of childhood psychosis. This can and does occur, however, if the attack (on System B) is consistent enough, or the child is in some other way predisposed.[8]

Unlike the effects of System A training, the effects of System B training show up as a rule during early life, the predisposing conditions being then most favourable; theoretically the effects could be expected to diminish in later life. And in actuality, it seems that the neurotic does sometimes grow out of his neurosis.

Speaking with some caution – since, as indicated, evolutionary and genetic considerations are also heavily involved, whose precise influence is extremely difficult to assess – one can suggest that one reason why primitive peoples have neurotic societies is because they have *not* been subjected to extreme (or, indeed, any at all to speak of) System A conditioning and training. A sample of children drawn from such a community, however, could no doubt be experimentally galvanized into something much closer

8. It is possible that the recent dramatic lowering of the age at which puberty is reached among Western children, both boys and girls, is partly due, paradoxically, to the ever-increasing encouragement System A receives. It may be that sexual development is not in fact being *favoured*, but on the contrary is under pressure. How might this be the case? By way of analogy, one observes for instance that a plant deprived of proper nutrient often burns up such resources as it has ('sacrifices' its leaves and so on) to hasten into a rapid flowering and seeding. The flower is often considerably smaller than it would be under normal conditions, though usually perfect.

While the reasoning here is not altogether straightforward, and while one has provided an analogy rather than a parallel (though it *may* be a genuine parallel), it is at least worth considering that the 'privation' of System B somehow produces an earlier puberty. There is of course the additional and vital question of the vigorousness of the pubertal change – as measured, say, by levels of hormone secretion. Is the change today in any sense a shallower one than it used to be? Is indeed the whole System B period somehow less of an experience, or an inadequate experience?

to the psychotic society. This in fact we appear to observe to an extent in certain primitive peoples brought into large-scale contact with the twentieth century. Conversely, one suggests that we ourselves tend to encourage a System A society by the way we treat our children – i.e. 'intellectually' and mechanically.

Be these considerations as they may, the facts of our present psychotic society are writ large in all forms of public media and public institutions, and many aspects of the nature of Western society – so large in fact that we tend even not to perceive them without some initial direction. Lake Erie is an example. This event is on such a scale that a genuine *effort* is required to appreciate its significance. To grasp that significance it is necessary first to build up in the imagination the industrial background, the nature of the social organization, the chains of circumstance and the whole infrastructure of events which *must* pre-exist before that event could exist – which, as it were, give birth to that event. Once one begins to make this effort of imagination, then one begins to appreciate the otherwise decontexted, out-of-focus event under discussion for what it is – the open sore on a diseased body.

These are fine phrases – if they mean anything. Since it is not only difficult to imagine that they do, but leads to so unpleasant a conclusion, that one would rather do almost anything than admit the implications of the kind of events described, the present book offers here only one other giant symptom of the deviant psyche of society.

The following extract is taken from the London *Evening Standard*, 8 October 1969.

Montreal, Wednesday – Three people were shot dead as a night of violence gripped Montreal when 4000 policemen walked out in a pay dispute . . .

Mobs rampaged down the city's main street smashing door windows and looting . . .

During the night fires, explosions, assaults, robberies and a full-pitched gun-battle kept most Montrealers huddled in their homes in

what was described as the worst reign of terror in the city's history.
The mob spread down the main streets, smashing shop windows and looting. . . . At one point they broke open a fire-arms store, hauling away rifles and other weapons in their cars.

And so the article continues.

What is one to *say* about this story? How is it that this *one* item of news does not bring every one of us to a halt in our tracks with the realization that something has gone unbelievably wrong with the twentieth century?

Deliberately, I have chosen not to use any of the myriad items of a similar nature which emerge almost daily from the United States, from gangster-ridden, materialistic New York or Chicago. Instead the item we have offered comes from former British Empire, almost lovable, Canada.

With this we adjust our lens focus.

The items which now follow attempt to turn the phrase 'the psychotic society' from a not uninteresting, but perhaps nevertheless fundamentally conversation-piece concept, into a literal reality. There are several points to bear in mind. The first is our earlier proposal that clinical psychosis and clinical neurosis do not differ from everyday behaviour so much in kind, as merely in degree. Second, that the word normal does not at all mean 'correct' or 'good', but simply 'that which approximates to the norm'. One asks two questions: (a) whether the present norm (the normality) of Western society approximates more closely to the norms of clinical *psychosis* than it does to the norms of clinical *neurosis*, and (b) whether that norm is in any case not static, but is noticeably moving in the direction of the extreme condition of full clinical psychosis.

The items following may seem an extremely slight platform on which to mount the rather grave charges we are preferring – indeed, these are deliberately arranged in order of 'triviality'. Moreover, one does not seek to deny the perfectly rational basis of some of them. One asks only whether the apparent rationality does not rather contain the seeds of *rationalization*.

The symptoms of our alleged patient, society, are these.

(1) The covering of domestic gardens with concrete to make a parking space for one's car.

(2) The replacement of telephone-exchange and postal-district names with numbers, and the numbering of personal bank accounts.

(3) The protective pre-wrapping of foodstuffs and the sterilization of milk for longer life (an interesting paradox).

(4) The fear which, for instance, many Americans going abroad have of catching a disease – as evidenced by long programmes of 'shots' prior to leaving.

(5) The high production rate and over-diversification of detergents, cleaning, cleansing and cosmetic products of all kinds.

(6) The over-production of motor-cars, and consumer durables of all kinds, and their triumph in importance, in many senses, over the human being.

(7) The city as a desert. (desert = 'a desolate and barren region', 'a place where nothing grows'.)

(8) The actions and reactions both of the British Labour Government, and still more of the Conservatives, towards permissiveness, pornography and marijuana.

(9) The marked increase in the use of technical and jargon terms based on Latin roots in everyday speech.

(10) The building of very high, square, transparent blocks of offices and flats, using non-organic materials – in respect of which the present book would like to coin the phrase 'the anti-womb'.[9]

(11) The denial of spontaneity and emotion at large by the institution of procedure, protocol, precedent, tradition, formalization, the pre-recording of radio and television broadcasts, and so on – and the general substitution of content by form, function by office.

(12) The shallowness of registered and expressed emotion,

9. One of the aims of these buildings appears to be to get as far away from the Earth as possible.

the inconsequentiality and triviality (as described by Henry Miller) of everyday life in the West.

(13) The trivialization on the one hand of woman as woman; and on the other the masculinization of woman.

(14) The techniques, of many kinds, of killing at a distance – in the most extreme forms of which the people killed are seen by the killer neither before nor after that event.

(15) Allied to this, the application of industrial techniques to the killing and disposal of human beings – as particularly in Nazi Germany, but also in the guillotine, the electric chair, and so on – making feasible the notion, and possible the phenomenon, of mass destruction, the counter-coin to mass production.

(16) The removal of feeling and emotion, morality and ethics, from the study of psychology. The replacement of a human-centred, or even an organism-centred, study of life by quantified, mechanistic and reductionist methods, and computer-based models.

Is this collection wholly nonsensical, or have we in it some hope of justifying what has been claimed?

It is *not* impossible to establish links between these behaviours of so-called normal society and the behaviours of hospitalized psychotics, and this linking is now undertaken. It will of course be appreciated, before we move on, that the list above does not by any means exhaust the dossier of possible charges against – the full complement of symptoms shown by – the psychotic society. One has, for instance, not even mentioned the education system, 'the devourer of souls'.

Let it be emphasized once again that one is fully prepared to acknowledge the role of reality, meaning here non-psychotic, factors, which have in all cases contributed to the position. For example, one is aware that the switching from telephone-exchange names to all-figure telephone numbers is part of a programme of rationalization[10] dictated by e.g. the wish to provide a world-

10. The double meaning of this word seems entirely to escape those who are so fond of using it in such connections.

wide direct-dialling system, whereby any person can ultimately dial any other person in any part of the world without the services of an operator.[11]

In hospitals up and down England there are psychotics who spend all their time attempting to stay 'clean'. Prior to breakdown and admission to hospital they may be taking as many as six baths a day, with as many changes of clothes, particularly of underwear. In later stages of the illness these people move on an endless cycle through the house, washing door-knobs, scrubbing already spotless floors, the insides of dustbins, and flushing the lavatory repeatedly on passing the lavatory door. In this connection we are reminded perhaps particularly of Joey, who thought the whole world a sea of excrement into which he would be dragged down.

A question the reader might like to consider is how many times a person would have to change his shirt every day before we would begin to consider him mad. Twenty-five times? Fifteen times? Eight times? Four times? Twice? Of course, these matters cannot simply be approached numerically. A further parameter is that of compulsion. An individual may like to change his shirt two or even three times a day – but he does not necessarily feel that he *has* to. He may do it when convenient – but when it is not (say, when travelling) he may shrug and dismiss the matter from his mind. It is the degree of uneasiness which the person demonstrates when prevented from acting in his chosen manner, from carrying out his 'normal' routine, which is one of the more important clues.

What, then, does one make of a whole nation or class preoccupied with cleanliness? One thinks here particularly of Americans who boil water before they drink it *in England*, who in some cases boil the water they wash in, and who feel

11. The writer would *still* claim that the exigencies of the situation are not accepted neutrally, let alone sadly, by the decision-makers, but are gleefully met half-way as a further step to de-personalization, one further point of no-contact between human beings. Whatever the intention in any case, the outcome is that one can no longer remember the telephone codes of one's friends.

physically unable to drink water that has not been boiled.[12] Compare here the *hippie* practice of passing a 'joint' from mouth to mouth, among strangers let alone friends.

Health is a logical extension of cleanliness. Some people take vitamin tablets and similar preparations daily. Others go through a programme of exercises when they get up or before they go to bed. Such practices are within the range of the normal. Once again, however, one must consider also, along with incidence, the degree of compulsion. Prior to actual psychotic breakdown some individuals are taking six, ten, twelve kinds of medicine daily, running twenty miles an evening, refusing all food except, say, boiled spinach, and are unable to tolerate being in a room unless a window is open.

Both of the matters just considered – cleanliness and health – are closely linked to ritualism. Obsessional psychotics perform many long, apparently meaningless rituals. They may walk up and down endlessly alongside one wall of a room, or sit making interminable ritual movements of the hands, the head or the whole body, or perhaps make and unmake a bed over and over again. The ritual in these cases is an end in itself and the content is non-existent (at least to the observer), or immaterial. An extremely interesting feature here is seen in the results of interference with the ritual by another person. When this occurs the patient, who till then has been relatively calm, and certainly controlled, will at that point exhibit an outburst of uncontrollable anger, become completely hysterical, break down and weep abjectly, or whatever. It is clear that the ritual has been exercising a strong containing or controlling influence.[13]

12. In connection with health practices generally, both national and individual, I would like it to go on record that I am perfectly aware of the realities of how disease is spread, and the need to encourage the washing of hands after using the lavatory, for stringent control of the maintenance and functioning of kitchens in restaurants, the need to enforce quarantine regulations for animals entering this country, and so on. What one is discussing are the beginnings of compulsion, *unreasonable* zeal and irrationality, in such connections.

13. Conceptually at least, as already noted, there seems to be a close link

The relevant question to ask of the psychotic society is at what point can (needful, or productive) *procedure*, *protocol* and *method* be said to become *ritual*. Again the beginnings of an answer would be sought in terms of frequency and compulsion. The writer suggests that the sheer bulk of protocol in Western society is alone already suspect. Could one really justify all of it? Does one not hear it commonly said that protocol holds up rather than facilitates progress? The pursuits of bureaucracy and procedure appear to have become self-justifying ends in themselves. The present writer is himself particularly concerned over the paramount importance which method and methodology have assumed in modern psychology. In this field of inquiry which is closest to the personal experience we term life, whose very subject-matter *is* (or should be) that experience, we find the strongest emphasis on a practice – the experimental method – which undeniably, and I suggest intentionally, interposes itself between the psychologist and his subject. The question really is whether the experimental method functions as a *necessary* filter, selector or range-finder, as is claimed – or as a shield, defence or Denial.

Allied to the objectification of the social sciences and the technicalization of their private terminology is the objectification of public language. So a man who works with his hands – a handworker – is a manual operative (which actually means precisely the same, deriving from Latin *manus* = hand and *opus* = work). Similarly a dustman is a refuse-disposal operative. One may

---

between this behaviour and both the controlling *and* psychologically satisfying (or comforting) aspects of instinctive behaviour discussed in the last chapter. Both involve control, both when interrupted produce psychological disturbance, both – in this book's view – are a function of the Ego. One thinks too in this general context of the controlling aspects of form in classical art, also discussed earlier, and the overall relation of form to System A. The classical Greek drama was governed by what look suspiciously like ritualistic and compulsive considerations (the insistence on five acts, the obligatory denouement, the nature and function of the chorus, and so on.) The origins of drama as ritual appears, actually, to be a fairly commonplace notion of the specialist literature of that field.

instantly multiply one's examples by opening any newspaper or magazine. Of far greater significance, however, is the escalation – here in the military context one finds oneself automatically using 'escalation' in preference to Anglo-Saxon 'growth' or 'build-up'[14] – of this same practice in connection with warfare. Even the word warfare itself begins to have an archaic ring – the as common, becoming more common, term is combat (Latin *combatuere* = to strike with). With this change of terminology the whole *affect*, the meaning in that sense, is lost – thus: 'In a successful medium-range air-strike the allied command today liquidated heavy concentrations of enemy forces in the Dortmund area.' What this actually means, in real terms, is something like as follows.

'Quite a lot of human beings from one group, sitting in machines, went over to another group of human beings where, using the machines, that first group blinded, smashed the chests of, and tore the limbs from a great many of the second group, only some of whom were soldiers. When the first group came away, everywhere in that place men, surrounded by parts of men, lay screaming and dying: some with eyes oozing from their skulls, others with flies clustered on their exposed, bleeding parts or on intestines, strung out sometimes yards from their stomachs. In other places away from that place, though they did not as yet know it, large numbers of children (indistinguishable except in terms of language and a few psychological attitudes from the children of the men of the first group) (and not the children who lay, their eyes similarly oozing, etc., among the dead and dying men) were now without fathers; many wives came to face the desert, the empty night, of the rest of their lives without their husbands.'

One need not pursue these points.

Suffice to say that the device of latinization, objectification and jargonizing of language resembles closely the complex intellec-

14. When one considers that the word 'growth' is from the same root as the words 'green' and 'grass' one sees again how close Anglo-Saxon is to natural reality.

tualizations and in particular the intellectual denial of emotion seen in many psychotics. Joey offered one striking instance of this.

A common and clinically widely accepted category of psychosis is that of paranoia. In paranoia the patient feels either mildly or desperately, diffusely or specifically, persecuted. The patient may, fairly specifically, imagine that people are talking about him, that the Government is spying on him, that an assassin has been hired to kill him. Or he may find certain aspects of the environment, such as dirt, or the presence of certain objects unbearable. He may remain calm and relatively normal until, say, a cat is brought into the room. Or he may tolerate the presence of visitors, even of strangers. But if anyone *touches* him he may become insane with rage or fear. Or he may selectively be able to tolerate men, but not women and the discussion of some topics, but not others – perhaps again the discussion of sex or body functions.[15]

Surely the attitudes of Western governments (the Establishment) to permissiveness, that is, to casual or frequent sex, 'daring' or way-out styles of dress or hairstyles, and so on, pornography and marijuana, resemble those of paranoia? They do so in two ways – first, in the selective nature of this response; second, in the exaggerated nature of the response in terms of the objective realities of what is objected to.

Let it be at once clear that I am not saying that we as yet know sufficient about the effects of, say, marijuana to give it an unqualified clearance certificate. The Establishment can, for

15. Because sex is so very often implicated in forms of paranoia one is tempted to say, and some have said, that these patients are exclusively persecuted by their own denied sexual urges. Joey, on the other hand, was apparently persecuted by excrement. It was when going to the lavatory that he had to hold on to the wall, would not sit on the seat and needed the supportive presence of other people. He at no time appeared to object to women as such, nor was he a stranger to such activities as masturbation ('master painting'). The present book would of course claim that the paranoiac is persecuted by his autonomic functions in general, by the autonomic *system* – that is, by System B.

example, and quite rightly does point to the fact that the increase in pot-smoking in the United States has been accompanied by a rise in the consumption of heroin and other hard drugs.[16]

One is on less controversial ground in respect of permissiveness, sex and pornography. Is there any evidence, for example, to show that a man works less efficiently in shorts and shirt sleeves, particularly on a hot day, than he works when wearing a suit? Yet lawyers, for instance, are not allowed to appear in court in shirt sleeves.

In the 1920s of the present century many women set out to look more like men. Hair was cropped short, breasts flattened, smoking became commonplace. While there was indignation in some quarters over this phenomenon, there was *never* any suggestion of any kind of legislation to prevent those aspects of women's activities. Indeed, this would have been seen to be wholly absurd, an unforgivable and ridiculous intrusion into the private affairs of the individual. In the last decade or so in our society, there has evolved a fashion movement which involves men coming to look more like women. Men grow their hair long, wear silks and bright colours, and cease such 'masculine' activities as consuming large quantities of alcohol. One has the definite impression that the conservative Establishment in particular would very much like to legislate against this trend – albeit somewhat indirectly – by, for example, re-introducing conscription, reducing National Assistance benefits, expelling from school pupils with long hair, and so on. If this *is* the case, it would seem that the Establishment is more threatened by the conversion of men to women than it was by the conversion of women to men.[17]

16. On the other hand official pronouncements on marijuana do not appear to say, or in fact say, the following: that for the present a total ban on marijuana will be enforced until its effects have been studied from all points of view; that such a large-scale, long-term study has already been mounted; and that in the event of the investigation returning a favourable report, or only a qualified negative, steps will be taken to legalize, or to control within such limits as seem reasonable, the use of this product.

17. Since this paragraph was written comes the news that a government in South America has banned by law the use of 'drug-culture' words in newspapers and magazines.

The paranoia is seen at its clearest in connection with sexual activity as such. Here the authorities – without appearing *too* ridiculous – can and do legislate. Obviously, the risk of appearing ridiculous is preferable to the risk of seeing nude bodies. In an earlier chapter it was proposed that sex in Western society is an underground institution. The extent to which this is true emerges only when a comparison is made with an open institution, such as aggression.

On any given day it is possible without difficulty, and with full public knowledge to see a display in which two men hit each other. There are no such displays of love-making. Latterly, but only latterly – and it is this about which the Government is so concerned – *adults* may see *pictorial* representations, or pseudo-representations, of this act. Pictorial displays of violence on the other hand are, of course, freely available to children of all ages and both sexes at every cinema and on every television screen.

From our cultural history emerge – with the possible exception of Rudolf Valentino – no heroes who were great lovers and nothing else. In the history of the whole of Western Europe only two names come to mind – Don Juan and Casanova. Even so the reputation of these men is only at best a somewhat tarnished one – and even so *none* of the three names mentioned, one observes, is Anglo-Saxon. We have, on the other hand many, many cultural heroes who were fighters.[18] Both in this country and America there exist also some extremely exclusive institutions, where over a period of several years one may study the history and practice of murder, with a view to becoming a qualified murderer oneself. There are no such institutions where one may study love-making.

If in this country a man walks up to a small girl aged, say,

18. In homes up and down England there are photographs of murderers on proud display: pictures of oneself, of one's ancestors, at the battle-front in their murderers' uniforms. Nowhere are there pictures of one's ancestors making love.

eight, and places his hand on her chest, several things will happen: he will lose his job; go to prison; and in prison be wholly rejected by the other inmates (unlike a murderer) and instead be physically beaten and tormented by them. If, however, this same man had struck the said child in the face, he would *perhaps* have lost his job – depending partly on whether the child's father was content to hit *him* several times, or whether the father brought a legal charge of assault. The man would in any case not go to prison, but be fined or bound over to keep the peace.

One is, of course, aware that the first act of the man placing his hand on the girl's chest might well have been the prelude to sexually assaulting and subsequently murdering the child. What we are concerned with, however, are the extremes of public reaction to this admittedly horrible possibility, let alone the act. It would be an equally horrible act to kill another man with a bayonet – or so one would think.

When one leaves a public lavatory in this country there are prominent signs with the instructions: 'Please adjust your dress before leaving' and 'Now wash your hands'. At the exit to a munitions factory there should be the notice: 'Now please cut off your hands'. There is, of course, no such instruction.

In our culture fuck, bugger, cunt and prick: and shit, piss, fart and turd are swear-words. Kill, maim, mutilate, beat, stab, hurt, destroy are not. In other words obscenity resides wholly in the sexual act and the acts of excretion: not in the act of killing.

While there is no evidence that frequent sexual intercourse is in itself and as such harmful (aside, that is, from the indirect and in principle totally avoidable consequences of pregnancy and venereal disease), we legislate against this act at every level, both public and private. There is on the other hand clear evidence that firing bullets into people kills them. In America, however, no legislation exists against the purchase and possession of firearms; or in this country against the possession of air-rifles.

In the light of the foregoing are there not some grounds for suggesting that the attitude of Western society to sex is paranoid?[19]

What of the shallowness of felt and expressed emotion, described by Henry Miller at the head of this chapter, and so much in evidence around us? Shallowness of emotional response is, one recalls, a feature of the personality of the hospitalized psychopath. He is not only incapable of deep emotion but of sustained emotion over any length of time. Thus he has no real or abiding interest, no convictions, no passion and no remorse. When he hits an old woman over the head and takes her money, it is with no intention of being cruel to her, or of an actual – except perhaps momentary – wish to hurt her. This man is not so much un-emotional as an-emotional.

Such is the way also of the news services, particularly of newspapers.

There remains still a very important feature of clinical psychosis which we have so far neglected. This is the moment of onset of true madness – the point at which the patient loses his perception of 'real' reality, and can no longer distinguish between it and the products of his fantasy, his compulsions and malfunctioning nervous system. Anyone who saw either of the two films, Kubrick's *Dr Strangelove* or Polanski's *Repulsion*, will have seen instances of this moment portrayed. In the first film it is when the general suddenly begins to speak of the 'precious bodily fluids' that one abruptly realizes that he has become

19. One or two general comments about the use and meaning of the word paranoia in this book should be made here. *Clinical* paranoid psychosis is, for this text, the persecution of System A by System B. However, the persecution of System B by System A is also possible – so that, for example, hippies genuinely and generally feel persecuted by the police.

In more psychological terms, clinical paranoid *psychosis* is viewed here as the capture of, or the fear of capture of, the Ego by the Self. Clinical paranoid *neurosis* is the fear of the capture or destruction of the Self by the Ego. Chairman Mao is presumably suffering from paranoid neurosis.

As at all points, we really need *two* terms to describe paranoia, which take account of the differences just described. One is continually in the position of having to use the same word for two situations – as in the case of the word 'romantic' in the last chapter.

(clinically) mad. In Polanski's film we are shown the increasing oddness of the heroine's behaviour – a tendency to isolation and the breaking off of social relationships – which remains, however, within the range of the normal. Then the point where a work-mate opens the girl's handbag and we see that it contains the decaying head of a rabbit. At this moment we realize that the balance of the personality has tipped over into madness.

It is such incidents as these – usually perhaps less dramatic, though by no means always – which the clinician looks out for in his interviews with those referred to him: the indicators of the loss of touch with reality.

Such a moment appears to have occurred in respect of a whole *society* at the end of the last war, when the world saw for the first time what had taken place in the Nazi concentration camps. Was there not evidence here that a whole nation had moved over the line into insanity? The majority of Germans, as might be the case with a psychotic patient, have been least able to accept this view. In general we all were, and are, reluctant to admit that such events could occur in our species on such a scale – hence, although the evidence is actually incontrovertible, the world has taken refuge in the notion that Nazi Germany constituted a special and self-contained case, which could in no way be applied to mankind or society at large.

I would like to propose that Lake Erie constitutes a second example of the phenomenon. Nor is it by any means the only candidate for that position.

If this is indeed so, then matters have reached a point where it is becoming too late to remedy them.

In speaking of *society* and not *individuals*, however, we must beware of the possibilities of mystification. Society is not an actual thing or object which can exist in the absence of human beings. It is a collection of human beings, and only a collection of human beings. The launching of the study known as Sociology in relative isolation from the study of the individual has aided the process of reification – the belief that there is such a *thing* as society. (To be accurate, the boot is probably on the other foot –

it is rather the impetus towards reification which has produced, among other things, Sociology.) Thus when one here speaks of a psychotic society one means one of two things, and only one of two things, there being no third, mystical possibility, namely: (1) that a majority of the individuals who make up a community has become psychotic, and (2) that not the majority, but only the controlling minority, has become psychotic (that is, those individuals who control the decision-making machinery, the means of production, or whatever).

Both the cases specified could apply to our present society. In neutral terms, what appears to be happening is that the behavioural norm of Western society is moving along the ascending growth curve away from an acceptable state of (System A) normality, towards clinical psychosis. The statements by the psychiatrists at the head of this chapter, together with those of many others, appear to bear this out.

To close this section one further word in connection with the study of psychology. The reason why the author attaches such importance to this branch of activity (apart from the fact that he is himself a psychologist) lies in the circumstance that in this study we have the best chance to see how our society regards life, and the value and meaning it places on it. The social sciences in general are *par excellence* the show-case of our own view of ourselves.

The example which follows here is drawn from a book already quoted at several points in these pages – Ian Oswald's book on *Sleep*. Ian Oswald is also a psychologist, a well-qualified and successful one. That I have already quoted from the work in question is, hopefully, some evidence of my generally non-partisan attitude to this at once learned and useful book. One could have drawn something very like what follows from almost any psychological text in publication at the present time. There is therefore no question of selecting this book as anything other than a wholly typical example of what is widespread throughout the whole profession. The two following passages are taken from widely separated points of the text, as indicated.

There are, unfortunately, people who conduct hysterical campaigns bolstered by misleading, sentimental stories which evade the basic issues, in order to further the cause of 'anti-vivisection' as they call it. These people, who remain blind to the millions suffering from disease, and who forget the other millions who are today alive and healthy instead of dead or sick, thanks to the advances of medical knowledge in the last thirty years (penicillin and anti-diptheria inoculation, for example), try to prevent all experiments, no matter how simple, which involve animals. They speak of 'torture' of animals by scientists even when referring, for instance, to small and harmless injections. Of course it hurts to have an injection. But it is not 'torture'.

In fact, the 'poor, dumb animals' upon which developing medical science depends, often delight in the extra attention they receive through experiments. Unless they are happy, they are not healthy. Only healthy animals are of use for experiments. (p. 25)

The presence of two, alternating kinds of sleep, clearly analogous to what is seen in the human, has now been discovered in numerous animal species, so that one can reasonably argue that experiments upon cat paradoxical sleep are relevant to human sleep with rapid eye movements. At Lyons, Jouvet made cuts through the brain-stem, separating the lower part wholly from the upper part, or even removing the latter completely, including the cerebral cortex. What might be called the rear-brained cat still went on alternating between two kinds of sleep, as far as could be judged by the appearance and disappearance of muscle tension and other bodily signs. Whenever the muscle tension vanished, as in the paradoxical sleep of the intact animal, a strong electric shock was applied to the leg, muscle tension returned and the animal appeared to be 'awake' for a few minutes. . . . In fact it was possible to deprive selectively the rear-brained animal of paradoxical sleep. . . . One could not possibly attribute dreaming to a rear-brained cat, *so one could not say this particular form of meat was 'dream deprived'*. Only a physiological explanation is possible. (p. 103, my italics)

A number of the comments one would like to make on these extracts would border on the libellous. These must therefore be left to the imagination of the reader. One can legitimately ask, however, whether Ian Oswald is not guilty of a certain amount of double-think. Or could one accept that the cat in the second extract 'delighted in the extra attention' it received during the experiment? One would hope so, because only a happy animal is of use to the experimenter.

Of very considerable interest in these extracts is the immediately obvious and absolute denial of the unconscious, a very common feature in the literature of academic and experimental psychology. The experimenter here is clearly totally unprepared to admit that any form of consciousness or awareness could be associated with the lower or extra-cerebral centres of the brain. 'Only a physiological explanation is possible.'[20]

In Chapter 6 it was suggested that one explanation for the insomnia suffered by clinical psychotics and pre-psychotics might be the desire to escape from dreaming and the unconscious.

The precise wording of the italicized sentence in the second extract quoted above gives particular cause for concern. To myself such statements are decisive links in the chain of evidence for the reality of the psychotic society, as defined by this chapter.

This aside, one must ask oneself in all non-sentimentality whether it is likely that any good for humanity can come out of any area of study (i.e. from experimental psychology) in which such attitudes are fundamental.

20. One has already noted elsewhere that psychologists are far more ready to assign awareness to a computer than they are, not only to a living organism such as a cat, but to a human being.

Part Six:

Evolutionary Prospect: The Three-Dimensional Personality

Art, Morality and Justice: The Triad in Eastern and Western Culture 14

The final chapters of a book are normally reserved in first place for the summing up of the central argument or a drawing together of threads: or such a chapter may contain suggestions for the further development of the main theme in relation to broader issues and future developments.[1]

The present book is untypical at least in that the major part of this chapter is devoted to broaching a further topic. Some attempt, certainly, is made at a summing-up – but the nature of the new topic itself largely rules out that possibility. As for the relating of the central theme to broader issues, hopefully this book has been concerned with little else throughout.

We turn essentially to a consideration of System C.

1. A moment's thought shows that an attempt to 'sum up' a living species still in the process of evolution would be a somewhat unreasonable undertaking. Such aspects as one perhaps could sum up are, by definition, *historical* features of the organism. The social sciences may indeed be indicted at this point precisely for choosing to concentrate on such historical issues – but worse, for assuring us that *these* are what is most meaningful. Of the more recently evolved (and in that sense more typical) higher behaviours of the human organism they have little to say that impresses. The social scientist seems to be almost solely concerned with the conditioned or robotic behaviours of the organism. As we shall see in the course of this chapter, the experimental demonstration of 'basic' behaviour is actually achieved only with organisms maintained either in states of unawareness, or subjected to forced choice. The free organism is a very different proposition.

How can it be that there is a third major system within the personality, or rather, how has one been able to avoid discussing it until now? This is not such a feat as it seems. Nearly every psychological (or religious, political, philosophical, or whatever) theory manages to sound most of the time as if it were speaking of the totality of human experience. It is partly a matter of skilful window-dressing, the effect of blind spots, and so on.

In connection with the proposed System C we shall be considering among other topics art, morality, justice and, not least, humour.

First, however, the *a priori* justification for the existence of this new System – the 'theoretical' considerations which demand, or at least allow of, its existence.

On the purely conceptual level one could propose that if Systems A and B exist, as claimed, within the one individual they could, taken *together*, be described as a position, or phenomenon, C. The manner in which the two interact might then perhaps already be grounds for speaking of a *System* C.

Diagrammatically this equation is:

$$B \times A = C$$

What are the possibilities, however, for justifying the following triadic model? Would this have a genuine relationship to some actuality?

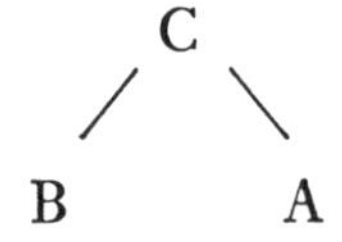

Possible initial support is provided first by the fact that the personality – as the present book has pictured it – is no stranger to triads. The following have for example, become familiar to us.

B* = e.g. emotion

B1 B2 love hate

However, let us for the moment cast our eyes further afield and ask what triads are in general use in fields of human activity – which phenomena, if widespread, might begin to look as if supported by some (unconsciously) perceived basis of personality, such as I believe we find in respect of the cross, the journey, the narrow gate, the old man of the sea, and many other archestructures.

There are indeed a great number of such triads. It is not easy to decide which of these to consider first. Additionally, let us admit straight away that the *content* of these triads often does not in every case accord well, at least not in all aspects, with the content of the personality triad we hope in due course to establish. Three of these already widely known triads follow.

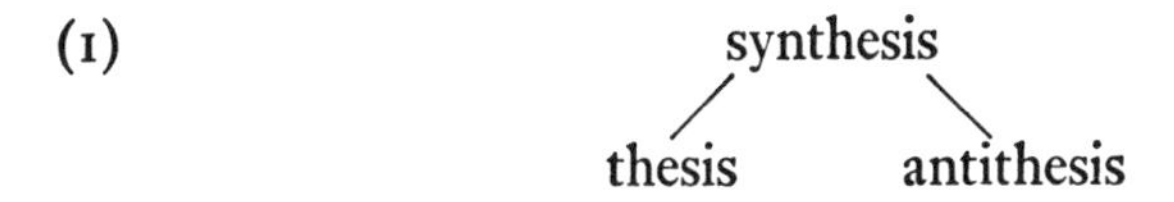

A description of the progress or development of argument in Greek philosophy.

(2)

reason

appetite passion

Plato's view of the structure of the soul. This triad is possibly a concertinaed or collapsed version of the pyramid of triads shown in the footnote on this page. That possibility is discussed later.

---

* Pre-empting the present discussion, we could produce a pyramid of triads, thus:

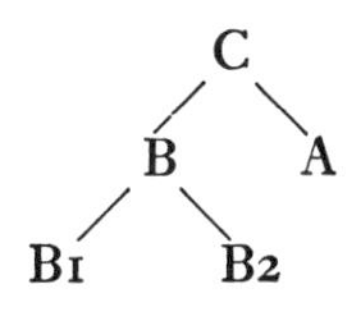

(3)

The 'third eye' of Shiva and mysticism. The eye of inward vision and clairvoyance.

Attention should be directed at the uppermost element of the triad in each case – to the qualities of synthesis, reason and inner vision.

Without drawing each out in the full diagrammatic form, two other extremely well-known threes in our cultural inheritance are Freud's id, ego and super-ego; and the Father, Son and Holy Ghost of the Christian faith. Freud's formulation, as it happens, fits well enough into the schema of B, A and C; that of Christianity less well.[2] Perhaps one should just point out in passing that the *trigrams* of the *I Ching* are also three-line figures.

Let us, however, for the moment not pursue the question of existing triads[3] and propose a few of our own, as follows.

---

2. One could go on at this point to consider the wide prevalence of the number three in superstition, magic and religion generally. The fairy grants three wishes; three Magi attended the child Jesus; accidents and deaths are said to happen in threes, and so on. The precise origins of these matters do not concern us.

Nevertheless, the considerable significance of three in human affairs has provoked much speculation. Psychoanalysis has proposed that the force of three derives from the male penis and the two testicles, or the two breasts and the vagina. One should, perhaps, not overlook the fact that our immediate section of the universe is made up of the sun, the moon and the earth.

As far as my own views are concerned, I believe that the wide occurrence (or concurrence) of a pattern or system – a positional relationship, a numerical quantity, or whatever – indicates the presence and influence of an arche-structure.

3. soul / \ body mind — Yet we cannot move on without mentioning another famous three, that of body, mind and soul. One has, it is true no particular grounds for setting them out as a triad. However, in ascending order of value or merit body would be held lowest,

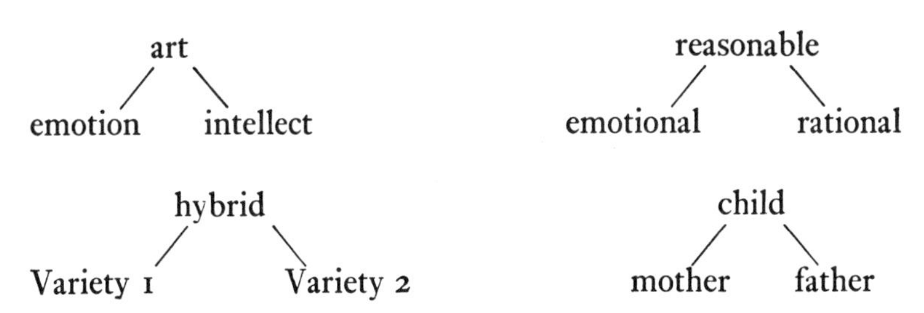

The full significance of these will emerge. For the moment, the point to bear in mind – as is most clearly and literally emphasized by the hybrid of animal and plant cross-breeding – is that the qualities of the resulting 'product' cannot be wholly predicted from a knowledge of the properties of the 'parental' elements. It is on this note that we can turn to a consideration of art, morality, justice and humour.

As I hope to demonstrate, these behaviours cannot be readily lodged under either System A or System B, as defined and described in this book. Certainly elements of both these systems can be discerned in the phenomena in question. Still other features, however, cannot. Since art has already been discussed elsewhere, we shall begin with this.

Art, as we have said, is a marriage of form and content. By form one means a number of things – the imposition of structure or control; the limiting or setting of boundaries; and also the fixing or capturing of what is otherwise fleeting or transient. By content one means inspirational content, not information. Though the content of art may happen to convey or include

---

and soul highest. Floods of ink have been spilled over the meaning of these terms. We, however, are concerned only with the fact that they exist – and that there are three of them.

There remains then still one further very well-known triad of great relevance to the present inquiry, which will be considered a little later.

certain factual information, this conveyance or inclusion as such is none of the purpose of art.[4]

The point is also not simply that aspects of both the A and B Systems are present. It is, first, that both *must* be present – even though in the various extreme expressions of romantic, classic, abstract or spontaneous art the admixture of the two may be far from equal. More importantly, the nature of these elements is changed. It is not altogether easy to convey how this is the case. Let us consider time. While a play, for example, confers a finite and definite temporal structure of one kind or another on some part or some view of life, there is no answer to the question 'how long is a play?' Certainly, one can ask, 'what interval of measured time elapses between the first raising and the final lowering of the curtain?' Yet (without speaking here of boredom) one of two plays of equal *objective* duration may, subjectively, appear to have occupied a considerably longer time than the other.[5] One might consider, too, the question of repeatability. Surely in *this* a play resembles a scientific experiment? To which the answer is that it does, and does not. It is very debatable whether even a play seen on two consecutive nights with the same cast can be said to be the same play on the two occasions. One has certainly not the same play when it is performed by a different cast with a different producer. What, however, of a painting? Surely this remains the same on every occasion? To which question one must again reply both yes and no. It is what one brings to the painting, as also to the play, of

4. Many might dispute this point. Art forms involving the written word, plays, novels in particular sometimes appear to be solely the vehicle for an explicit, cognitive message. However, whereas it is true that when the *informational* content of a work is increased beyond a certain point, that work ceases to be a work of art – so a telephone directory is not art, nor, usually, is a Government pamphlet about the filling of sandbags – it is possible to *reduce* the informational content (as well as syntax and whatever else) of a written work almost indefinitely without destroying it as art. So perhaps James Joyce, Dylan Thomas, Harold Pinter, E. E. Cummings and many others.

5. Perhaps one is merely saying here that the question of duration is intrinsically irrelevant to the play's value as art.

course, that is always different. *We* are never the same person twice. One of the features of art with which System A – the intellect – cannot readily cope is the unchanging ability of such works to form new and meaningful relationships with us at every point in our continually changing lives.[6]

Art performs many miracles. It captures the uncapturable moment – returns it to us ever and again, always new; makes permanent without fixing, cages without limits, repeats without boredom. Particularly, it 'captures' timelessness: so that the cave paintings of 20,000 years ago are today as immediate and direct in their appeal as when they were first created. Would not such achievements seem beyond the scope of either System A or System B alone?

Morality is as vast a subject as any discussed in this volume, and cannot therefore be dealt with in any but the most general terms. However, it seems to me that *any* conception of morality must be somewhere based on a freedom of choice – the free choice between two or more alternatives. Since how could a man, compelled to perform an action by powers greater than his power of resistance, be considered to be acting immorally? In another context, a person suffering from, for example, kleptomania is clearly (to use the legal phrase) in a state of diminished responsibility. Indeed, the establishment of diminished responsibility, by reason of a *psychiatric* condition, is now an accepted, albeit sometimes reluctantly accepted, basis for a defence plea in English courts. Here it is recognized that a man suffering, say, from paranoid delusions is from an internal point of view in much the same position as is, from an external point of view, a young child forced to steal by his father. Both of these individuals are not free to choose, but are 'possessed' by some force, inner or outer, greater than their resistance to it.

With the use of the word possession one is now able to link the present discussion with our considerations in Chapter 1,

6. Where is the scientific experiment or the mathematical computation which we could bear to repeat or see repeated even ten times – let alone again and again throughout our lives?

We have in fact been discussing no other subject throughout the whole of the book.

The possessed man is not a free man, and is not at that point immoral because he is not, and cannot be held to be, responsible for his actions. The immorality of this person may, however, *precede* the actual state of possession. The crime or immorality can reside in allowing oneself to be possessed in the first place – if prior to that point one was free.[7]

All legends and stories connected with any aspect of possession place great and specific emphasis on the crucial nature of the initial point of choice. Thus Faust initially *seeks* the pact with the Devil: and until he actually puts his signature to the pact he remains a free man. The signing is a symbolic or stylized representation of the actual choice point. So the *afterwards* helpless victim must first *invite* the vampire to cross the threshold. So the wilful sister in *Goblin Market* chooses to eat of the goblin fruit. And thus Eve and then Adam, long before her, chose to do the same. *Until* they ate of the fruit they were free. Here in the Genesis story one sees the paramount importance accorded by Christianity to the moment of temptation. And so it is, too, with Dr Jekyll and with all the magicians, characters and situations examined in the earlier part of this book. He who touches pitch shall be defiled – but he does not *have* to touch it in the first place.[8]

What does experimental psychology have to say on these

7. From here on, for the purposes of the discussion, this prior freedom is assumed to have been in fact the case. It is assumed that one is at some stage preceding possession free to choose, free to reject the offer, free to remove oneself from the situation.

8. Christianity, of course, recognizes that the mere *presence* of the tempting object may prove overwhelming: that once one has even entered the situation of temptation, one is already lost. So it counsels the avoidance of the tempting situation. The choice point has then been pushed further back. It does not reside in the moment of temptation (for then the cause is already lost) but at the moment when the individual – at that point still away from the situation and still free – decides or chooses to put himself in the way of it. cf. 'I'll go into the Casino but I won't actually play' or 'I'll have just one drink and then I'll stop', and so on.

matters? It is probably not selling the experimental psychologist short to say that he views this whole problem as one of conflicting reinforcement schedules. Not that he is overmuch concerned with morality as such – but he is extremely interested in situations where choice is made between alternative courses of action, and the parameters which favour or predetermine the emergence of a particular choice over others.

It can and has been repeatedly shown that a naïve rat, meaning a rat not previously the subject of experiment or other experiences unknown to the experimenter, will make whatever choice he has been conditioned to make by the schedules of rewards and failures provided by the experimenter – in terms of such variables as the frequency, recency and total volume of reward (or, less straightforwardly, of punishment). A rat who finds food behind a trapdoor bearing a cross, but never behind a trapdoor marked with a circle, will always, given a choice of two doors – out of a maze, or whatever – take the door marked with a cross.[9] The very eagerness or strength with which the rat pushes the door can be predicted (and verified by instruments) from a detailed knowledge of his reinforcement history.

One has no wish to challenge these findings as such. They are facts, in that sense meaningful, and certainly of great interest. The only question which occurs to one is *why should the rat do anything else*? Why should a rat, particularly a hungry rat, *bother* to do anything else?[10] What occurs when he *is* 'bothered'?

It looks in fact, as we shall see, as if the laws of conditioning operate only in a psychologically – as it were, clinically – sterile situation, that is, when all other things are equal. One can say that the findings of conditioning are the laws of the robot organism (perhaps even of the 'possessed' organism). To the point

9. This means in practice something rather less than 100% – but certainly a very high percentage of trials indeed. As will be shown there are, nevertheless, grounds for challenging the interpretation even of this result.

10. The rat in experiments is usually a hungry rat: failing this he must be a frightened rat, a randy rat, or whatever. Otherwise he tends to do nothing at all – that is, he rests, or preens, or potters.

which, and on the occasions when, he does *not* resist his conditioning (in practice something like 100% of the time), the rat's behaviour follows the rather precise lines demonstrated.

Digressing very slightly, one feels strongly that the error of applying the findings of such conditioning studies as these to human beings lies in ignoring the question of what happens when the conditioning is resisted: lies in the denial of the fact that there *can be any resistance* – except in terms of another conflicting reinforcement schedule. To be fair, the psychologist is not simplex on these points. He would agree that the interactions of the very numerous reinforcement schedules to which every organism (and particularly the human being) is subjected in the course of its life are extremely complex, and not readily forecast.

Though one does not usually see specific statements to this effect, it appears however that the academic psychologist tacitly and wholly denies any idea of an 'independent' or overruling volition (what used to be called free will) in the human organism. If an individual does not respond as conditioned or predicted by the reinforcement schedule administered, the explanation is that some other, stronger conditioning, some other set of conditioned responses, have been brought into play which were till that moment somehow in abeyance. If the individual does nothing at all, for instance, it is suggested that his general level of motivation is low – implying that if he *were* to do something, it would be what the reinforcement schedule pre-ordains.

If the foregoing were true, of course, then no one would be *responsible* for anything. On this view we are the products, the simple victims, of our conditioning (our past experience, in that sense). And without responsibility there can, of course, be no such thing as morality.

Even the observed behaviour of the experimental rat, however, calls this view and the allegedly solely dominant role of the conditioning processes into question. The following brief considerations must, unfortunately, constitute the whole of the attack we have space in which to mount on the conditionist position.

If the rat is presented with a choice situation where failure is painful, and/or where the execution of the choice made requires a considerable effort, perhaps even a degree of danger, rather unusual response-behaviours are observed. Let us suppose the rat is placed on a stand high above the ground, from which he can escape only by jumping through one of two doors some distance away. The correct door yields instantly to pressure and deposits the rat on a cushion. The wrong door is solid, the rat receives a painful bump on the nose and falls a rather long way to a safety net. In such choice situations the rat indulges in what psychologists have labelled 'vte' (vicarious trial and error). That is, he examines both doors carefully, leans out towards them, makes little movements as if jumping (though clearly not actually intending to jump at that moment), as if trying out the two alternatives in a kind of hypothesized reality. *Something* would, at any rate, appear to be going on in his head. Let us anthropomorphize, and perhaps sentimentalize, the situation completely. It is *as if* the rat is saying to himself: 'Now, wait. There's a door with a cross and a door with a circle. Up till now it's always been the cross. Hasn't it? Never yet had a good result with a circle. O.K., the door with the cross it is.' And he jumps.

If the rat has undergone some reinforcement with each door, his vicarious trial and error behaviour is more marked and lasts longer. Here he must 'weigh' degrees of conflicting experience.

If the experimenter at some point randomizes the correct solution – that is, offers no correct solution, it being sometimes one door and sometimes the other for no actual reason – a further interesting development occurs. The rat settles for one door or the other (the door with the cross, the door on the left, or whatever) and thereafter jumps for that door every time, regardless of the outcome. This stereotyped response persists long after the experimenter re-regularizes the experiment so that a correct solution is again possible. During this lengthy period a rat will time after time monotonously bump his nose and fall into the net.

These responses are not easy to account for in straightforward conditioning terms. Would one not expect the rat's behaviour to remain as random (as disorganized) as the schedule itself? Moreover, it is possible to regard the rat's response as extremely stupid *or* extremely shrewd. The stereotyped response might be a breakdown of all attempts to cope with the situation. Or the rat may have decided, like some gamblers, that since nothing he does seems to have any effect, in what seems a game of pure chance, he might just as well do the same thing every time.

The demonstration of vte and stereotyping already perhaps creates certain difficulties for the conditionist position. Already there it begins to be as meaningful to speak of 'drawing on one's experience' as of 'response to conditioning'. Certainly the outcome is, in effect, in many (though perhaps not in all) cases the same. The first phrase, however – 'drawing on one's experience' – unmistakably *implies* (whether justifiably or not) some kind of choice.

The difficulty of challenging the strict conditionist position at the level of animal experiment lies in 'persuading' the organism to act against its conditioning, and by means other than introducing or postulating another reinforcement schedule. The question one is seeking to resolve is, whether an organism (in particular perhaps the human organism) can act against its conditioning (its experience, its learning) – and if so whether merely under the counter-influence of some sort of counter-conditioning, or whether by means of some part of the personality which exists in *some kind* of independence (either total or relative) of the conditioning processes.

As it happens, evidence for this postulate exists, and as it happens the evidence is actually from experimental sources. One finds it only in occasional or passing references in the literature, so that it is clear that the psychological establishment fails to grasp (or does not wish to grasp) the significance of this phenomenon for the structure and functioning of personality. Instead, as the extracts which follow admit, experimenters by and large prefer to concentrate their attention on *animals*, where these 'difficulties' do not arise.[11] The extracts are from a discussion of conditioning principles, mainly in connection with the responses of human subjects to various conditioned stimuli (nonsense syllables, flashing lights, and so on).

Razran assumed that these differences among subjects were due to their attitudes. . . . So he tried giving subjects positive and negative instructions at different times. For example, if he told them to try to associate the nonsense syllable with eating pretzels he usually got positive results, but if he told them to *avoid* forming the association, zero or negative conditioning was apt to result. Razran explained his results by assuming that a conditioned response in adults is dependent on two factors. . . . The first factor is the one that operates in animals, and is the one to which the laws of conditioning apply directly. The second factor develops in the child around the age of three to five and obscures the functioning of the first factor. . . . The ultimate explanation of this [second] system may be in terms of principles like those of conditioning, but at the descriptive level it has quite different laws.[12]

. . . Razran presented a series of conditioned stimuli during a long eating period, having misinformed the subjects as to the purpose of the experiment. For example, he gave 40 flashes of coloured lights (the conditioned stimulus) during a two-minute period of eating pretzels (the unconditioned stimulus), having told his subjects that he was trying to 'find out the effects of eye fatigue on digestion'. With this technique, none of many subjects, during several hundred total hours of experimentation, gave any sign of thinking he was expected to secrete saliva in response to [the light]. As a result, the data show striking confirmation of Pavlov's main behavioural findings in acquisition, extinction, spontaneous recovery, generalization, etc.[13]

11. One suggests that they *would* and could arise were it not for the fact that the experimental animal in most circumstances has no interest in, or motivation towards, anything but the next most obvious step. We could say that a rat 'falls' every time he is 'tempted' because he has no reason not to fall.

12. G. H. S. Razran: *Conditioned Responses: An Experimental Study and a Theoretical Analysis*, Archives of Psychology No. 191, 1935.

13. G. H. S. Razran: 'The Effects of Subjects' Attitudes and Task-Sets upon Configural Conditioning', *J. Exp. Psychology*, 24, 95–105, 1939.

In considering these results one is immediately faced with the question as to how attitudes can control a response like salivary secretion, which is involuntary . . .

The attitudinal factor that Razran demonstrated so clearly probably complicates all studies of the conditioned response in man. Consider the psycho-galvanic [skin] reflex – not only is it involuntary, but most people do not even know they have one. Yet the conditioned psycho-galvanic response will often drop out entirely if the experimenter merely says 'O.K., no more shocks. I just want to try the light (the conditioned stimulus) a few more times.' If the subject really believes the experimenter and ceases to expect a shock after the light, the psycho-galvanic response conditioned to the light drops out almost immediately.' R. S. Woodworth and H. Schlosberg, *Experimental Psychology*.

In brief, the foregoing says as follows: that *normal* human subjects respond to classical conditioning in accordance with Pavlovian laws when they are in a state of unawareness of what is happening or supposed to be happening; when they are aware, and negatively instructed or orientated, the conditioned response is *not* usually established – even when the response is one over which human beings have in fact no conscious control (or even know exists): when they *are* aware and *positively* instructed or oriented, conditioning usually does take place, again in line with the known laws of classical conditioning.

There are many points of great interest in the above extracts. Though they do not use the term, are not these experimenters in fact discussing human (free) will – the exercise of choice and control in the management of the psychological economy? 'The second factor develops in the child around the age of three to five and obscures the functioning of the first factor.'[14] This statement suggests that up to that age the child is more or less the helpless victim of whatever (here classical) conditioning he experiences. Thereafter he apparently begins to be able to exercise some kind of volitional supervision over these processes.

14. What is it that begins to emerge at this age – System A? Or may we already begin to speak of a rudimentary System C?

Why, one wonders, is this statement not printed on the front of every psychology textbook?

Something along the lines discussed is on the other hand assumed by many *non-experimental* psychologies and philosophies, as well as religions.

Woodworth and Schlosberg report that 'If the subject really believes the experimenter . . . the conditioned psycho-galvanic response drops out almost immediately'. May one alter 'believes' into 'trusts' the experimenter? What one would very much like to know is whether this same result would be obtained with clinically neurotic and psychotic subjects. One suspects not, though for the moment this remains only an assumption. One *should* bear in mind, however, that in these reported experiments one is dealing with normal, balanced subjects.

It is indeed remarkable (as the writers themselves comment) that a function of the autonomic nervous system – a function in this case that the subject does not even know he has – can be regulated by a conscious attitude and a conscious instruction, that is, by the central nervous system (System A). The theme of the present book has after all throughout been that the two major personality systems are antagonistic, have different aims, do not work in conjunction, influence each other as a rule only occasionally and to small extents, and then more often than not interact disruptively and negatively.

In particular, it will moreover be obvious that the result described by Woodworth and Schlosberg is *precisely* the opposite one to that obtaining in connection with the lie-detector test, or with the tachistoscopic projection of taboo words. In the lie-detector, the presentation of words associated with the crime (which operate as conditioned stimuli) produces a psycho-galvanic response, *despite* the absolute determination of the conscious mind not to allow this to occur. Yet in the experiments quoted above, the autonomic system readily obliges the merest hint or whim of the conscious personality, and allows itself to be switched off. Autonomic functions are outside the direct control and often even the awareness of consciousness

(System A), as we all know. Yet here we have clear evidence of such control. Whence and why this sudden, miraculous co-operation?

In seeking an answer to this question we now, very reluctantly, terminate this brief excursion into experimental findings and turn instead, first to a consideration of the word 'balance'. Balance (meaning literally 'two dishes') is used not only of a stable personality but in one of our most important triads – that of justice. The figure of Justice does not at first thought appear to be a triad, but on examination is found to be such. The figure holds in her hand a pair of scales (in which in Egyptian and other mythologies the souls or hearts of the dead were weighed). The outcome of the balance of the two pans is justice, fairness and impartiality.

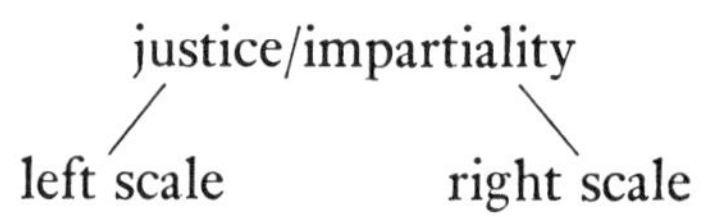

When one speaks of a balanced personality, does one mean one in which the *two* parts of the personality are 'in balance'? Or is the literal meaning of the word not to be considered? In connection with the mature or adult person(ality) we use also the following expressions: integrated, level-headed, even-tempered, impartial; we say of such a person that he is not one-sided, biased, partisan, not impaired in his judgement.

'Integrated' means literally whole or complete (Latin *integer* = undiminished, unimpaired, whole, complete). 'Impartial' means not divided, separated (Latin *partire* = to separate). 'Impaired' literally means unequal (Latin *paria* = equal). It is very interesting also that the mathematical sign for 'equals' is *two* lines. In English, of course, the word pair comes to mean two.

One sees, therefore, that the concepts of 'two-ness' and 'not two-ness' are fundamental to many of these words and concepts. In the integrated (and desired) situation 'two-ness' is replaced by 'one-ness'. We are concerned in this chapter particularly with this one-ness and to show, more especially (a) that this one-ness is the *product* of two-ness and (b) that the qualities or attributes of one-ness are other than, or are additional to, the attributes of the two contributing parts.

Let us revert momentarily to the experimental situations described by Woodworth and Schlosberg and the opposed findings of lie-detector experiments. Once again we wholly anthropomorphize the position. In the former case is it not as if the two parts of the personality, the central nervous system and the autonomic nervous system, are working together as two colleagues engaged in a combined project? System A calls over 'I say, Bert, would you mind switching off that psycho-galvanic reflex!' 'Right-o!' shouts Bert, switching it off. The situation is one of ease and harmony (Greek *armonia* = a fitting or joining together). In the case of the – guilt-ridden – individual and the lie-detector one has a very different situation. System A screams 'For God's *sake* switch off that reflex, you bastard!' And System B returns 'I'll see you in hell first!' It is indeed into hell that the personality has moved at this point.

In the situation of harmony the miracle is achieved of the two antagonistic Systems working together as if there were one System, an outcome, once again, that could hardly have been predicted. There is no *manipulation* involved here: rather we have a situation of coexistence and benign mutual influence[15] – an appropriate, undisputed division of labour – and, incidentally, an extended range of achievement potential. Is not this, too, an excellent description of a good marriage? Or rather, one should say that this is the *possibility*. One should refrain from speaking euphorically of this situation as if it were more than an *occasional* actuality.

In summary, we can perhaps claim that a first case has been made out for the reality of System C. That it is, moreover, not merely a conceptual framework within which hypothetically to contain the actually existent Systems A and B. System C

15. The lion, perhaps, lying down with the lamb, and so on.

seems *actually* to exist – though as it were as a beginner, a guest lecturer, an occasional student, not as yet a permanent feature of our psychological landscape.

The manifest attributes of this third system seem to be wholly desirable: in respect of it we find in use such terms as balance, justice, reason, integration, synthesis, harmony, inner vision, clairvoyance and so on.[16] However, let us descend from these somewhat heady heights to more mundane considerations, which nevertheless perhaps carry rather more weight.

When the two Systems A and B are working in harmony or balance a freedom is conferred upon the personality, or if preferred, a whole personality is created, which the person does not possess when either system is favoured at the expense of the other, or attempts to function as if it were itself sole arbiter of the individual's destiny. The integration of the two systems either creates or permits to emerge, or both, a – potential – third system, System C. One is driven to postulate this third system partly by reason of the fact that the 'liberated' personality evidences qualities and abilities not found in either of the component Systems.

At this point we may perhaps attempt to produce a descriptive expression or label which can be used to summarize and convey the nature of System C as we perceive or experience it in others or ourselves – in terms, that is, not of theoretical concepts but of the life experience and of actual people.

There exists in Yiddish the expression '*mensch*', this being a straight derivation from the German *Mensch*. The German word means 'a human being'. Much more than the English expression,

16. Such, essentially, is F. Schiller's view of the aesthetic experience. Though his terminology is quite different, his view of the personality when liberated by the artist is substantially that expressed here. See F. Schiller's *On the Aesthetic Education of Man* (1793), transl. E. M. Wilkinson and L. A. Willoughby, O.U.P., 1967.

however, *Mensch* conveys the qualities which are only sometimes indicated by the phrase in English – namely the qualities of sympathy, decency and humanity. In Yiddish in particular these aspects of the term are wholly paramount: thus in that language *mensch* means a good guy, a straight person, a regular person, a decent sort, a vulnerable and feeling, though by no means soft or stupid, individual; the kind of man or woman one can turn to, and is glad, even proud, to know. There is no comparable term in standard English. However, among English-speaking Jews, who, deliberately or otherwise, tend over generations to produce English equivalents of such Yiddish expressions, the word 'person' now carries the value of *mensch*.[17]

Since 'person' both exists as an English word and can without too much forcing be made to embrace precisely those behaviours which we have postulated for System C, let us adopt the term 'the Person' for our third major personality content, conferring on it the necessary status by capitalization. The over-all personality triad thus emerges as:

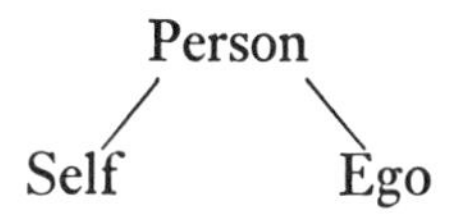

Possibly the most striking of the attributes of the third system is humour. One does not intend here to attempt to analyse humour or to define it. Whatever the reader understands by humour will serve for present purposes. However – turning specifically for a moment to laughter – one is thinking of the relaxed response suggested in the phrase 'helpless with laughter'. One excludes forced laughter, and hysterical or manic laughter. In forced laughter the individual concerned is almost certainly not experiencing the inner state which is the correlate of true

17. An example of its usage occurs in the film *Funny Girl*, where in the spoof ballet-scene Barbra Streisand asks the male lead who has approached her with a series of traditional ballet leaps: 'Can't you walk across the stage like a person?'

laughter: though for reasons of his own he wishes us to think that such is the case. Hysterical or manic laughter is best viewed as a (minor) state of 'possession'. True laughter, on the contrary, is best described as a *release from possession.*[18] Genuine humour may be further considered to lack as an *intention* all cruelty, spite, contempt, derision, and so on. These comments on the subject must suffice, otherwise we shall find ourselves, despite the earlier disclaimer, engaged in the business of definitions.

The question of humour is raised here principally in order for us to discuss its absence. For the hallmark *par excellence* of the pure System B and System A statement is the lack of humour. There is not one intentional smile, much less a laugh, in the whole of the Old and New Testaments, the *I Ching*, of Marx's *Capital*, of *Mein Kampf*, of the Conservative party's Manifesto, of any textbook of Physics or Chemistry, of Woodworth and Schlosberg's *Experimental Psychology* – or indeed *anywhere* except in self-confessedly humorous works. Ignoring for the moment the gross evidence of compartmentalization on view here, let us consider instead the question that most people would certainly voice, namely, 'Why *should* there be any humour in those books?' The question implies that humour would constitute an intrusion in such contexts, be out of place, unseemly and irrelevant.

Can we see, however, that this objection arises, first, from an unquestioned belief in the essential rightness of compartmentalization? I am not concerned here with whether or not, or for what reasons, it might be a good idea, say, to have a joke printed at the foot of each page of a physics textbook – or for that matter included at intervals in the text. One is asking why it is apparently *self-evidently* true for the majority of individuals that this should not be so.

For the present book the key to the situation lies ultimately with the archetypes and archetypal behaviours of the Ego and the

18. The notion of 'release from possession' is actually a very useful and accurate description of the state of balance that arises from the integration of Systems A and B.

Self; and with the essential estrangement of System A from System B that we see in all societies both of East and West. At its broadest the division consists of the, as far as possible, complete split into the two camps of 'Caesar' and 'not Caesar'. Societies as a whole may be viewed as occupying one or other of these two camps – but the two camps, of course, also exist within each single individual, as repeatedly emphasized. By virtue of the felt need to guard the frontiers of the camp against the constant threat of infiltration by the enemy (the forces of the opposing system) there arise the two concepts and mental states of Self-importance and Ego-importance.[19] These two ideas are normally, and from the point of view of this text incorrectly, covered by the one term, self-importance. For immediate purposes, a small 's' will be used for that common term, and the capitalized version, Self-importance, when the importance of the Self as opposed to the Ego is at point.

Ego-importance (which is what is normally called self-importance) is the easier of our two to identify. It is demonstrated by the love of status and the trappings of status, by titles, honours, uniforms, medals, by the existence of, and adherence to, rigid and complex rule-systems and procedures, hierarchical structures, by emphasis on precedent, precedence and tradition, by competition, by praise and spoils for the victor and contempt for the loser, and so on. One need hardly continue this list.

Self-importance, as defined by this book, is harder to demonstrate – though it is exposed shrewdly enough at many points for instance in the Old and New Testaments. One may speak, for example, of the pride of humility, of the holier-than-thou or 'I have suffered more than you' syndrome. The subject of the sin

19. Additionally, within the area of the Ego we have a further phenomenon of compartmentalization which arises from the nature of the Ego already discussed. The 'method' or style of the Ego is to divide, analyse, separate, and it is this approach which has so dramatically advanced our knowledge and control of the physical universe. The style of the Self (an opposite, as always) is that of association and unification.

of actually seeking after suffering, of more than willing martyrdom, is also allied here. Such are discussed, for example, in T. S. Eliot's *Murder in the Cathedral* and Robert Bolt's *A Man for All Seasons*. There are, too, a number of versions, in various communities of the East and the Americas, of the pride in 'winning' by *giving away* most. The most generous – often therefore finally the poorest – man has the most merit. This is perhaps a variant of the 'inverted snobbery' recognized in our culture.

Leaving that context altogether, one notes the Self-satisfaction which many women (but fewer men) derive from fecundity – the high production-rate of children – which is another instance of the vaunting of the Self.

Thus, in this short space, it proves after all not too difficult to provide instances of Self-importance or Self-pride.

One may now see clearly how equally dangerous the notion or presence of humour is to *both* Self-esteem and Ego-esteem. Such essentially vulnerable structures, built as they are on ultimately untenable premises, must at all costs be protected from the corrosive properties of the amused or detached view.[20]

These considerations lead naturally to the questions of reason and reasonableness.

It was earlier suggested that Plato's famous triad of appetite, passion and reason was a collapsed form of the fuller paradigm given in the footnote on page 481. Actually, there are two possibilities. If by reason Plato meant rationality or rationalism, then his triad is really something like B1, B2, A, thus:

If, however, by reason one understands that which is reasonable – reasonability or reasonableness, one has rather:

20. Perhaps I should point out that I am not objecting to people taking themselves seriously – indeed, I would prefer them to take themselves tragically. The objection is to taking oneself importantly.

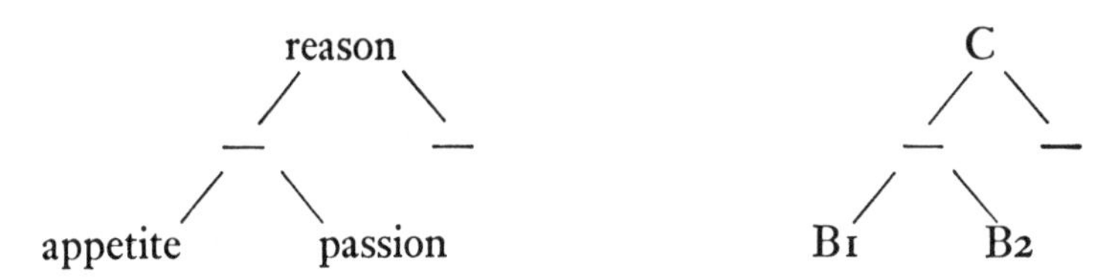

The present text, at any rate, wishes to distinguish sharply between rationality and reason. Reason is then understood to refer to what is contained in the two words reasonable and reasonableness. Rationality is the abstraction from rational and rationalist. The rational can be defined for present purposes as logical but cold; the reasonable as less rigid, and while in no way irrational, certainly containing a measure of the warmth and yielding qualities of System B. Reasonableness is seen as an attribute of the Person, rationality as a characteristic of the Ego.

Again, however, we are more concerned to discuss the absence rather than the presence of reasonableness. Both System A and System B are notably, and characteristically, *un*reasonable. They are extreme and absolute: in political terms, totalitarian. It will not be difficult for the reader to generate examples for himself. Only one is offered of each kind of extreme view from the two systems under discussion, with one other from the third system for comparative purposes.

System B: The sins of the fathers shall be visited upon the children to the third and fourth generations.

System A: If the peasants have no bread, then let them eat cake.

System C: Whatever turns you on.

The third statement here emphasizes what is so sadly absent in the two earlier statements – tolerance.[21] The nature of intolerance is familiar to all of us, and need not be discussed here. One may, however, perhaps distinguish between active and

21. Essentially the willingness to concede that there are many ways of approaching any one matter, among them those which suit some people and those which suit others, and not one of which can necessarily be called 'best', except perhaps in the case of a particular individual.

passive intolerance. Active intolerance is where an alternative viewpoint or method is so intolerable that the opponents of it are moved to destructive action to wipe out the 'abuse' and the anathema. Passive intolerance – which is as undesirable since it can always become active – is where a rival concern is suffered to exist, but without the concession that it has every *right* to exist. That right is conceded in perhaps the most famous – if slightly unreasonable! – statement of tolerance of all: 'I disapprove of what you say, but I will defend to the death your right to say it.'[22]

It is a saddening observation that the majority of individuals in our society – as in all present societies – appear content to live out their lives under what we must call the dominance of archetypal symbols and situations – the largest archetype being that of the part-personality, Ego and Self.[23] The sadness arises from the fact that all these situations are merely hang-overs, an unwelcome legacy, left-overs in fact from the historical past, by which one means the evolutionary history of our species, and no longer essential to our survival in the literal sense they once were. Paradoxically they are now more likely to prove the agents of our destruction. As long as we tolerate and seek to encourage their continued existence we are living as it were in the past, or as in a prison. As prisoners we do, certainly, enjoy the advantages of prison – a kind of security, perhaps even certainty, a regularity – and moreover have some timetable of future events. Nothing particularly good is going to happen for us, of course, but neither then is anything particularly bad. This is not actually true; *that* hope is illusory. Like all prisoners, nevertheless, while avoiding certain of the dangers at large in the world (largely in the present case those high-lighted by

22. Ascribed to Voltaire.

23. These initial tendencies are, as already more than once stated, reinforced by conditioning practices.

abnormal psychology – i.e. personal rejection, disorientation, un-belonging, and so on), we forego also our freedom, and the hope of the new.

In the last decade we have seen a violent movement towards change in the young people of the West, particularly among university and college students. Since something of this kind has happened with almost every young generation that ever was, there might seem little to make an issue of in this phenomenon – apart, perhaps, on this occasion from the unusually high degree of extremism which has accompanied the rebellion. To myself as also to other commentators,[24] however, it seems possible that *some* aspects of this movement do contain the seeds of real change – while the *majority* of its aspects do not, as we shall see.

One could make out a case to show that what these young people are primarily in rebellion against is System A. They are against privilege, the tyranny of status, the law and order of the law-books, the dead hand of tradition, the system of formal examinations, uniforms of all kinds and people in uniform (policemen and soldiers). They demand sexual freedom, equality with their teachers, the abolition of the class structure, of academic methods in teaching and practice, and so forth. We are all now familiar with the items on the list.

Already, however, in the items cited we perceive the dismal truth. *The very large majority of these students are intent on breaking away from the archetypes of the Ego, in order only to place themselves under the direction of the archetypes of the Self.* In political terms, they are bent on exchanging the chains of the Right for the chains of the Left. Their very intolerance of their opponents illustrates quite sufficiently that they have not escaped from the A – B pendulum.[25]

This holds true not only for the radical students, but for the hippies and all other members and sub-members of the current protest movement. 'Doing your own thing' turns out to mean

24. e.g. Charles Reich, *The Greening of America*, Allen Lane The Penguin Press, 1971.

25. To say nothing of their frequently unrelieved Self-importance.

'doing the *collective* thing'. How could it be that if all these people were doing their *own* thing they should all turn out to be doing the *same* thing? This is too much of a coincidence.

Of great interest too is the fact that the protesters are able to see the essential mediocrity, drabness, pointlessness and hypocrisy of the Establishment, without catching any glimpse of such features in themselves – just as the Establishment in turn readily perceives these qualities in the protesting factions, without being able to turn the same clairvoyant regard upon itself.

Timothy Leary's *The Politics of Ecstasy* contains many incisive and original insights into fundamental sicknesses and degradations of established Western society. For these we must be very grateful. Yet what is his main alternative – a return to System B. The straightforward recommendation of a return to religion, to the collective experience, even indeed to the tribal experience.

All versions of the radical movement – and in particular the Socialist movement – operate under the delusion that they are new. In fact, as we have shown, their basis is extremely ancient, pre-dating the rise to dominance of System A. One clue, perhaps, is in the word 'radical' itself. It is taken from Latin *radix* meaning a root. These movements, then, derive their impetus from the roots or origins of the personality. The official view, however, is that radical somehow means progressive. A better description would actually be *re*gressive – for this, in personality terms, is what is involved.

Certainly the slogan 'expansion of consciousness', the cry of the hallucinogenic divisions of the radical (the people's) army, means the admission of the unconscious into consciousness. To this event as such one has, of course, no objection – unless the intention then is to allow these aspects of the personality to *dominate* waking consciousness. This, alas, turns out to be what *is* meant. Thus there is no real merging of the unconscious and conscious potential, which could lead to genuine liberation – that is, to the emergence of System C.

The students of Essex University upset the tables and scattered the papers of the official Commission sent to investigate them, with the shout of 'You are irrelevant'. These are heartening words in themselves, for indeed the Establishment (society as it is at present constituted) is probably irrelevant to the future as such, and the future of our species in the long term. Equally irrelevant in this connection are, however, the views of the Left. It would seem to be *politics* as such which are irrelevant.[26] There is no point in replacing the tedium and the half-truths of the Right with the half-truths and tedium of the Left.

In a recent television interview Michael Foot said that he did not believe 'that people want to live grabbing and fighting and competing with each other'.[27]

Conservative spokesmen for their part are frequently heard to say that they do not believe 'that people want to be cossetted and coddled and made into milk-sops, but want a chance to show what they can do'.

These men of the Left and Right appear genuinely incapable of understanding each other. The present author, without wishing to appear to set up in business as superman, is himself perfectly able to understand both points of view. While, certainly, the more peaceable and other-oriented nature of the Left is in itself most welcome in our present society, one is unable to support the belief that a viable society can emerge on the sole basis of System B1.

At this point we must change tack and once again raise the matter of consciousness. It will however be necessary first to attempt some further justification and explanation of this term.

It seems to me that the problem of the nature of consciousness above all else should be occupying the attention of mankind. This one must repeat, is very much a minority view among psychologists. The academic or experimental psychologist particularly is

26. Compare here Leary's 'the soft chuckle which comes neither from the left nor the right, but some centre within'.

27. *Cameron Country*, B.B.C.2, 14 February 1970.

adamant in the view that if we avoid looking at this matter long enough, it will go away. Consciousness is, he generally maintains, a product of our fantasy or of loose semantics – at best an unfortunate label surviving from the past for something much more tangible and less mysterious, though he omits to say exactly what. And yet without consciousness and awareness – the *fact* of awareness, that is, not the *concept* – how could there be anything at all of what we term the life-experience? The world, while it may in some way exist in our absence, exists in our presence only because we are aware that it exists: just as we ourselves exist only in that we are aware that we do.

This last statement would perhaps not be challenged as such by the experimentalist. He, however, by no means wrongly, asks the question 'how do we become aware?' The answer – his answer – is by means of the senses, by the use of external receptors such as our eyes, ears, touch and so on.[28] As we well know, a man without eyes not only does not see, but is not aware of light. Therefore the experimental psychologist is extremely concerned to study the eye and its mechanisms and how its functioning relates both to the eye's past experience of stimulation, or particular kinds of present stimulation, and so on. Nor, I

28. The usually unexpressed assumption is that if we removed all sources of stimulation – as it were amputated all the senses – there would be nothing left. Thus the human organism turns out to be a vacuum surrounded by information. That, perhaps, is slightly overstating what the academic psychologist maintains. He is rather maintaining something along these lines – that there is a central sense organ (the brain) which does not of itself sense anything directly. It can, however, receive messages from the other senses – the receptors outside itself – and these it sorts, codes and records in accordance with its own built-in methods and terms of reference. I, however, am still concerned to know how it is that this central organism is *self-aware* of its functions. (The eye, the hand, the ear, etc., are not self-aware, that is, when connections to the brain are severed.) *How does the central organ read its own output tapes?* – for this is another way of talking about consciousness. Consciousness is so to speak the 'little man' who sits at the consul of this vast computer termed the nervous system, looking at the output. If one denies that there is such a 'little man', then how on earth *can* we conceptualize the fact of our awareness? Because a fact it *certainly* is.

reiterate, is this task in any sense a misguided or pointless undertaking.

Let us, however, ask two questions. First, what of the depot or receiving station to which the signals from the various receptors are passed? Does this depot in any way differ from the receptors themselves, or is it simply an extension of them – the other end of the bundles of nerve fibres (whose functions, incidentally, are well understood)?

The answer appears to be, yes, the depot is very different. For at some stage, somehow, these signals *pass into awareness. We* at that point have a conscious experience. As far as we can tell what we call a conscious experience can only occur in the central depot. The eye, the ear, the skin – *none* of these organs – have this property. Thus if a nerve-ending is stimulated so that a message (detectable with electronic apparatus) starts on its path to the central depot, but at the same time an electronic or other block is introduced at some point along that path, even quite close to the terminus point, the message is not delivered and we feel *nothing*. Although the skin, or whatever, has registered and transmitted a message, we do not experience that fact.

This was our first question. The second is more crucial. What on earth does a psychologist mean when he says, for example, *that he is studying the mechanism of the eye*? *Who* is studying the mechanism of the eye? The position is yet worse if the psychologist says that 'at college he made the decision to concentrate on eye-mechanisms'. What business does he have using these terms? Is not a (conscious) decision in strict psychological terms a non-event involving a non-existent attribute of the nervous system? Is the fact of the matter actually that the psychologist uses one frame of reference as a living person and another, ludicrously inadequate, frame of reference as a psychologist?

One finds oneself somewhat, though in fact more so, in the situation of the hero in H. G. Wells's *Country of the Blind*. This seeing man is unable to explain what seeing is to the people of that country, who are all born blind. The present situation, where one tries to explain or define consciousness to the psychologist, is far worse, and diabolically so. For the psychologist, while stating that he does not know what is being discussed, is *only pretending to be blind*. He in fact knows perfectly well what awareness is. It is as much a fact of his experience as anyone else's. At worst, then, he is denying what he knows to be true; at best, without realizing it, he puts on a pair of blinkers along with his psychologist's hat.

These are harsh words in a professional context. It is not actually inevitable that the matter be approached in this head-on, argumentative manner. The objections may be outflanked, perhaps, by allowing consciousness to stand as a hypothetical construct, to see whether thereby any closer understanding of any objective event or events is achieved. Such an outcome tends to back-justify one's original assumption. Let us then see whether any aspect of the *behaviour of consciousness* may be described, such that – because the behaviour is seen to exist – one may be encouraged to believe that consciousness itself exists.

The present book has proposed that consciousness is a movable commodity. This notion was broached in connection with the Journey, in Chapter 8. Let us for the moment tentatively describe consciousness as a state of arousal (of an area, or of a level) of the cortex (or of course extra-cortex or subcortex). And let us postulate that this state of arousal tends to occur most often in connection with evolving or recently evolved parts of the nervous system; and less often in connection with parts of the nervous system where evolution, to all intents and purposes, appears to have ceased. We do know, of course, that certain types of electrical activity are displayed by the cortex when an individual is observably conscious, which are not in evidence when he is observably not conscious. (One would not, however, wish necessarily to say that this electrical activity *is* consciousness, though it certainly appears to be a physiological accompaniment of that condition.) If one can show that such electrical activity at any time moves its location (and even if the electrical activity at the new site is not identical with the electrical activity at the first site) at a time when a marked change is observed or reported in the subject's inner psychological state,[29] one has a *tentative* case for the statement that 'consciousness' has moved.[30]

When a person ceases to perform a cognitive task – *which produces a known pattern of electrical cortical activity* – closes his eyes and allows his mind to drift, or specifically begins to day-dream – a *new* electrical activity is observed from the back of the brain (the occipital region), the so-called alpha rhythms. Of course, one swallow does not make a summer. One sees, however, that the beginnings of a case already exist.

If consciousness *is* associated principally with the most recently developed parts of the brain – let us, incidentally, not forget that the cerebellum has undergone considerable recent evolutionary development along with the cerebral cortex[31] – then one looks to find consciousness more connected with the *surface* of the cortex than with the subcortex. The sites for the archetypes of the Ego would probably prove to be imbedded *within* the cortex and not at the surface of the cortex. While the present book has not gone into that issue, in order not to overcomplicate the discussion, there *are* recognizable *layers* in the cortex. In other words, there is a *thickness*, and an interior of the cortex as well as a surface and a surface area.

Ignoring for the moment the question of a physiological site for System C, on the basis of our present theories we would expect 'most consciousness' – whatever that means – to be in operation when System C is functioning; rather less consciousness to be involved when System A is operating; and System B to involve least consciousness. (Of course, in many ways System A is still a going concern, with System C as it were still to some extent only the claimant or pretender to the throne.) The latter

29. Fortunately we *can* ask a human subject what seems to be going on in his head.

30. Perhaps too at this point one might remind the reader of the various spatial metaphors noted in Chapter 9 in respect of mental activities – to cast one's mind back, to be withdrawn, to be absent-minded, and so on.

31. It is to be hoped that experimenters will attempt to ascertain whether the electrical activity of the *cerebellar* cortex, or of other brain structures, changes e.g. during dreaming.

part of the hypothesis is perhaps seen to be borne out in the elusiveness or dimness of most dreams, and by the fact that dreaming time is normally some 11% of waking time.

As to the actual site of System C, one is inclined for a variety of reasons to think of parts of the frontal lobes. A piece of evidence of somewhat bizarre antecedents, supporting this view, is offered at the close of this chapter. On the other hand, System C might involve the *whole* of the cortex – perhaps the 'silent areas' in particular. This would then allow System *A* to be sited rather closer to the (motor) cortex, concerned with actual doing.

The notion of the movement of consciousness helps in understanding a well-known and widespread phenomenon, and in turn itself receives support as an idea from the existence of this phenomenon. It is suggested that the idea of the migration of souls, of the soul ascending to heaven after death, of the belief of many mystics that their souls can pass out of their body during sleep and trance and travel to other places (indeed, the very notion of the soul *as such*) may arise from the fact of a moving consciousness. This movement would then be itself an archestructure (not an archetype) – that is, an unconsciously perceived or experienced property of the nervous system – manifesting itself as a basic ingredient or element of legend, religion, and whatever.

While one must reluctantly forego here a general discussion of what the mystic considers consciousness to be, one or two aspects of the occult are of great relevance. The first of these is automatic writing.

Automatic writing is a phenomenon which anyone can experience without too much difficulty. What happens, as an end-product, is that one's hand, provided with a pen or pencil, writes of its own accord. Anyone prepared to spend a few minutes each day sitting relaxed in a preferably darkened and silent room,[32] with one hand resting on a writing-pad on the arm of a

32. The darkness which, for instance, the spiritualist medium insists on we may now see as a step towards reducing System A and conscious, cognitive activity.

chair or on a low table (as long as one is comfortable and relaxed) holding a pencil, will find that his hand twitches occasionally of its own accord. It is necessary to avoid thinking about the hand – which does not mean that one need be wholly unaware of it (one should above all not be *trying* to make the hand do anything). In time, perhaps even on the first occasion, one's hand will scribble meaningless patterns on the paper. In further time it will write actual words. As an end-product one can have long, in fact rather endless and rambling, conversations with one's own hand.[33]

Many would find the mere prospect, let alone the actual experience, of automatic writing rather unnerving – no doubt imagining that they were becoming, or at least running the risk of becoming, insane. It is precisely this kind of fear, of course, which causes System A so totally to deny System B.

As far as my present theories are concerned, during automatic writing one allows System B to function robotically (i.e. without consciousness being transferred to System B) and to use an aspect of conscious motor control usually reserved for, and by, System A – especially during waking consciousness.

Automatic writing may be thought of as the first stage of the full mediumistic trance, as demonstrated in the spiritualist medium. This full trance condition requires for its achievement a rather greater willingness to allow events to take their course than the majority of individuals in our society are probably able to bring to the situation.[34]

33. We shall not here attempt to consider the content of automatic writing, beyond saying that it is frequently ill-tempered, sometimes abusive, often childish. At other times, however, the subject-matter may be lofty and inspired, if not always particularly original. However, it is possible to develop beyond these levels. Some people are said to have written books in this manner, as also in full trance. Joan Grant, for instance (*Lord of the Horizon*, etc.) claims that some of her novels are written in this way.

34. Many of R. D. Laing's comments are relevant here.

'The process of entering into *the other* world from this world, and returning to *this* world from the other world, is as natural as death and giving birth and being born. But in our present world, that is both so terrified and so un-

When I myself experienced the full mediumistic trance for the first time, the subjective concomitants were, in this order, a wind rushing through the head, a sense of being engulfed by something, with awareness dwindling rapidly to a pin-point, followed by unconsciousness. The onset of the condition was very sudden, giving the impression of some (damming) barrier having been broken through. I was reminded in retrospect of the accounts of being 'filled with the Holy Ghost', and of similar narrated experiences of the Old and New Testaments. The second and subsequent occasions of entrancement were less dramatic, and without loss of consciousness.[35]

The state of mediumistic trance, from one's own subjective point of view, can only be described as a state of possession. The sensation is absolutely literally as if some other person had entered one's body and were using it entirely for his or her own purposes. 'Someone else' walks about the room, talks with other people – using one's own body. One hears oneself speak with the voice of a child or of a girl, quite naturally and unmistakeably, with no trace of falsetto. Somehow one has 'raised a devil', a series of devils, out of one's own personality.

This condition can be seen as a logical extension, an extreme form, of the position on view in automatic writing. There, one proposes, a robotic System B has taken over one small portion of the conscious nervous system and is using it for its own pur-

---

conscious of the other world, it is not surprising that when 'reality', the fabric of this world, bursts and a person enters the other world, he is completely lost and terrified and meets only incomprehension in others. . . . It makes far more sense to me as a valid project – indeed as a desperately urgently required project for our time, to explore the inner space and time of consciousness. . . . We are so out of touch with this realm that many people can now argue seriously that it does not exist. It is very small wonder that it is perilous indeed to explore such a lost realm.' (*The Politics of Experience*)

35. Experienced mediums usually counsel against allowing oneself to enter the condition of complete unconsciousness of so-called deep trance – for clearly one is then in no position to do anything about what might occur.

poses.[36] In the full mediumistic trance System B has robotically taken over certainly the whole of the conscious motor system – perhaps, indeed, the whole of System A, for the 'personalities' that arise in the medium are very capable not only of action, but of thinking for themselves.

Where, in the meantime, is one's own consciousness? Somehow or other, and certainly subjectively, it appears to have withdrawn from the scene of operations, out of the way. One suggests that it has perhaps withdrawn completely into System C (the 'silent' association or areas of the cortex?), leaving System A vacant (the motor and immediately adjacent areas of cortex, perhaps) for their temporary occupation by the Self or System B.

It will be clear, therefore, that there is no suggestion that the medium is possessed by the spirits of the dead. That the spiritualists should believe this is, however, perfectly understandable – for such is the most obvious explanation. The subjective experience of possession readily leads to the conclusion that an agency has taken one over from *outside*. The fact of the matter seems to be that an agency has taken over from *within* – though because the possession arises in System B it is, in that sense, outside System A, while nevertheless still inside the total person.

The fact that the medium in trance acquires information telepathically about others present also helps to support the notion of 'a voice from beyond'. So also does the fact that the medium may utter sentences in a language that he or she ostensibly does not know. One says ostensibly, because it is obviously not always easy to know how much experience of such a thing as a foreign language a person may or may not have had during his life. Experiments have shown that the nervous system is capable of retaining very large amounts of information that the subject is not consciously aware of having even heard.

36. There would seem to be, incidentally, many clear parallels between the trance condition and the hysterical conditions of severe neurosis – with the difference that in the latter case the 'possession' remains, during the illness, an unconscious one.

Before leaving the question of consciousness, three further points may be mentioned.

First, an analogy by means of which the suggested 'moving' action of consciousness may be grasped without mystification. If one has a room containing a variety of lighting systems – say, concealed lighting along one wall, a central chandelier and a small lamp on a low table hidden by a settee – one may light that room by any of the devices mentioned. Each, however, will give the room a different character, and be respectively more suitable for particular occasions. One knows from one's own experience that lighting can create or dispel shadows, or give prominence to certain aspects of a room at the sacrifice of yet others. Nonetheless, it is of course always the same room. Not only that, but the various lamps each operate from the same source of current and on the same general principles.

If the room is thought of as the nervous system – or the total physiological personality – and the lights as consciousness, one can appreciate that the schema of personality functioning proposed need contain no element of mysticism.

The second point concerns the puzzle of why the (narrowly) experimental psychologist (especially in view of his general keenness to reify the non-existent) should so steadfastly seek to destroy the notion of consciousness. This puzzle is perhaps best solved by the fact of the movement of consciousness of which, if true, we are all at some level aware. The safest way of avoiding the whole question of the movement of consciousness *into the so-called unconscious* is by denying the very existence of consciousness in the first place – in something, perhaps, of the way that a murderer might feel it safer to deny ever having been to the place where a crime occurred, than to say that he was there, but engaged in something else. The outright denial of the *un*-conscious as such, which we have noted at many points, is in fact then only the experimental psychologist's *second* line of defence.

Third of our three points is the very vexed question of whether, since it can move within the nervous system, consciousness can move *outside* the physical personality. Can it exist, as it were, anywhere except in the presence of the physical brain and in association with it? One has no answer at present to this question, but this book suggests that one should be prepared to retain an open mind to the problem. It is undoubtedly the most controversial topic in the whole study of organisms and behaviour. The phenomena of telepathy, among other matters, suggests that the solution to the problem is not likely to be a simple one.[37]

The remaining points are either final summaries of positions taken up elsewhere, or they are additional points which it proved difficult to accommodate in the argument as it in fact developed.

First in Figure 28 a summary diagram is found of the male and female personalities in terms of the triads discussed in this chapter. The overall impression obtained from this first part of the figure is that nature, in the organism, appears to work in terms of pairs of antagonistic functions, which when in harmony yield in fact a third situation not expressed by either opposite.[38]

37. As far as I am personally concerned, however, I am quite sure that we must regard awareness and consciousness as phenomena which are wholly discontinuous from the phenomena of the physical universe. This is not tantamount to saying that the former can exist in the *absence* of the latter. It *is* tantamount to saying that the former are instances of a different order of events.

38. On the physical level, even the latest phenomenon of evolution, the cerebral cortex, is in two paired sections. The vast majority of our organs in fact demonstrate this bilateral symmetry, as it is termed, and indeed the large majority of organisms throughout nature. Why should this not also be true at the psychological level, and in terms of personality? The symmetry described is, as we noted, general rather than particular. Our right lung has, for example, three ventricles against the left lung's two. Similarly, the right and left hemispheres of the cerebral cortex do not have identical functions.

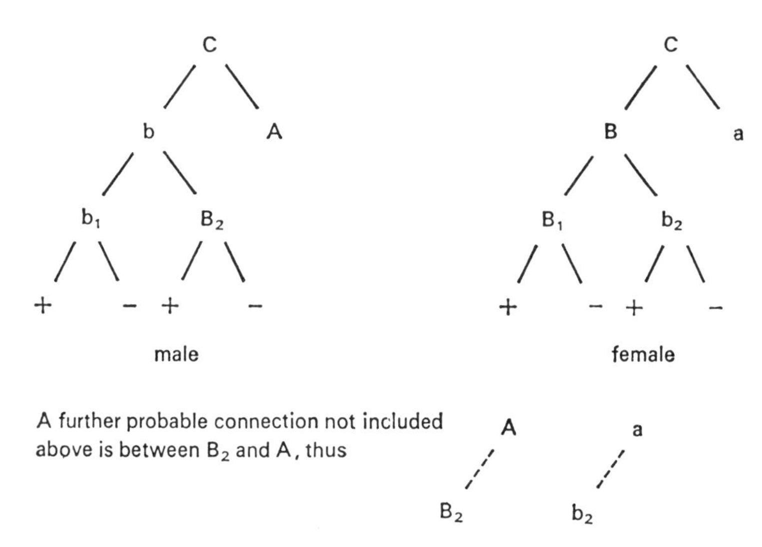

Figure 28. Full triadic representation of male and female personality.

In this section of the figure capital and small letters indicate the greater and lesser strength of the component.[39]

In order not to complicate these diagrams over-much the

39. The implication has been throughout this text that these units are independently inherited, and this implied contention is now made wholly explicit. I believe that the behavioural clusters identified, and which are designated here with letters of the alphabet, are separately encoded in blocks within the chromosome – and that at some later stage of knowledge geneticists will be able to show these. It would therefore follow that their inheritance is relatively discrete or absolute. While psychoanalysts, for example, always assume that when a quality is in strong evidence its opposite must have been repressed and in fact is seeking to 'get out', this does not always appear to be the case. Not every kind person, it seems to me, is secretly a hateful person, nor is every gentle person denying his aggression.

Former colleagues and other writers with whom I have tried to discuss these issues have usually been indignant at the notion that personality traits could be so grossly and crudely inherited. Yet, as one knows, for example, from any given population of mice we can breed selectively a 'tame' and a 'wild' strain. Dogs, too, are not merely bred for their physical appearance – but for qualities such as loyalty, determination, tenacity (cf. the husky and the bulldog), aggression, docility with children, and so on. These behaviours are bred true in particular varieties, and always appear in their offspring.

probable connection between B2 and A is not included in the main figure. If A did evolve principally out of B2, with which the male is more endowed, this would neatly account for the apparently greater capacity for logical thought in men over

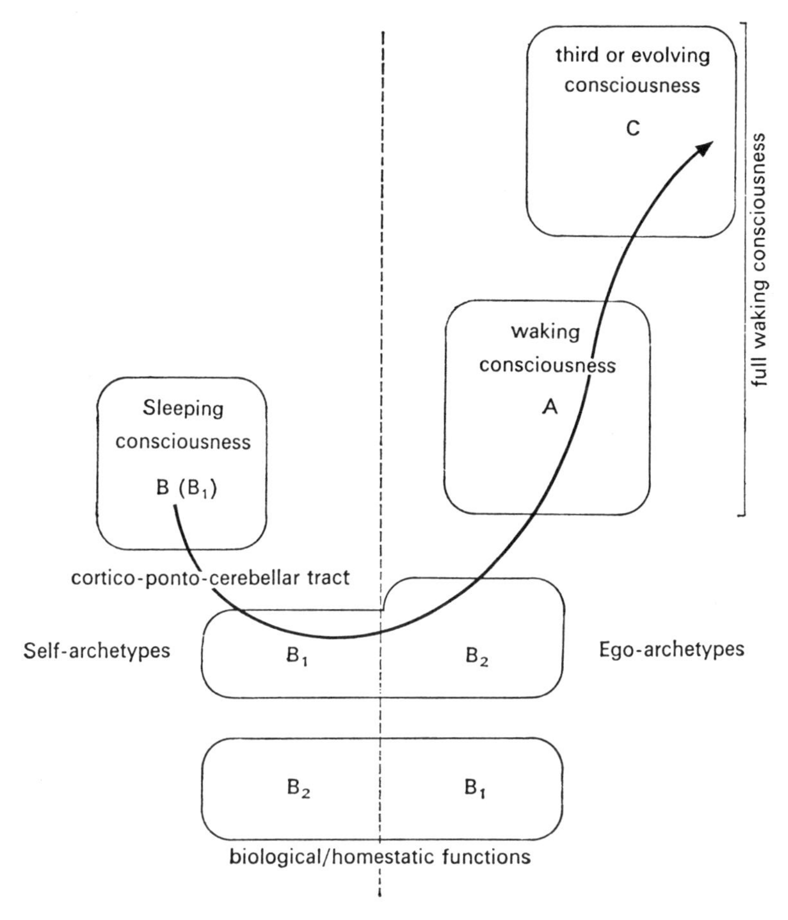

Figure 29. Schematic representation of full composite personality (male and female).

women. A further brief word on the connection between musculature – aggression in that sense – and cognitive thought is added later in this chapter.

Figure 29 will be seen, first, not to differ in essential principles from the pyramidal structure of Figure 28, although appearing rather different at first glance. This second figure is a composite of the two parts of the earlier Figure 26 of Chapter 12, with the addition of System C. It would be neater, and perhaps therefore initially more convincing, if we could place B2 ahead of B1 in the lowest level of the figure. However, the evidence is that, for instance, the forepart of the hypothalamus is concerned with quiescence and the rear with arousal. Nevertheless, such a crossing over of influence (from point of origin to point of effect) is actually a commonplace of the nervous system as a whole – seen at its most striking in the crossing of the nerve fibres from the right side of the body to reach the opposite, left-hand cerebral hemisphere.

Further reference will be made to the second part of the latter figure. For the moment we turn back to the pyramidal models of (a).

The plus and minus signs at the foot of the figures appear to be new, but actually they are not. It will be recalled that I drew attention to the fact (in Chapter 3) that 'aggression' contained both 'fight' and 'flight'. These are the positive and negative poles of that particular behaviour. It may also be recalled that we found two opposites for the word 'love' (or trust), namely 'hate' and 'fear'. The wider dimensions of these two, as it were, nuclear concepts are shown below. For the moment let us say that the true opposite of love is hate; while fear is more properly a sensation generated not from a state of anti-love, but from the state of (readiness for) flight. Very briefly, the position is this:

When one thinks about it carefully, one realizes that although aggression is in fact often accompanied by hate, it need not be. Let us consider the case of a farmer who, during a time of food shortage following the passage of a war across the countryside calmly, dispassionately, even with pity, shoots the man he finds in his barn – knowing that there is no food to spare for strangers. Or the case of the psychopath who kills a man in order to steal his car, as we might remove a pile of books from a chair in order to sit on it, and with as much emotion. Instances of hate without aggression are also not impossible to come by. A miser may hate people who have more money than himself, or who waste money, without the notion of wanting to perform aggressive acts against them. It is perfectly possible, too, for a physical coward (or a woman) to hate very intensely, but nonetheless passively.

One is not trying to play with words here, and it is certainly true that psychologists, psychoanalysts in particular, do sometimes speak – far too loosely, in my opinion – of the 'aggression of the miser', and so forth. This usage arises partly from their predetermined view of personality (i.e. that the opposite of what is on view is always at work down below), briefly discussed in the footnote on p. 516. The concept of, and the word 'aggression' might usefully be reserved exclusively for the physical act of attack or for the obvious surrogate of such attack – and some other word such as resentment or rejection be employed in respect of what are essentially states of 'shutting out'.

The view that hate and aggression may not be directly, and in that sense not very strongly connected, leads to some interesting considerations. One would like, however, to repeat at this point the warning that it is an error to think of aspects of the personality as existing in watertight compartments. There are what we might call centres or nucleii in respect of certain activities, but these always relate, as it were, at the edges to other centres, which they affect and are affected by. In brief, it is possible to envisage a structure of personality, still essentially pyramidal, where however B2 is not a major component of System B. Hitherto we have considered it as a part of that system. The alternative model would be somewhat as follows.

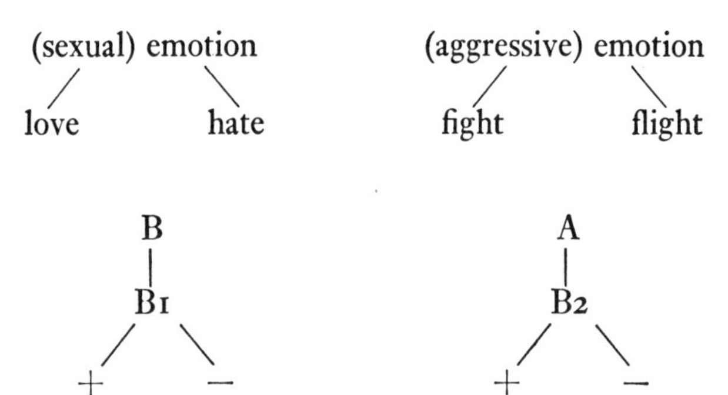

This alternative arrangement has much to recommend it. If fear or anxiety arises exclusively in connection with flight (the negative pole of aggression), this would account for the close relationship of anxiety with the Ego – noted and commented on by Freud. It would also explain, for instance, why we find it difficult or inappropriate to apply the word 'coward' to women. The Self, one could say, does not know cowardice, while the Ego does. *Hence* saints and martyrs go to their awful deaths with more composure than many heroes. While the Ego gives itself medals for bravery – which would make no sense if bravery were a matter of course.

Bearing these alternative proposals for the structure of personality in mind, let us consider what we have termed centres or nucleii of behaviour. These are seen as starting points from which quite wide and ramified conceptual structures emanate or irradiate (throughout consciousness), becoming very different in that process, but still possessing on closer inspection a certain basal referent in common.[40]

In the course of writing this I attempted to assign further actual words to the positive and negative poles of B1 and B2. The results of this attempt follow, in no particular order. It will be clear that one is tapping some of the most basic aspects of our emotional life.

40. Perhaps it was the felt archestructure of this fact of our being which also lay behind Goethe's attempts to show e.g. that all parts of a plant had basically the same structure (*Urform*) as the leaf.

B1		B2	
+	—	+	—
accept	reject	fight	fear
yield	close	kill	anxiety
love	deny	hate	worry
want	forbid	destroy	shame
permit	prevent	attack	anguish
		confront	torment
		assault	

As always, these lists are in no sense exhaustive. It is a sad testimony to my, and perhaps others', psychological make-up, however, that words connected with aggression occurred to me more readily – hence they are here in the majority. What one suggests in connection with the above is that if the various states described by the groups of words in question could (physiologically) be traced down, or back, they would be found to emanate from a single point or area of the nervous system – and not from as many points, for example, as there are words. One hesitates to say it, but many of the terms of B1, + and —, appear to have a great deal in common with bowel movements. Freud, of course, proposed this long ago – and pointed out that, for instance, mean people tend to suffer from constipation, hence the whole concept of 'anal retention' in Freudian psychology. Others of the words of the B1 lists have also equally clear associations with eating and hunger.

One of the tests of the validity of this thinking is perhaps to examine whether or not we as a culture have produced stock figures (in a sense, archetypes) on these bases – figures who sum up the areas under discussion. One then comes up with

B1 + = the saint
B1 — = the miser
B2 + = the warrior
B2 — = the coward

It would be an extremely interesting exercise to take all the nouns of the English language referring to emotions or attitudes, and arrange them into groups on a common-sense basis, to see how many nuclear starting-points one ended up with. Possibly such an exercise could tell us most of what we require to know about the basic structure of personality.

There follows next in Table 13 a summary of the attributes of the three major Systems A, B and C. The intention is here merely to remind the reader of the main areas covered and the main conclusions reached.

TABLE 13

B – SELF	A – EGO	*C* equivalents
female	male	whole
being	doing	alive
unconscious	conscious	
feeling	thinking	reasoning
associative	reductionist	creative
cyclic	linear	
(curved)	(straight)	
Pavlovian	Skinnerian	
visceral	muscular	
neurosis	psychosis	sanity/choice*
species	individual	

Some of the emotions and mental states associated with our three major personality systems are:

SELF	EGO	PERSON
effacement	duty	well-being
submission	right	relaxation
sympathy	will	cheerfulness
compassion	justification	constructiveness
unity	pride	empathy
	contempt	

* Catholics and other Christians, Communists, fascists, members of the Labour party, members of the Conservative party, psychoanalysts, learning theorists, Pavlovians, and so on have one thing in common – there are certain choices they may not make and still retain their group membership. The fact that many of them feel no wish to make those choices does not alter the truth of this. These are psychological 'sets', that is, they pre-determine which phenomena are allowed to exist.

'Hero'-figures associated with the Self are Christ, Buddha, the martyrs, etc.; with the Ego, all warriors, warrior-kings and rulers. And the overall *styles* of the Self and the Ego respectively are Being and Doing.[41]

In the System B society, being replaces doing as far as possible – so, for instance, the emphasis on meditation and withdrawal (both metaphorically, and literally – e.g. into monasteries) or 'dropping out' – the withdrawal from the world of activity and striving. In the System A society, doing replaces being as far as possible. Thus, instead of relaxing one takes a tranquillizer, for sleep one takes a sleeping pill. In the System A society the deed is taken for the wish. (This appears, incidentally, to be Krishnamurti's distinction between experience and experiencing.)

Thought is evolved muscle. Emotion is evolved viscera.

Some of the historical and archetypal figures mentioned have come close to being the whole man, as we see below. True System C man does not, perhaps, as yet exist. This appears to be the Superman of whom Nietzsche was speaking. We associate him, however, with the totality of life. His being alive is, or will be, fuller and richer. (Christ, for instance, said: 'I came that they might have life, and have it more abundantly.') The overall style of System C is therefore Life or Living.

Christ, though essentially a Self figure, was close to the Ego: just as the benevolent ruler, though an embodiment of Ego, may be close to the Self. One may conceptualize the position thus:

B		A	
Self-mother	Self-father	Ego-mother	Ego-father
Nature	Christ	Benevolent Despot	Tyrant

In earlier chapters I have spoken of the radical differences in the personalities of man and woman, and the activities for which they are and are not ideally suited.

41. Two recent films glorified respectively the Self and the Ego – *Easy Rider* and *Butch Cassidy and the Sundance Kid*. (In the first film, one *action* opens the way to a lifetime of being; in the second a lifetime of doing procures brief moments of *being*.)

The figure which follows (looking somewhat like the opening of a Hollywood film of a former era) attempts to summarize the position. However, the figure makes no claim at all to be exhaustive, nor is it to be taken this time as a literal model of the nervous system.

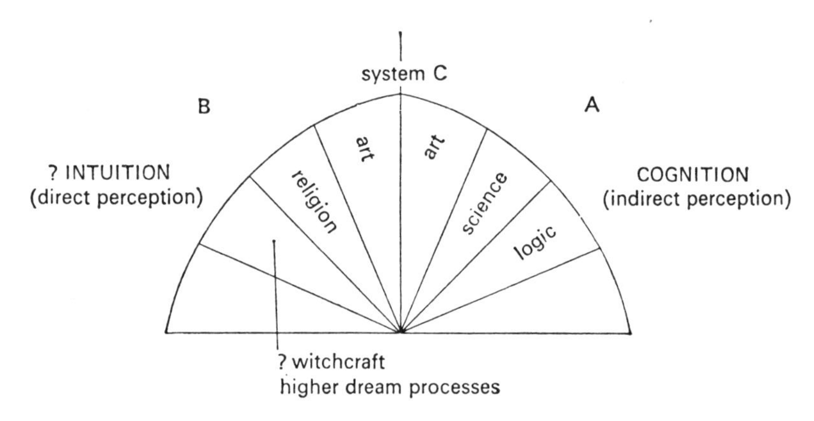

Figure 30

The gaps on the left-hand side of the figure, where 'witchcraft' has been tentatively indicated, await the as yet unrealized talents of woman. Hopefully, and presumably, these activities will embrace such matters as telepathy and precognition.[42]

Our society desperately needs to find a way back to System B – not in order to revive the dominance of the Self, as little to be desired as the dominance of the Ego – but in order to make a

42. It is of interest that in connection with Art (which the diagram here places nearest the dividing or joining line between man and woman, male and female) one notes that a number, probably the majority, of male artists possess feminine qualities. This is such a known phenomenon that it is perhaps hardly necessary to define one's terms of reference. It is less often realized that many female artists similarly display male characteristics. This view is expressed, for instance, by Walter Sorell in *The Story of the Human Hand*, and is a view with which my own observation agrees. There is, incidentally, no suggestion at all by the present text that this represents some kind of norm at which women should aim, assuming that it were in their power to do so. One is referring essentially to the genetic make-up of a part-sample of an, in any case, small number of women.

balanced integration of the two systems possible.[43] It is likely that the present dramatic increase in drug-taking of all kinds is a rather desperate attempt on the part of many individuals in our society to find that way back. A very interesting point is that it is precisely in *America* the stronghold of System A, that the swing towards System B is most marked – a point which, in his own terms, Charles Reich makes in *The Greening of America.* Drug-taking and 'living for kicks' are perhaps the only slightly less desperate attempts of the 'normal' individual to achieve what Laing sees some of his hospitalized patients as trying to achieve.[44]

We end on a speculative note. Gods, we have said, are frequently found to have feet of clay. The son of the Earth lost his strength when Hercules raised him from the ground; and Icarus, too, met with disaster when he severed his connections with terra firma. The sign of the labyrinth, the underground system of pathways, found inlaid on the floor of some medieval cathedrals is a further description of the same situation (the same continuum) – i.e. that man is composed of earth at one end and stars at the other – and that we attempt to sever the connection at our peril.

The labyrinth (built below ground), the pyramid (squat, com-

43. Clearly, women are in very many ways in a position to play the central role in this reorientating of society and its ethos.

44. Thus: 'A further attempt to experience real alive feelings may be made by subjecting oneself to intense pain or terror. Thus one schizophrenic woman who was in the habit of stubbing out her cigarettes on the back of her hand, pressing her thumbs hard against her eyeballs, slowly tearing out her hair, etc., explained that she did such things in order to experience something 'real'. It is most important to understand that this woman was not courting masochistic gratification; nor was she anaesthetic. . . . She could feel everything except being alive and real. . . . The . . . person may 'go for kicks', court extreme thrills, push himself into extreme risks in order to 'scare some life into himself', as one patient put it.

In the progressive loss of the real presence of the other, and hence loss of the sense of me-and-you-together, of we-ness, *women may become more remote and threatening than men* [my italics]. The last hope of a point of breakthrough . . . may be through a homosexual attachment, or the last loving bond may be with the other as child or animal.' (*The Divided Self*)

pact, windowless) and the cathedral (tall, soaring, the light blazing through its great windows) together make an archestructure of man.

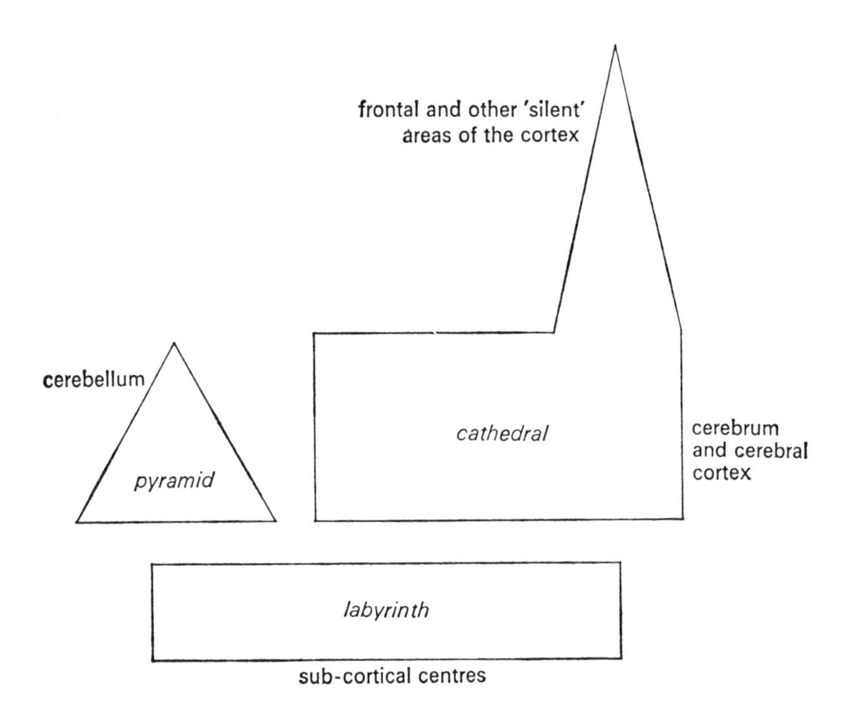

Figure 31. Labyrinth, Pyramid and Cathedral as a composite archestructure of the human brain.[45]

This last figure (Figure 31) should be compared with Figure 29, from which it does not essentially differ, as with the human brain, which it symbolically represents.

45. One notes that the pyramid and the spire of the cathedral are, in effect, and in section, triads.

Conclusion

A question put by an academic referee who read this book at the manuscript stage was, did I feel that I had said all there was to say about the personality? At various points in the text, and now again below, I have tried to indicate the developing nature of my proposals. These closing remarks which follow are intended to meet a few of the criticisms and possible misunderstandings which this book may produce.

Every theory in every line of inquiry is provisional. There are no exceptions to this statement. The apparently unshakeable Newtonian explanation of the universe proved to be provisional. Einstein's views will suffer the same fate.

The turnover in theories of psychology, however, is still more rapid. This does not imply this time a criticism of the psychologist; it simply shows the occupational risk. The turnover arises among other things, from the complex and peculiarly nebulous nature of psychological material, and not least from our own inevitable involvements with it.

The personality is in some ways like the half mythical animals of which hunters bring back reports, but which always evade capture or final identification. Or the angel with whom Jacob wrestled, but who nevertheless always escapes before daybreak. There is this to add, however: each time we wrestle with the angel we discover something new about him. Each successive time we meet him, we fight a little longer and a little more surely.

One of the tests of the validity of a theory of personality, apart from its ability to help us understand certain phenomena, is the quantity of data which it succeeds in handling. The theory I have proposed here may fairly claim to handle a considerable amount.

Nevertheless, I am well aware of the theory's inadequacies and the difficulties which arise for its internal structure (quite aside from the matter of its external validity). In Chapter 12, for example, I briefly touched upon the question of possible differences between the male and female Ego; and between the male and female Self. Of course, it is not simply a matter of inventing plausible (or plausible-sounding) terms. We can all do that till the cows come home. It is a matter of *requiring* terms to account for observed and described behavioural phenomena. The remark was made at one point that philosophers collect explanations for which they have no facts. This is another of the professional hazards which face those who undertake to study the human psyche, from whatever standpoint. I am not as yet in fact sure, by way of example, whether the division into the male and female Ego, and the male and female Self, is wholly justified.

I hope that, in particular, the previous chapter has further suggested the developing nature of some of my views. There I propose, for example, that there may be a greater disunity between the B1 and B2 sub-structures than I have in general suggested. Since completing the book I have in fact become in creasingly drawn to the idea of four poles (two sets of positive and negative terminals) to the basic 'emotional' battery: Love (+) / hate (—); and fight (+) / flight (—). These would then constitute the poles of the Self and the poles of the Ego respectively.

These poles may be arranged on the perimeter of a circle, like the points of a compass. One then has something very reminiscent of some of Jung's basic representations of the personality.[1]

1. The *I Ching*, too, makes great use of the four points of the compass.

I am of the opinion that the compass, whatever its reality value in the external world, is also an archestructure. I do not wish to open up this by no means simple issue at this stage – but if we imagine the cerebrum to be the

Matters need not be left there. Four (archetypal?) figures which correspond to the four personality poles – as suggested in the previous chapter – are perhaps: the Saint; the Miser; the Warrior; the Coward. Let us take these and attempt, on a purely chance-your-arm basis, to add the dimension of sexuality. This produces:

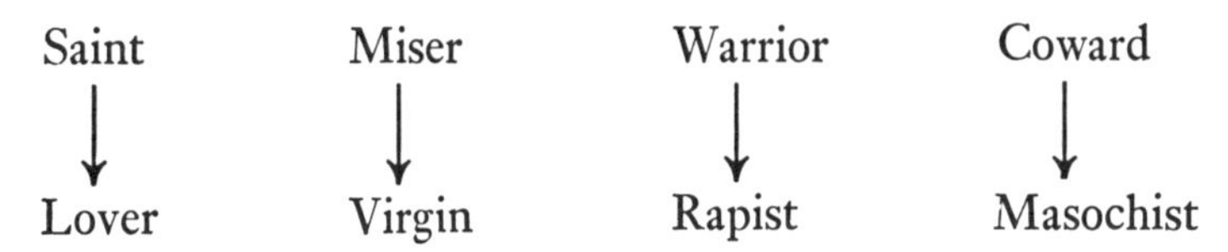

Then let us add the further dimension of male and female – again on a purely rough-and-ready, undefined basis. This gives us:

Female Saint	Male Saint	Female Miser	Male Miser	Brunhild	Hector	Female Coward	Male Coward
Courtesan	Libertine	Spinster	Voyeur (?)	Dominant (butch) Lover	Dominant (masculine) Lover	Harpy	Torturer

One sees, then, that other kinds of models of the personality are possible. This one, on which we have here made a tentative beginning, resembles more a child's house of building blocks.

---

north and the cerebellum the south (the cold and warm qualities of these being quite appropriate), there is then the further possibility of considering the two hemispheres of the cerebrum *or* the cerebellum or indeed *both* as West and East respectively. I believe that with a closer inspection of this model one might be in a position to elucidate more precisely the relationship of East-Lucky-Right-Left in various ancient and modern cultures.

A left-handed boxer, incidentally, is called a south-paw. Both the compass and 'animal' (i.e. paw) references are of great interest.

Some would feel, perhaps, that it has about as much value, but suppose we said that each of these figures might be the perceived or personified mean of certain inherited characteristics, normally distributed in the population? That would be somewhat less literary – if that is the aim.

Are we then with this latest draft model suggesting that the general theory of this book, which one has argued out in such detail, is really after all to be scrapped? Not at all. For we note that the square is the basis of a pyramid: just as the circle (on which the compass points are arranged) is the base of a cone. And that finally, the vertical section both of the pyramid and the cone are triads – or, at least, triangles. I believe, indeed, very much that with such manipulations of the geometric shapes in question, plus the additional symbolics of the positive and negative terminals of the electrical battery, and the magnetic poles of the geographical compass, one comes very close to the real structure and functioning of the human personality.

Perhaps this is an appropriate moment to apologize to the experimental psychologist.

I have been described by one in other respects generous critic as something of a crank in my attitude to experimental psychology. I was distressed to hear this view. For while it is true I have made, and do make, no secret of my disapproval of some aspects of the experimentalist approach, my comments on it have not been intended as a destructive criticism in quite that sweeping sense. I would rest my defence not simply on the fact that I have quoted extensively from the writings of experimentalists; but on the fact that I have been at pains to suggest throughout how my own theories could be put to test experimentally – even sometimes to the point of underselling the theory. I have, then, no objection whatsoever to the experimental testing of hypotheses, *wherever possible*. By which last I mean, however, quite definitely and unreservedly to say that it is not always possible.

There is a further important aspect to this matter. This has nothing to do with what experimental psychology may or may not be able to achieve. It has to do with what experimental psychology itself represents 'emotionally' or psychically to those who engage in it, and its meaning as a cultural and behavioural phenomenon of our time. In this context, and this sense, my views are quite unequivocal. Here I see the academic psychologist spear-heading a movement towards a society and a way of life under which, despite the use of the term 'way of life', it will not be possible for our species to continue to exist. I am not here speaking of hydrogen bombs or biological warfare, though these are symptomatic. Of what I precisely mean I have tried to say something in the chapter on the psychotic society.[2] I cannot take time to enlarge on this further here. I would, however, like to cite one more very small example of what is currently taking place at the heart (the non-heart) of the civilized world. The incident which is described was reported in *Newsweek* magazine, 27 July 1970. An American television company set up a fictitious emergency to ascertain whether allegations concerning (certain?) medical ambulance services in America were well grounded. The subjects of the drama were only acting. The ambulance men were real.

> He lay writhing on the floor, a shabbily dressed middle-aged man, his eyes closed in pain, his mouth sucking desperately for air. Impassively the ambulance driver and the attendant who had been summoned looked on. 'He's gotta have at least thirty-eight bucks or we don't take him', one of them snapped to the stricken man's room-mate. Pointing to two one-dollar bills on the kitchen table, the room-mate pleaded: 'That's all I can find. But he's got a job, he's good for the money.'
>
> But this was not good enough for the Mid-America Ambulance Company. Visibly annoyed, the attendants helped the room-mate prop the victim in a kitchen chair. Then they departed – but not before one of them pocketed the two dollars from the table.

Is it *likely* that such an incident can take place in isolation from

2. That chapter has ended up in the final draft as Chapter 13. The 'moment' (in this case this book) is seen again to contain all the information one needs for the understanding of (in this case my) attitudes and motivation.

the fabric of society or that such attitudes can grow up overnight? Would one not be inclined to suspect that such incidents and attitudes might tend to be typical, to be symptomatic of a general mode of behaviour, of no less than a way of life?

I have the feeling that some of my readers and former colleagues will take me to task for introducing human values and the wider context into what purports to be a scientific or objective examination of the personality. To this I would say only that I cannot for one minute imagine how any book on any scientific subject whatsoever could fail at some point to be concerned with human values. (There are, after all, no *other* values.) But I cannot conceive for a micro-second how any book on the social sciences could, essentially, be concerned with anything else. For if the social scientist is not to be the guardian of human values and the psychological health of the human community – then who, today, is? It is for the gross dereliction of this particular function that I would here indict the academic psychologist. One cannot stand by in silence when the psychologist degenerates into a man whose sole function is to produce habits out of a rat.

I have, lastly, been asked whether as a consequence of my views on the structure of personality I consider war between the races (between the gene-pools), and 'war' between the sexes, as inevitable. One's first comment is that if a difference is in fact genetically and not merely culturally produced it cannot, whatever else, be wished away or otherwise treated as if it did not exist. One sees such differences, then, as irreducible in the absolute sense. It is just as well that this is so, for it seems clear that the new emerges continually, and perhaps only, from the meetings of difference – that is, from thesis meeting antithesis. To say, however, that these phenomena are irreducible is not actually to say that they are unmodifiable, or their expression incapable of any control. The first indispensable step in control, nevertheless, must be the identification and precise understanding of the phenomena. To this end the present book is offered as a contribution. Already in such a process of description, means of control or remedy suggest themselves. However

I dislike the overtones of the word control – and remedy suggests a disease or illness. I prefer some such word as channelling. At all events, this control or channelling must in no sense be understood to imply any kind of *manipulation* of one group by another.

I do not, then, even if it were possible, wish for uniformity. Neither, on the other hand, do I wish for competition, in the usual sense of that term. Competitive interaction (despite Conservative views on the matter) is in essence and by nature negative and destructive. That is to say, any initial credit balance will in the long run always be offset by a larger (though perhaps 'invisible') deficit. The interaction of differences – which is what I am speaking of – need, however, not necessarily be harmful.

To repeat an earlier remark it seems to me, whatever else, that to believe that the behavioural and psychological differences between men and women – and, to a lesser extent, the behavioural and psychological differences between races and varieties of men – are not at least partly genetically based, is to be wholly naive. It goes without saying of course that 'different' means neither superior nor inferior.

Leaving with this matters on which others have commented or may comment, if there is one matter on which I would take *myself* to task, it is on having concentrated too much on the more easily grasped and too little on the less readily grasped. I have, that is, a sense of having made rather a contribution to System A and Knowledge I – which does not urgently require support – than to System B and Knowledge II, which does. One comforts oneself to an extent with the thought that had one written a treatise solely with reference to, say, the *I Ching*, no social scientist would have bothered to read it. Ideally, of course, any contribution should be neither to System A nor System B, but to System C.

In conclusion I would like it to go on record that I am very aware of many significant contributions to the understanding of personality that I have not even found space to mention. Some would run counter to my own views, others are supportive.

Contributions for which I have considerable respect include those of some experimentalists – C. L. Hull, D. O. Hebb, and others.[3] My excuse for omissions – apart from the fact that cognitive theories, at least, already receive adequate attention from our cognitive society – is the limitations imposed by space, in turn imposed by the desire to present my views in the compass of a single volume. I hope to remedy at least some of the omissions in a later volume.

In connection with the work of current social scientists, I would like to make one point and to draw one last distinction between the bulk of that work and my own. I suggest that the majority are *horizontal* (two-dimensional) studies of the organism; while the present – and those of Freud, Jung and some others – are *vertical* studies: *ideally* three-dimensional, but in fact perhaps also at least sometimes two-dimensional – though then still in respect of the vertical, not the horizontal, axis. The essential 'flatness' of the academic and experimental approaches are perhaps glimpsed in the fact that some experimentalist's idea of a depth interview is one which lasts an hour instead of ten minutes. I believe as a generalization that the horizontal or lateral approach arises, apart from the question of one's own individual constitutional leanings, and out of a fear and a denial of the unconscious. The two-dimensional *vertical* approach (such as I believe Freud's to be) arises perhaps also from a constitutional predisposition, and again from a fear: but on this occasion of the annihilating, Medusa eye of consciousness.

3. The views of the two psychologists named are normally reckoned under theories of learning, not theories of personality. My own very wide definition of personality, however, embraces all mental contents.

Select Bibliography

CHAPTER 1

Amphitryon: Plautus
Doctor Faustus: Christopher Marlowe
Dr Jekyll and Mr Hyde: R. L. Stevenson
Dracula: Bram Stoker
Faust, Parts I and II: J. W. v. Goethe
Frankenstein: Mary Shelley
Love in Infant Monkeys: H. F. Harlow, *Scientific American*, June, 1959.
New Maps of Hell: Kingsley Amis, Science Fiction Book Club, 1962.
The Interpretation of Dreams: Sigmund Freud, Allen & Unwin, 1954.

CHAPTER 2

Larousse World Mythology: Hamlyn, 1965.
Magic, Supernaturalism and Religion (originally published as *The History of Magic*): Kurt Seligmann, Allen Lane The Penguin Press, 1971
New Larousse Encyclopedia of Mythology (*2nd Edition*): Hamlyn, 1968.
The Black Arts: Richard Cavendish, Routledge & Kegan Paul, 1967.

CHAPTER 3

A Comprehensive Etymological Dictionary of the English Language: E. Klein, Elsevier, 1967.
An Etymological Dictionary of the English Language: W. W. Skeat, Oxford University Press, 1946.
Cold Comfort Farm: Stella Gibbons, Longmans, 1932.

Language: Its Nature, Development and Origin: Otto Jesperson, Allen & Unwin, 1922.
Straight and Crooked Thinking (2nd Edition): R. H. Thouless, Pan Books, 1953.

CHAPTER 4

A Textbook of Physiological Psychology: S. P. Grossman, Wiley, 1967
Physiological Psychology (2nd Edition): C. T. Morgan and E. Stellar McGraw Hill, 1950.

CHAPTER 5

Conditioned Reflexes: I. P. Pavlov, Oxford University Press, 1927.
Experimental Psychology: R. S. Woodworth and H. Schlosberg, Methuen, 1955.
Introductory Lectures on Psycho-Analysis (2nd Edition): Sigmund Freud, Allen & Unwin, 1933.
Joey: A 'Mechanical Boy': Bruno Bettelheim, *Scientific American*, March 1959.
Learning to Think: H. F. and M. K. Harlow, *Scientific American*, August 1949.
Science and Human Behaviour: B. F. Skinner, Macmillan, 1953.
Studies in Word Association: C. G. Jung, *Collected Works, 2nd Edition, Vol. 2*, Routledge & Kegan Paul, 1969.
The Dynamics of Personality and Hysteria: H. J. Eysenck, Routledge & Kegan Paul, 1957.
The Divided Self: R. D. Laing, Tavistock Publications, 1959.
The Politics of Experience and the Bird of Paradise: R. D. Laing, Penguin Books, 1967.
The Psycho-Pathology of Everyday Life: Sigmund Freud, *Standard Edition of Complete Works, Vol. VI*, Hogarth, 1966.

CHAPTER 6

Sexual Behaviour in the Human Female: A. C. Kinsey, Saundars, 1953.
Sexual Behaviour in the Human Male: A. C. Kinsey, Saundars, 1949.
Sleep: Ian Oswald, Penguin Books, 1966.
The Nervous System: Peter Nathan, Penguin Books, 1969.

CHAPTER 7

The Politics of Experience and the Bird of Paradise: R. D. Laing, Penguin Books, 1967.

CHAPTER 8

A Textbook of Physiological Psychology: S. P. Grossman, Wiley, 1967.
Physiological Psychology (2nd Edition): C. T. Morgan and E. Stellar, McGraw-Hill, 1950.
Sleep: Ian Oswald, Penguin Books, 1966.
The Ambidextrous Universe: Martin Gardner, Allen Lane The Penguin Press, 1967.
The Left-Handed Book: Michael Barsley, Souvenir Press, 1966.
The Story of the Human Hand: Walter Sorell, Weidenfeld & Nicolson, 1968.

CHAPTER 10

A Dictionary of the Bible: J. Hastings, T. & T. Clark, 1898.
Fossil Man: Michael H. Day, Hamlyn, 1969.
History of the Primates (9th Edition): W. E. Le Gros Clark, British Museum (Natural History) 1965.
Man and the Vertebrates: A. S. Romer, Penguin Books, 1954.
Prehistoric Art: P. M. Grand, Studio Vista, 1967.
Prehistoric Art in Europe: N. K. Sandars, Penguin Books, 1968.
Shanidar : The First Flower People: Ralph S. Solecki, Knopf, 1971; published as *Shanidar : The Humanity of Neanderthal Man*, Allen Lane The Penguin Press, 1972.
The Ambidextrous Universe: Martin Gardner, Allen Lane The Penguin Press, 1967.
The Archetypes and the Collective Unconscious: C. G. Jung, *Collected Works, 2nd Edition, Vol. 9, Part I*, Routledge & Kegan Paul, 1969.
The Naked Ape: Desmond Morris, Cape, 1967.
The Origin of Species (1859): Charles Darwin, Penguin Books, 1968.
The Stone Age Hunters: Graham Clark, Thames & Hudson, 1967.

CHAPTER 11

I Ching or The Book of Changes: transl. Richard Wilhelm and Cary F. Baynes, with a *Foreword* by C. G. Jung, *3rd Edition*, Routledge & Kegan Paul, 1968.
Synchronicity: An A-Causal Connecting Principle: C. G. Jung, *Collected Works*, *2nd Edition*, *Vol. 8*, Routledge & Kegan Paul, 1969.

CHAPTER 12

King Solomon's Ring: Konrad Lorenz, Methuen, 1952.
Love and Hate: Irenäus Eibl-Eibesfeldt, Methuen, 1971.
The Archetypes and the Collective Unconscious: C. G. Jung, *Collected Works*, *2nd Edition*, *Vol. 9*, *Part I*, Routledge & Kegan Paul, 1969.
The Study of Instinct: N. Tinbergen, Oxford University Press, 1951.

CHAPTER 13

The Betrayal of the Body: Alexander Lowen, Macmillan, 1967.
The Divided Self: R. D. Laing, Tavistock Publications, 1959.
The Will to Meaning: Victor Frankl, World Publication Co., 1969.

CHAPTER 14

Experimental Psychology: R. S. Woodworth and H. Schlosberg, Methuen, 1955.
On the Aesthetic Education of Man (1793): Friedrich Schiller, transl. E. M. Wilkinson and L. A. Willoughby, Oxford University Press, 1967.
The Politics of Ecstasy: Timothy Leary, MacGibbon & Kee, 1970.

Index